FUNDAMENTALS OF TURBULENCE MODELING

Combustion: An International Series
Norman Chigier, *Editor*

Bayvel and Orzechowski, Liquid Atomization
Chen and Jaw, Fundamentals of Turbulence Modeling
Chigier, Combustion Measurements
Kuznetsov and Sabel'nikov, Turbulence and Combustion
Lefebvre, Atomization and Sprays
Li, Applied Thermodynamics: Availability Method and Energy Conversion
Libby, Introduction to Turbulence
Roy, Propulsion Combustion: Fuels to Emissions

FUNDAMENTALS OF TURBULENCE MODELING

Ching-Jen Chen
Dean of Engineering
Professor of Mechanical Engineering
Florida A&M University and
Florida State University
Tallahassee, Florida
U.S.A.

Shenq-Yuh Jaw
Associate Professor
Department of Naval Architecture
National Taiwan Ocean University
Keelung, Taiwan
Republic of China

USA	Publishing Office:	Taylor & Francis 1101 Vermont Avenue, NW, Suite 200 Washington, DC 20005-3521 Tel: (202) 289-2174 Fax: (202) 289-3665
	Distribution Center:	Taylor & Francis 1900 Frost Road, Suite 101 Bristol, PA 19007-1598 Tel: (215) 785-5800 Fax: (215) 785-5515
UK		Taylor & Francis Ltd. 1 Gunpowder Square London EC4A 3DE Tel: 0171 583 0490 Fax: 0171 583 0581

FUNDAMENTALS OF TURBULENCE MODELING

1 2 3 4 5 6 7 8 9 0 EBEB 9 0 9 8 7

This book was set in Times Roman. Editorial and Composition Services by TechBooks. The acquisitions editor was Lisa Ehmer. Cover design by Precision Graphics.

A CIP catalog record for this book is available from the British Library.
⊗ The paper in this publication meets the requirements of the ANSI Standard Z39.48-1984 (Permanence of Paper)

Library of Congress Cataloging-in-Publication Data

Chen, Ching Jen, 1936–
Fundamentals of turbulence modeling / Ching-Jen Chen, Shenq-Yuh Jaw.
p. cm. — (Combustion: an international series)
ISBN 1-56032-405-8 (alk. paper)
1. Turbulence—Mathematical models. I. Jaw. Shenq-Yuh. II. Title. III. Series: Combustion (New York, N.Y. : 1989)
TA357.5.T87C45 1997
620.1′064—dc21 97-18161
CIP

ISBN 1-56032-405-8 (cloth)

CONTENTS

PREFACE

This book is the outgrowth of the research interest and lecturing by one of us (C.-J.C.) at the University of Iowa, Florida A&M University, Florida State University, and many other research institutes and universities in the United States, Japan, China, and Taiwan. S. Y. Jaw revised the manuscript and introduced the most recent works, including his own at the final phase of the book. The intent of this book is to summarize the progress made in turbulence modeling for colleagues, engineers, graduate students, and friends.

Turbulent flow motions occur in many engineering problems and in our environment. Although the Navier–Stokes equations can properly describe the turbulent flow, it is too costly and often unnecessary for engineers to obtain such a complex and detailed solution. Instead, averaged Navier–Stokes equations are often sufficient and practical to describe the turbulent motion confronted in engineering problems. To recover the information lost during the averaging process, a turbulence model must be introduced. At present, no such unified model exists. Oftentimes, turbulent flow modeling is criticized as "postdiction" rather than "prediction."

The present-day state of turbulence modeling is such that some prediction capability has been achieved. The most promising and practical model for engineering problems is the second-order turbulence closure model. Therefore, the present book emphasizes the second-order closure model and its application to engineering problems.

The book is divided into six parts: 1., Introduction to Turbulence, which defines the phenomenon of turbulent motion and the averaging process to handle irregular turbulent fluctuation; 2., Second-Order Closure Turbulence Models, which derives and constructs various turbulence models and details the determination of turbulent coefficients or constants; 3., Discussion of Turbulence Models, which verifies and tests the turbulence models and discusses various levels of approximation; 4., Near-Wall Turbulence, which derives and discusses the turbulence model near the wall and the treatment needed in the prediction of near-wall turbulence; 5., Applications of Turbulence Models, which documents applications of turbulence models to many problems from simple to complex and from two-dimensional to three-dimensional situations; and 6., Turbulent Buoyant Flows, which addresses the additional modeling required that is due to the appearance of gravity and density difference.

We are indebted to many colleagues, graduate students, and friends who have been interested in our work and who interacted with us on research in turbulence modeling. I (C.-J. C.) thank A. Wada and N. Tanaka of the Central Research Institute of the Electric Power Industry for their help and intrest in turbulence modeling. We also owe special thanks to V. Pai and Y. Haik for the typing and additional editing of the final manuscript.

We will be happy to receive any discussion and criticism on the ideas presented in the book. While we make no claim to originality of most of the contents of the book, we have attempted to present the turbulence modeling in a systematic and logical fashion, bringing in the historical view and utilizing a set of turbulent model postulations.

Ching-Jen Chen
Sheng-Yuh Jaw
July 1997

CHAPTER

ONE

INTRODUCTION TO TURBULENCE

It is important to recognize that the turbulent flow motion is always three-dimensional, unsteady, rotational, and, most important, irregular. The irregularity of turbulent motion is due to the inherent nonlinear nature of the Navier–Stokes equations when the Reynolds number is beyond the critical value. Thus, contrary to laminar flow, which is regular and deterministic, turbulent flow is stochastic and chaotic. In order to predict the gross or average behavior of turbulent flow, a mathematical model must be established. This book presents the fundamentals of turbulence modeling.

1.1 HISTORICAL VIEW

The first turbulence modeling may be traced back to the drawings of Leonardo da Vinci in the fifteenth century. As shown in Fig. 1.1 (circa 1495 A.D.), he gave the following description: "The clouds scattered and torn. Sand blown up from the seashore. Trees and plants must be bent." In the present day language of turbulence, one may interpret this in the following way: The clouds' motions are irregular and random, and the turbulent eddies are cascading. The sand particles on the seashore are entrained. Trees and plants are bent, subject to large turbulent shearing forces near the ground. The subsequent major events of historical development of the modeling of fluid motions are summarized in the following periods of time.

Fifteenth and sixteenth centuries. L. da Vinci (1452–1519) used visual and descriptive models to describe turbulent flow in his drawing. Many drawings are compiled in Richter (1970)[131]. However, no mathematical model was available to describe the flow motions.

Figure 1.1 Leonardo da Vinci's turbulence model. (From Ritchter (1970).)

Seventeenth and eighteenth centuries. I. Newton (1643–1727), L. Euler (1707–1783), D. Bernoulli (1700–1782), and J. d'Alembert (1717–1783) were among those cited by Struik (1948)[170] as having contributed to create a mathematical model for fluid motions on the basis of the assumption of continuum, inviscid, and by obeying Newton's law:

$$\rho \frac{DU_i}{Dt} = \rho G_i - \frac{\partial P}{\partial x_i}. \tag{1.1}$$

No mathematical model was available to describe viscous flow.

Nineteenth century. L. M. H. Navier (1785–1836)[114], J. B. Fourier (1768–1830), B. de St. Venant (1797–1886), and G. G. Stokes (1819–1903) are cited by Struik (1948)[170] for their contributions in creating a mathematical model for flow motion on the basis of the assumptions of continuum, viscous flow, and obedience to Stokes' (1845)[169] postulations for stress, τ_{ij}, and Fourier's (1955)[47] postulations for heat conduction, q_i.

A viscous fluid model can be shown as

$$\rho \frac{DU_i}{Dt} = \rho G_i - \frac{\partial P}{\partial x_i} + \frac{\partial \tau_{ij}}{\partial x_j}. \tag{1.2}$$

A viscous flow, thermal conduction model can be shown as

$$\rho C_p \frac{DT}{Dt} = \tau_{ij} \frac{\partial U_j}{\partial x_j} + \frac{DP}{Dt} + \frac{\partial q_i}{\partial x_i}. \tag{1.3}$$

No mathematical model was available to describe the turbulent flow and heat transfer.

Nineteenth and twentieth centuries. O. Reynolds (1842–1912)[128], L. Prandtl (1875–1953)[124, 125], T. von Kármán (1881–1963)[180, 181], and Sir G. I. Taylor (1880–1975), see Schlichting (1979)[149], contributed to developing the mathematical model for turbulent fluid motion on the basis of the assumptions of continuum flow, averaged flow, viscous flow, and obedience to a set of turbulent postulations.

An averaged viscous fluid model can be shown as

$$\rho \frac{DU_i}{Dt} = \rho G_i - \frac{\partial P}{\partial x_i} + \frac{\partial \tau_{ij}}{\partial x_j} + \frac{\partial \tau_{ij}^t}{\partial x_j}. \tag{1.4}$$

An averaged viscous flow, thermal conduction model can be shown as

$$\rho C_p \frac{DT}{Dt} = \tau_{ij} \frac{\partial U_j}{\partial x_j} + \frac{DP}{Dt} + \frac{\partial q_i}{\partial x_i} + \frac{\partial q_i^t}{\partial x_i} + \Phi, \tag{1.5}$$

where τ_{ij}^t and q_i^t are the turbulent Reynolds stress and heat flux, respectively. The development of a turbulence model is still not complete.

1.2 NAVIER–STOKES (N-S) EQUATIONS: VALIDITY FOR TURBULENCE

The foundation of turbulence modeling is the Navier–Stokes (N-S) equations. Therefore, it is necessary to understand them and to validate that they are capable of describing turbulent fluid motion.

1.2.1 Stokes's Postulations for a Viscous Fluid Model

Consider the postulations used in deriving the N-S equations:

1. The fluid motion can be considered to be a continuous medium (continuum).
2. The viscous diffusion of U_i (or ρU_i) is proportional to the gradient of U_i (rate of strain).
3. The fluid is isotropic.
4. The fluid is homogeneous; that is, $\tau_{ij} = F(U_l, P, \rho, T)$.
5. When the fluid is at rest, the stress is hydrostatic.
6. When the flow is a pure dilatation, the average stress is equal to the pressure (Stokes's hypothesis).
7. Viscous fluid model constants (or coefficients) require experimental determination, namely, ρ, μ, μ_2 (second viscosity).

There is nothing said in these postulations about excluding turbulent flows. As long as the flow does not violate the postulations, the N-S equations should be valid for turbulent flows.

For the ρ and μ constants, the N-S equations and the energy equation are

$$\frac{\partial U_i}{\partial X_i} = 0, \tag{1.6}$$

$$\rho \frac{DU_i}{Dt} = \rho G_i - \frac{\partial P}{\partial X_i} + \mu \frac{\partial^2 U_i}{\partial X_j \partial X_j}, \tag{1.7}$$

and

$$\rho C_p \frac{DT}{Dt} = k \frac{\partial^2 T}{\partial x_j \partial x_j} + \mu \Phi, \tag{1.8}$$

where

$$\Phi = \left(\frac{\partial U_i}{\partial x_j} + \frac{\partial U_j}{\partial x_i} \right) \frac{\partial U_i}{\partial x_j}.$$

One may be concerned that the continuous cascade of turbulent eddies may violate the continuum assumption. Let us examine the limiting eddy and vortex stretching in turbulent flows. For example, in turbulent flows the patterns in Fig. 1.2 are often seen. The eddies tend to be smaller for larger Reynolds number (*Re*). The turbulent eddies are always irregular, time dependent, and three-dimensional with the associated vortices always being stretched. One may ask, will the cascade of an eddy evolve such that the eddies become indefinitely small?

1.2.2 Limits in Vortex Stretching

Let us isolate a vortex and examine its stretching (see Fig. 1.3). Consider the curl of the N-S equations. Then we have the vorticity equation, where the vorticity (ω) is the curl

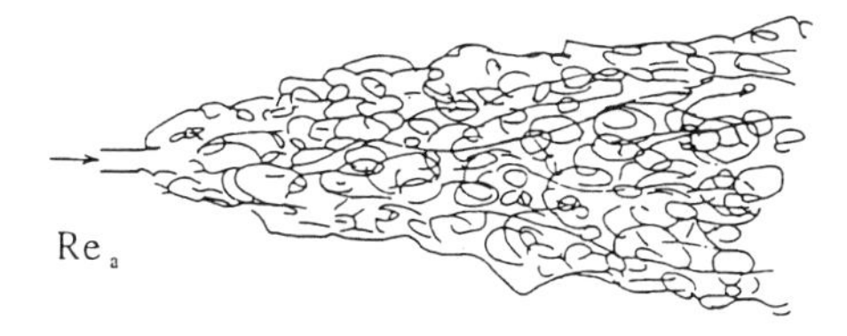

Re $_a$ < Re $_b$

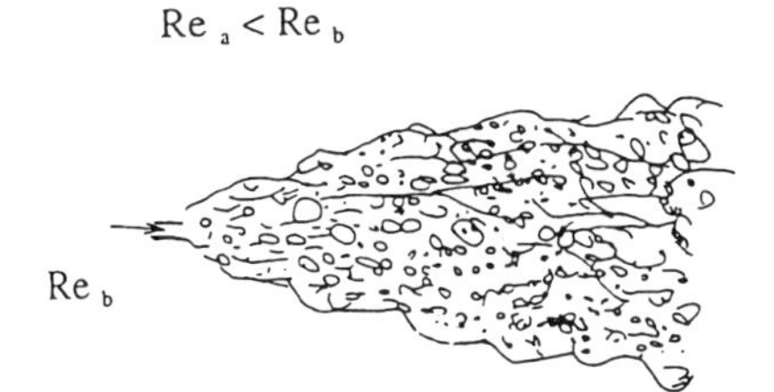

Figure 1.2 Eddy patterns in jet flows.

of the velocity and is given as

$$\frac{D\boldsymbol{\omega}}{Dt} = \frac{\partial\boldsymbol{\omega}}{\partial t} + (\boldsymbol{v}\cdot\nabla)\boldsymbol{\omega} = (\boldsymbol{\omega}\cdot\nabla)\boldsymbol{v} + \nu\nabla^2\boldsymbol{\omega}, \tag{1.9}$$

where

$$\boldsymbol{\omega} = \nabla\times\boldsymbol{v}. \tag{1.10}$$

We see that when $D\omega/Dt = 0$, the vortex stretching will stop. From an order-of-magnitude analysis of Eq. 1.9, balancing the right-hand side will occur when $\omega u/l \doteq \nu\omega/l^2$ or $ul/\nu \doteq 1$.

Here, ω is the limiting vorticity, u is the eddy velocity, and l is the eddy size. We find that when ul/ν (the eddy Re) is $O(1)$, $D\omega Dt \Rightarrow 0$, or, in other words, there will be no more vortex stretching. Thus, the smallest turbulent eddies should be at $ul/\nu = 1$.

Let us consider this from another point of view. The Kolmogorov scaling (see Hinze, 1975 [60]) (ν, ϵ), which refers to the small eddies, gives $u = (\nu\epsilon)^{1/4}$, $l = (\nu^3/\epsilon)^{1/4}$, and

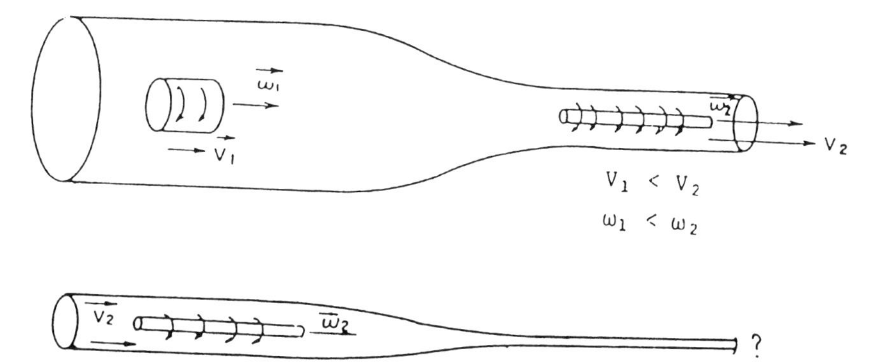

Figure 1.3 Examining the stretching of a vortex.

$t = (\nu/\epsilon)^{1/2}$, where ϵ refers to the rate of dissipation of turbulent kinetic energy. From the Kolmogorov scale, we also find $ul/\nu = 1$.

For example, consider turbulent flow in air. If $u = 10$ m/s and $\nu = 10^{-5}$ m^2/s, then $l = 10^{-6}$ m is the size of the small eddy. There are 10^{29} molecules/m^3 under normal one atmosphere conditions. In this case, there are 10^{11} air molecules/l^3 (where $1\,l^3 = 10^{-18}$ m^3). For an eddy with 10^{11} molecules, the continuum assumption is valid. We therefore conclude that the N-S equations are valid for turbulence.

This has also been proved by direct numerical simulation (DNS, Moser and Moin, 1984 [111] ; Spalart, 1986 [164]; Kim, Moin, and Moser, 1987 [74]; Mansour, Kim, and Moin, 1988 [102]) and by large eddy simulation (LES) (Deardorff, 1970 [38]; Schumann, 1975 [150]; Moin and Kim, 1982 [111]) of low *Re* turbulent flows. DNS predicts turbulent flows by solving directly the instantaneous N-S equations, whereas LES uses a filter model for the small-eddy behavior and computes the three-dimensional, time-dependent large-eddy structures in turbulent flows. These predictions are validated by experiments. However, at present, both approaches require much computer memory and central-processing-unit time and are not practical in engineering applications. For instance, in order to capture the fine scale turbulent eddies, DNS requires the grid size to be smaller than the finest turbulent eddies. In general, the grid number required by DNS is about 9/4th the power of the turbulent *Re* of the flow. For a flow with turbulent *Re*, (R_T), equal to 10^5, which is often encountered in engineering applications, the grid number required would be higher than 10^{11}, which is far beyond the capability of computers currently available.

What an engineer always would like to know is the mean effect of the turbulent quantities and not that of the instantaneous fluctuation quantities. Thus, a more practical approach to describing turbulent flows would be to model the averaged turbulent transport quantities. In this approach, models are created to simulate the unknown transport quantities that are formed from the averaging process. This approach is adopted in this book.

1.3 AVERAGING PROCESSES

1.3.1 Methods of Description

Statistical analysis. This type of analysis is used to study turbulent structures by various correlations ($\bar{u}^2$, etc.) and by probability density functions ($B(u)$, etc.). For example, let u^* be the instantaneous velocity. The time history of the velocity component u^* is shown in Fig. 1.4. To avoid looking at the history of instantaneous flow fluctuations every time, we replace or pack the same information into a probability distribution function $B(u^*)$. Here $B(u^*)\Delta u^*$ is the probability of finding u^* in the range Δu^*, or by definition,

$$B(u^*)\Delta u^* = \lim_{T\to\infty} \frac{1}{T} \sum_i (\Delta t_i).$$

Here Δt_i is the time interval that u^* is in Δu^*. In order to construct $B(u^*)$, one may measure a variable, f, as

$$\bar{f} = \lim_{T\to\infty} \frac{1}{T} \int_{t_0}^{t_0+T} f(u^*)dt = \int_{-\infty}^{+\infty} f(u^*)B(u^*)du^* \text{ (moments).}$$

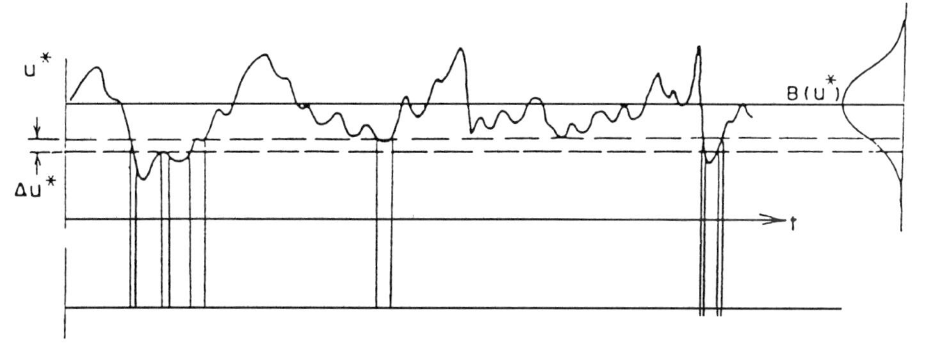

Figure 1.4 Time history of velocity component u^*.

For example,

$$U = \int u^* B(u^*) du^* = \frac{1}{T} \int_0^T u^* dt$$

$$\sigma^2 = \int u^{*2} B(u^*) du^* = \bar{u}^2 = \frac{1}{T} \int_0^T u^{*2} dt$$

$$S\sigma^3 = \int u^{*3} B(u^*) du^* = \bar{u}^3 = \frac{1}{T} \int_0^T u^{*3} dt$$

$$k\sigma^4 = \int u^{*4} B(u^*) du^* = \bar{u}^4 = \frac{1}{T} \int_0^T u^{*4} dt$$

where

U = mean
σ = variance
S = skewness
K = Kurtosis or flatness

From the measurements of these moments, one may construct $B(u^*)$, which appears as the kernel in the integral. A properly and accurately constructed probability density function should contain almost all turbulence information in the statistical sense. Some examples of probability density functions are given in Fig. 1.5.

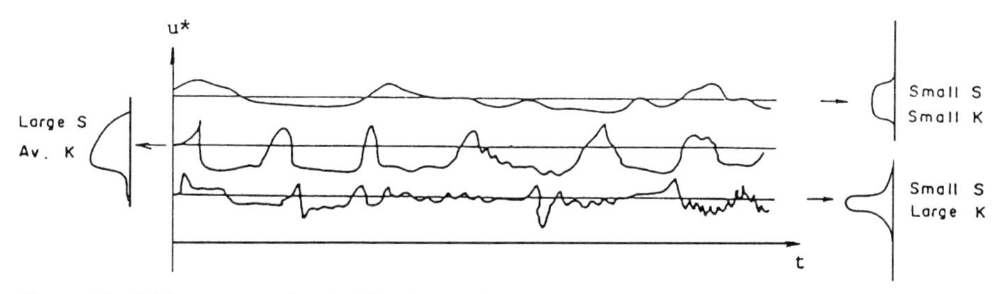

Figure 1.5 Different types of probability density functions.

A similar procedure may be applied to the two- and three-dimensional cases. However, the analysis is more complicated. Hence, although statistical analysis may reveal details of turbulent structures, it is cumbersome to use for solving engineering problems.

Phenomenological analysis. In this approach, turbulence model postulations are made. The analysis loses some details of the turbulence physics, but it can provide solutions to engineering problems. The starting point of phenomenological analysis is the construction of some process of averaging.

1.3.2 Reynolds Averaging

Let $u_i^* = U_i + u_i$, where u_i^* is the total or instantaneous value, U_i is the mean value, and u_i is the fluctuating value.

Long-time averaging ($T \rightarrow \infty$):

$$\bar{u}_i^* = U_i = \frac{1}{T} \int_0^T u_i^* dt.$$

In this case, U_i is not a function of time. See Fig. 1.6.

u_i^* = total value

U_i = mean value

u_i = flactuating value

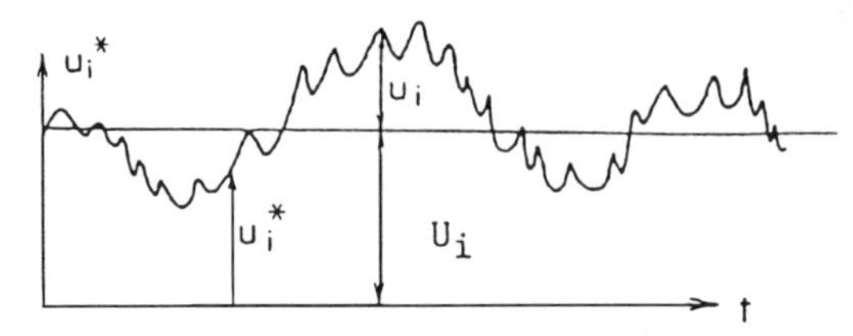

Figure 1.6 Reynolds long-time averaging.

Short-time averaging (ΔT short):

$$\bar{u}_i^* = U_i(t) = \frac{1}{\Delta T} \int_{-\frac{\Delta T}{2}}^{+\frac{\Delta T}{2}} u_i^*(t + t')dt'.$$

Here t' is the time variable in ΔT duration near t. In this case, U_i is a function of time and ΔT. If ΔT becomes large, the short-time average recovers the original Reynolds long-time average. A short-time average is somewhat similar to a signal that has been smoothed or has had a high-frequency component filtered out. It should also be noted that $\overline{u_i} = 0$ and $\overline{u_i u_j} \neq 0$ in general. See Fig. 1.7

1.3.3 Ensemble Averaging (Phase Averaging)

In a short-time averaging process, the experiment needs to be performed only once. However, the averaging depends on the filter time interval, ΔT, which is not known in

u_i^* = total value

U_i = mean value

u_i = flactuating value

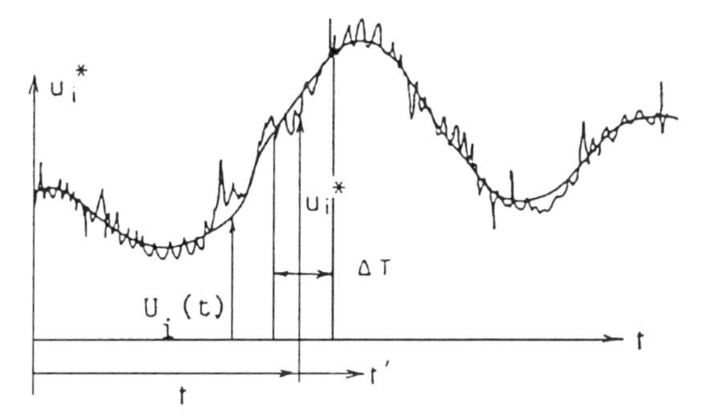

Figure 1.7 Reynolds short-time averaging.

advance. An alternative way of constructing an average for a time-dependent flow is by obtaining an ensemble average, which requires one to perform the experiment repeatedly for N times, and then the averaging is taken over the entire range of experiments over N experiments, as indicated in Fig. 1.8.

$$\overline{u_i^*} = U_i(t) = \lim_{N \to \infty} \frac{1}{N} \sum_{n=1}^{N} u_i^*(t, n), \tag{1.11}$$

where N = number of experiments.

Also, note that

1. $\overline{Au_i^*} = A\bar{u}_i^* = AU_i$
2. $\overline{u_i^* + u_j^*} = U_i + U_j$
3. $\overline{\partial u_i^*/\partial x_j} = \partial U_i^*/\partial x_j$
4. $\overline{u_i^* u_j^*} = U_i U_j + \overline{u_i u_j}$.

Equation 1.11 includes the short-time Reynolds average and is potentially more general. However, it is more difficult to produce because it requires N experiments. N should be large enough so that the ensemble average is no longer dependent of N.

1.3.4 Density-Weighted Averaging (Compressible Fluids)

The use of the Reynolds average for compressible fluids is cumbersome. For example,

$$\overline{\rho^* u^* v^*} = \overline{(\rho + \rho')(U + u)(V + v)} = \cdots$$

is a messy operation. Alternatively, a density-weighted average (or mass-weighted average) may be used.

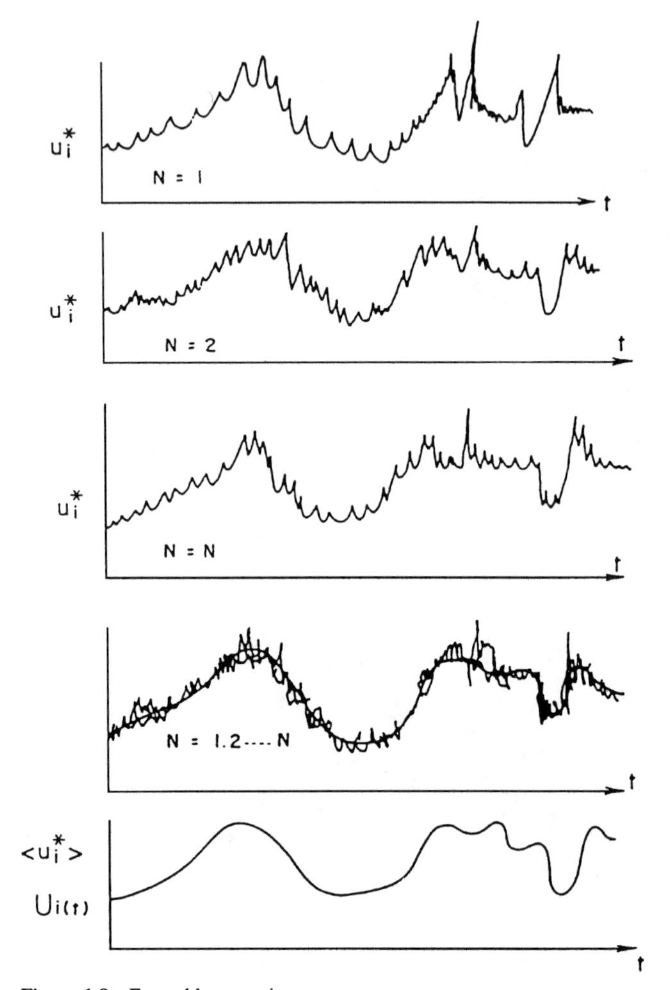

Figure 1.8 Ensemble averaging.

Letting $\tilde{f}$ be the density-weighted average for f, we define

$$\tilde{u}_i^* = \tilde{U}_i = \frac{\int_0^T \rho^* u_i^* dt}{\int_0^T \rho^* dt} = \frac{\overline{\rho^* u_i^*}}{\bar{\rho}}. \tag{1.12}$$

Then, instead of

$$u_i^* = U_i + u_i \quad \text{(Reynolds average)},$$

we have

$$u_i^* = \tilde{U}_i + u_i'' \quad \text{(density-weighted average)}$$
$$\rho^* = \bar{\rho} + \rho' \quad \text{(Reynolds average)}.$$

From Eq. 1.12, we have

$$\overline{\rho^* u_i^*} = \bar{\rho}\tilde{U}_i.$$

But,

$$\overline{\rho^* u_i^*} = \overline{\rho^*\left(\tilde{U}_i + u_i''\right)} = \overline{\rho^*\tilde{U}_i} + \overline{\rho^* u_i''} = \bar{\rho}^*\tilde{U}_i + \overline{\rho^* u_i''}.$$

Thus,

$$\overline{\rho^* u_i''} = 0,$$

or

$$\overline{(\bar{\rho} + \rho')u_i''} = 0.$$

That is,

$$\overline{\bar{\rho} u_i''} + \overline{\rho' u_i''} = 0,$$

or

$$\bar{u}_i'' = -\overline{\rho' u_i''}/\bar{\rho} \;\neq 0.$$

In general,

$$\bar{u}_i'' \neq 0.$$

For example, the mass conservation equation (Eq. 1.6) under density-weighted averaging becomes

$$\overline{\frac{\partial \rho^*}{\partial t} + \frac{\partial \rho^* u_i^*}{\partial X_i}} = 0,$$

or

$$\frac{\partial(\bar{\rho} + \bar{\rho}')}{\partial t} + \frac{\partial \overline{\rho^* u_i^*}}{\partial X_i} = 0.$$

Hence,

$$\frac{\partial \bar{\rho}}{\partial t} + \frac{\partial \bar{\rho}\tilde{U}_i}{\partial X_i} = 0,$$

which is simpler than the continuity equation with conventional Reynolds averaging:

$$\frac{\partial \bar{\rho}}{\partial t} + \frac{\partial \bar{\rho}\tilde{U}_i}{\partial X_i} + \frac{\partial \overline{u_i \rho'}}{\partial X_i} = 0.$$

1.3.5 Conditional Averaging (Sampling Average)

In some problems such as intermittent phenomena or thermal spikes, a special averaging process may reveal more physics of turbulence. Let l_i be the duration of the laminar flow and t_i be the turbulent flow duration. Then, we can define the intermittency, γ, as the ratio of the turbulent flow duration to the total flow duration, that is,

$$\gamma = \frac{\sum_i t_i}{\sum_j l_j + \sum_i t_i}.$$

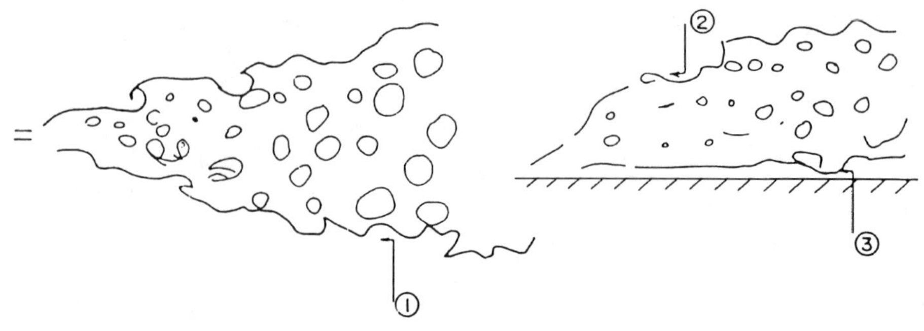

Figure 1.9 Signal sampling at three different locations in flow.

For example, the signals at Locations 1, 2, or 3 in Fig. 1.9 may be represented as shown in Fig. 1.10. In digital recording of total N data, we may define the intermittency as

$$\sigma = \sum_{i=1}^{N} I/N,$$

where, $I = 1$, when the data are turbulent, and $I = 0$, when the data are not turbulent. Also, one must remember that $\overline{u_{j,t}^*} \neq \overline{u_{j,l}^*}$ because

$$\overline{u_{j,t}^*} = \lim_{N\to\infty} \sum_{i=1}^{N} I u_j^* \Big/ \sum_{i=1}^{N} I,$$

whereas

$$\overline{u_{j,l}^*} = \lim_{N\to\infty} \sum_{i=1}^{N} (1 - I) u_j^* \Big/ \sum_{i=1}^{N} (1 - I),$$

wherein the subscripts t and l refer to turbulent and laminar signal, respectively (see Fig. 1.11).

For a thermal spike signal,

$$\bar{T}_t^* = \lim_{N\to\infty} \sum_{i=1}^{N} I T^* \Big/ \sum_{i=1}^{N} I.$$

We see that T_t^* can be far greater than T_l^*, as shown in Fig. 1.12, in which case a conventional Reynolds averaging will not capture the peak of the thermal spike.

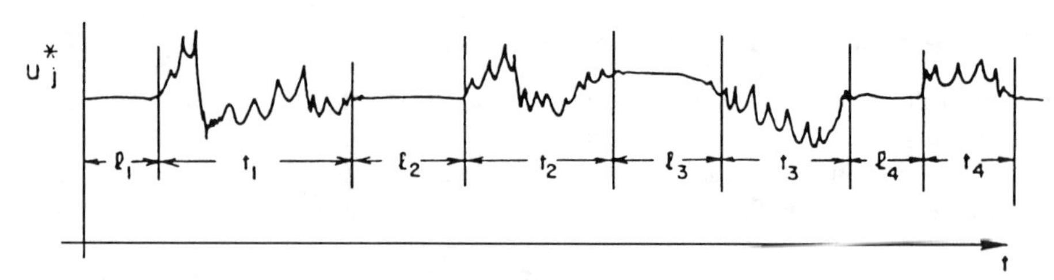

Figure 1.10 Representation of sampled signals.

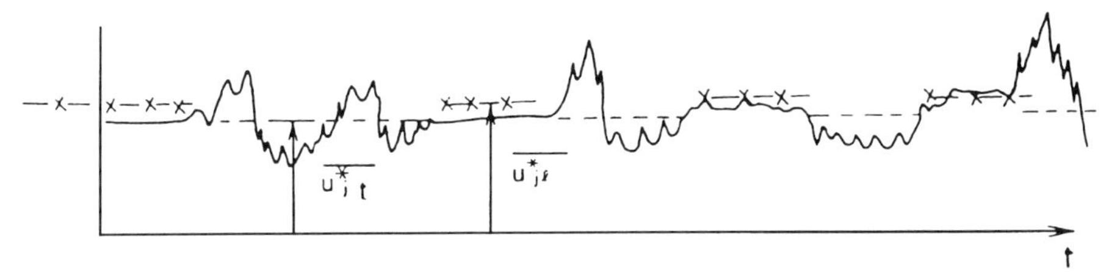

Figure 1.11 Intermittent sampling.

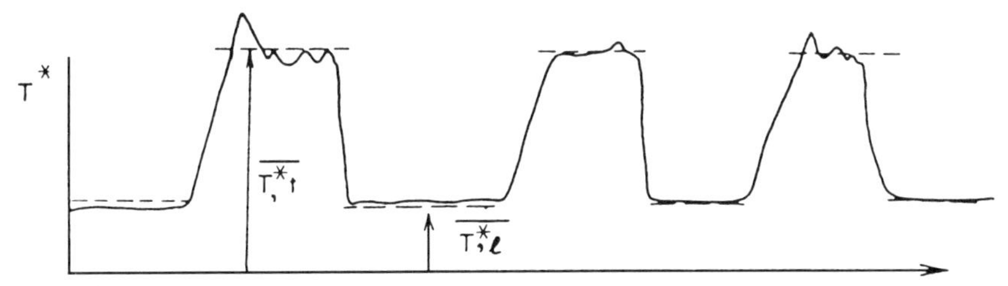

Figure 1.12 Thermal spikes.

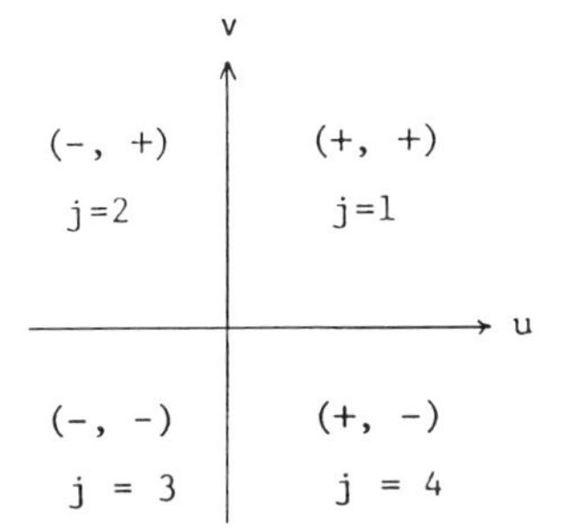

Figure 1.13 Quadrant conditional averaging.

Quadrant conditional averaging (sampling). This is used to monitor or investigate the details of the turbulent directional characteristics (u, v). Let j be the jth quadrant at the time t (see Fig. 1.13). Then

$$H_j(t_i) = 1 \quad \text{if } (u, v) \text{ is in } j\text{th quadrant at } t = t_i$$
$$= 0 \quad \text{if not.}$$

With $u^* = U + u$ and $v^* = V + v$, we can define the jth-quadrant conditional average (see Fig. 1.14) as

$$\langle \overline{uv} \rangle_j = \lim_{N \to \infty} \sum_{i=1}^{N} H_j(t_i)(uv)_i \Big/ \sum_{i=1}^{N} H_j(t_i).$$

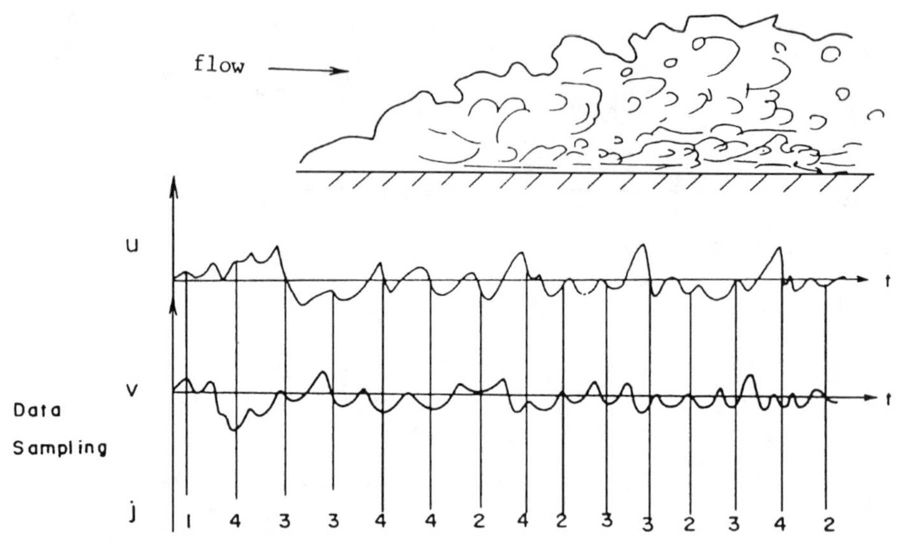

Figure 1.14 Data sampling in conditional averaging.

1.4 AVERAGED INCOMPRESSIBLE TURBULENCE EQUATIONS

1.4.1 Navier–Stokes and Energy Equations

We start out by defining the turbulent quantity to be equal to the sum of the mean value of the quantity plus its fluctuating value. Thus, $P^* = P + p$, $T^* = T + \theta$, $u_i^* = U_i + u_i$, and $\tau_{ij}^* = \tau_{ij} + \tau_{ij}'$. Taking ensemble average of instantaneous incompressible N-S and energy equations, Eqs. 1.6–1.8, we have the following equations.

Continuity equation:

$$\frac{\partial U_i}{\partial X_i} = 0. \tag{1.13}$$

Momentum equation:

$$\rho \frac{DU_i}{Dt} = \rho G_i - \frac{\partial P}{\partial X_i} + \frac{\partial \tau_{ij}}{\partial X_j} + \frac{\partial \tau_{ij}^t}{\partial X_j}, \tag{1.14}$$

wherein

$$\tau_{ij} = \mu \left(\frac{\partial U_i}{\partial X_j} + \frac{\partial U_j}{\partial X_i} \right)$$

and

$$\tau_{ij}^t = -\rho \overline{u_i u_j}.$$

Here, τ_{ij} is modeled from the viscous fluid model, whereas τ_{ij}^t needs to be modeled from the turbulence model.

Energy equation ($C_p = C_v$):

$$\rho C_p \frac{DT}{Dt} = \tau_{ij} \frac{\partial U_i}{\partial X_j} - \frac{\partial q_i}{\partial X_i} + \frac{\partial q_i^t}{\partial X_i} + \Phi^t, \tag{1.15}$$

where

$$q_i = -\kappa \partial T / \partial X_i$$
$$q_i^t = -C_p \rho \overline{u_i \theta}$$
$$\Phi^t = \overline{\tau'_{ij} \frac{\partial u_i}{\partial X_j}}$$
$$\tau'_{ij} = \mu \left(\frac{\partial u_i}{\partial X_j} + \frac{\partial u_j}{\partial X_i} \right).$$

Here, q_i is modeled from the Fourier postulations, whereas q_i^t and Φ^t need to be modeled from the turbulence model. Equations 1.7–1.9 give five equations for the five variables P, U, V, W, and T, whereas there are no equations for $\overline{u_i u_j}$, $\overline{u_i \theta}$, and Φ^t. Therefore, the set of equations is not closed. The problem of closing the turbulence mathematical model is called the *turbulence closure problem.*

Before the closure problem can be addressed, the ensemble-average quantities of $\overline{u_i u_j}$, $\overline{u_i \theta}$, and Φ^t must be derived. These additional unknowns, a total of ten (six from $\overline{u_i u_j}$, three from $\overline{u_i \theta}$, and one from Φ^t) can be derived from the fluctuation equations for u_i and θ. The fluctuation equations for u_i and θ are obtained from

$$[\text{N-S}] - \overline{[\text{N-S}]} \rightarrow \text{Equation for } u_i,$$

or

$$\rho \left(\frac{\partial u_i}{\partial t} + U_l \frac{\partial u_i}{\partial X_l} + u_l \frac{\partial U_i}{\partial X_l} + u_l \frac{\partial u_i}{\partial X_l} \right) = -\frac{\partial p}{\partial X_i} + \mu \left(\frac{\partial^2 u_i}{\partial X_l \partial X_l} \right) + \frac{\partial \overline{\rho u_i u_l}}{\partial X_l} \tag{1.16}$$

$$[\text{Energy}] - \overline{[\text{Energy}]} \rightarrow \text{Equation for } \theta,$$

or

$$\rho C_p \left(\frac{\partial \theta}{\partial t} + U_l \frac{\partial \theta}{\partial X_l} + u_l \frac{\partial T}{\partial X_l} + u_l \frac{\partial \theta}{\partial X_l} \right) = k \frac{\partial^2 \theta}{\partial X_l \partial X_l} + \frac{\partial \overline{C_p \rho u_l \theta}}{\partial X_l} - \Phi^t. \tag{1.17}$$

1.4.2 Turbulence Transport Equations

In order to close the problem, we will derive $\overline{u_i u_j}$, $\overline{u_i \theta}$, and so forth from Eqs. 1.16 and 1.17. These are the second-order, one-point correlation-based transport equations.

Reynolds-stress transport equations. The equation for $\overline{u_i u_j}$ is obtained by the following method:

- Multiply Eq. 1.16, for the ith variable, by u_j.
- Multiply Eq. 1.16, for the jth variable, by u_i.
- Add the results of the above two operations.
- Take the average of the total addition.

By following the above-mentioned procedure, the equations for $\overline{u_i u_j}$ obtained are

$$\frac{D\overline{u_i u_j}}{Dt} = \frac{\partial}{\partial X_l}\left(-\overline{u_i u_j u_l} - \overline{\frac{p}{\rho}(\delta_{jl}u_i + \delta_{il}u_j)} + \nu\frac{\partial\overline{u_i u_j}}{\partial X_l}\right)$$
$$- \left(\overline{u_i u_l}\frac{\partial U_j}{\partial X_l} + \overline{u_j u_l}\frac{\partial U_i}{\partial X_l}\right) - 2\nu\overline{\frac{\partial u_i}{\partial X_l}\frac{\partial u_j}{\partial X_l}} + \overline{\frac{p}{\rho}\left(\frac{\partial u_i}{\partial X_j} + \frac{\partial u_j}{\partial X_i}\right)}. \qquad (1.18)$$

For understanding the physics of this equation, let us consider what the seven terms on the right-hand side of this equation represent.

- The first two terms on the right represent the turbulent diffusion of the momentum.
- The third term represents the molecular diffusion of the momentum and is usually negligible compared with the first two terms. However, it is the term with the highest derivative. Hence, it must be retained.
- The fourth and the fifth terms represent the production of stresses, $\overline{u_i u_j}$, by the interaction of Reynolds stresses $\overline{u_i u_j}$ and the gradient of the mean values.
- The sixth term represents the viscous dissipation of the Reynolds stresses, $\overline{u_i u_j}$.
- The seventh term represents the pressure-strain (PS) of the flow, which tends to restore isotropicity of the flow.

Turbulent kinetic energy (k) equation. From Eq. 1.18 for $\overline{u_i u_j}$, we can easily obtain the turbulent kinetic energy equation by setting $i = j$. The mean turbulent kinetic energy is denoted by $k = \overline{u_i u_i}/2$, whereas the fluctuating turbulent kinetic energy is denoted by $k' = u_i u_i/2$. Then, the turbulent kinetic energy (k) equation can be obtained as

$$\frac{Dk}{Dt} = \frac{\partial}{\partial X_l}\left(-\overline{k' u_l} - \frac{\overline{p u_l}}{\rho} + \nu\frac{\partial k}{\partial X_l}\right) - \overline{u_i u_l}\frac{\partial U_i}{\partial X_l} - \epsilon + 0. \qquad (1.19)$$

Note that the PS term goes to zero because of the continuity equation. The physics hidden in the equation is revealed when each of the terms on the right-hand side is studied carefully.

- The first two terms on the right represent the diffusion of the kinetic energy from the high intensity to the low intensity that is due to turbulent fluctuating motions.
- The third term represents the molecular diffusion of the turbulent kinetic energy.
- The fourth term represents the production of the turbulent kinetic energy that is due to the interaction of turbulent stress and the gradient of the mean-flow velocity.
- The fifth term, ϵ, represents the rate of dissipation of the turbulent kinetic energy that tends to occur at the small-eddy scale. ϵ is always positive because

$$\epsilon = \mu\overline{\frac{\partial u_i}{\partial X_l}\frac{\partial u_i}{\partial X_l}}.$$

- The PS term of the Reynolds stress transport equation in the last term vanishes, and, hence, its net contribution to the turbulent kinetic energy transfer rate is zero.

Rate of dissipation equation. The dissipation rate of turbulent kinetic energy,

$$\epsilon = \nu \overline{\frac{\partial u_i}{\partial X_l} \frac{\partial u_i}{\partial X_l}},$$

directly affects the growth rate of k. It is thus one of the most important turbulent quantities. It appears naturally in the derivation of the k equation. The equation for the rate of dissipation of the turbulent kinetic energy, ϵ, is obtained by the following approach:

- Perform the operation $\partial/\partial X_l$ on Eq. 1.16 for the ith variable.
- Multiply the resulting equation by $\partial u_i/\partial X_l$.
- Take the ensemble average of the product and multiply by two.

By going through these steps, the ϵ equation is obtained as follows:

$$\begin{aligned} \frac{D\epsilon}{Dt} = & \frac{\partial}{\partial X_l}\left(-\overline{\epsilon' u_l} - \frac{2\nu}{\rho}\overline{\frac{\partial u_l}{\partial X_j}\frac{\partial p}{\partial X_j}} + \nu\frac{\partial \epsilon}{\partial X_l}\right) \\ & - 2\nu \overline{u_l \frac{\partial u_i}{\partial X_j}} \frac{\partial^2 U_i}{\partial X_l \partial X_j} - 2\nu \frac{\partial U_i}{\partial X_j}\left(\overline{\frac{\partial u_l}{\partial X_i}\frac{\partial u_l}{\partial X_j}} + \overline{\frac{\partial u_i}{\partial X_l}\frac{\partial u_j}{\partial X_l}}\right) \\ & - 2\nu \overline{\frac{\partial u_i}{\partial X_j}\frac{\partial u_i}{\partial X_l}\frac{\partial u_j}{\partial X_l}} - 2\overline{\left(\nu\frac{\partial^2 u_i}{\partial X_l \partial X_l}\right)^2}. \end{aligned} \tag{1.20}$$

Here the fluctuation or instantaneous quantity of ϵ is given by

$$\epsilon' = \nu \frac{\partial u_i}{\partial X_l}\frac{\partial u_i}{\partial X_l}.$$

In order to understand the ϵ equation, let us examine the various terms on the right-hand side of the equation.

- The first two terms again represent the turbulent diffusion of ϵ.
- The third term represents the molecular diffusion of ϵ.
- The fourth and fifth terms represent the production of ϵ.
- The last two terms represent the destruction (source or sink) of the dissipation rate of the turbulent kinetic energy.

Reynolds turbulent heat flux equation. If we define the fluctuating viscous dissipation term as

$$\Phi' = \tau'_{ij}\frac{\partial u_i}{\partial X_j} = \tau'_{nm}\frac{\partial u_n}{\partial X_m},$$

then the equation for the Reynolds turbulent heat flux, $\overline{u_i\theta}$, can be obtained by using the following approach:

- Multiply Eq. 1.16, for the ith variable, by θ.
- Multiply Eq. 1.17 by u_i.

- Add the two products.
- Take the ensemble average of the resultant sum.

Thus, the $\overline{u_i\theta}$ equation can be written as

$$\frac{D\overline{u_i\theta}}{Dt} = \frac{\partial}{\partial X_l}\left(-\overline{u_i u_l \theta} - \delta_{il}\frac{\overline{p\theta}}{\rho} + \alpha\overline{u_i\frac{\partial\theta}{\partial X_l}} + \overline{\nu\theta\frac{\partial u_i}{\partial X_l}}\right)$$

$$-\left(\overline{u_i u_l}\frac{\partial\theta}{\partial X_l} + \overline{u_l\theta}\frac{\partial U_i}{\partial X_l}\right) - (\alpha+\nu)\overline{\frac{\partial u_i}{\partial X_l}\frac{\partial\theta}{\partial X_l}} + \overline{\frac{p\partial\theta}{\rho\partial X_i}} + \overline{\Phi' u_i}. \qquad (1.21)$$

Here α and ν are the thermal diffusivity and the kinematic viscosity, respectively.

1.5 TURBULENCE CLOSURE PROBLEM

The averaged N-S equations, as indicated by Eq. 1.14, are given as

$$\frac{\partial U_i}{\partial X_i} = 0$$

$$\rho\frac{DU_i}{Dt} = \rho G_i - \frac{\partial P}{\partial X_i} + \frac{\partial \tau_{ij}}{\partial X_j} + \frac{\partial \tau_{ij}^t}{\partial X_j},$$

wherein

$$\tau_{ij} = \mu\left(\frac{\partial U_i}{\partial X_j} + \frac{\partial U_j}{\partial X_i}\right)$$

and

$$\tau_{ij}^t = -\rho\overline{u_i u_j}.$$

The above-mentioned four equations are for U_i and p. The problem of specifying unknowns, $\overline{u_i u_j}$, is the turbulence closure problem. Historically the first attempt to model the Reynolds stress, τ_{ij}^t, is to model it as a function of the mean-flow equations. Because the $\overline{u_i u_j}$ term is modeled directly without any additional differential equation, this model is also called *the zero-equation model or the first-order closure model.* For example,

1. $-\overline{uv} = \nu_t \partial U/\partial Y$. This is the Boussinesq eddy viscosity model.
2. $-\overline{uv} = l^2\,|dU/dY|\,\partial U/\partial Y$. This is the Prandtl mixing length model.
3. $-\overline{uv} = \kappa X \partial U/\partial Y$. This is the Prandtl wake mixing length model.
4. $-\overline{uv} = \kappa^2|(dU/dY)/(d^2U/dY^2)|^2\partial U/\partial Y$. This is the von Kármán mixing length model.

In the early twentieth century, because of the lack of computing machines, the above-indicated turbulence models were very popular because they are relatively simple and intuitive. However, models were very limited in the sense that the model could apply only to a particular problem or problems with similar geometry and were inappropriate for predicting other types of flows with different geometries. These simple models, in general, have a low-prediction capability. Nevertheless, they possess the advantage of simplicity and are capable of modeling turbulence with similar flow conditions and geometrical shapes.

1.6 SUMMARY

In recent years the zero-equation model that imitates the laminar viscous stress

$$\frac{\tau_{ij}}{\rho} = \nu\left(\frac{\partial U_i}{\partial X_j} + \frac{\partial U_j}{\partial X_i}\right)$$

has ad-hoc-ly been generalized to model the turbulent Reynolds stress as

$$\frac{\tau^t_{ij}}{\rho} = \nu_t\left(\frac{\partial U_i}{\partial X_j} + \frac{\partial U_j}{\partial X_i}\right) - (2/3)\delta_{ij}k.$$

Here, the eddy viscosity, ν_t, must be modeled. Prandtl (1925) and Kolmogorov (1941) have shown that if the turbulent kinetic energy k and its dissipation rate ϵ are known, then by dimensional analysis, the eddy viscosity can be obtained by

$$\nu_t = C_\mu k^2/\epsilon.$$

For simple flows without separation, such as boundary layer flow and pipe flow, ν_t can be modeled with a simpler model of mixing length with some degree of success. Otherwise, for more general and complex flow careful modelling of k and ϵ must be made. This leads the second-order closure problem that will be discussed in Chapter 2.

With the advance of high-speed computing after World War II, turbulence modeling also advanced. From the late 1950s to the present, turbulence modeling in engineering practice has progressed to the second-order turbulence closure model in which the second-order turbulence transport quantities $\overline{u_i u_j}$ and $\overline{u_i \theta}$ are modeled. In the second-order closure, the $\overline{u_i u_j}$, $\overline{u_i \theta}$, k, and ϵ equations contain many additional unknowns, such as $\overline{u_i u_j u_l}$, $\overline{p u_i/\rho}$, $\overline{p/\rho \partial u_i/\partial x_j}$, $\overline{\nu \partial u_i/\partial x_l \partial u_j/\partial x_l}$, and so forth. These unknowns must, hence, be modeled.

Although complicated and tedious, these models can be potentially more useful and less problem dependent. The second-order closure models provide a higher degree of prediction capability and reduce the postdiction. Moreover, the solution of the second-order closure model is within the capacity of currently available computers.

CHAPTER

TWO

SECOND-ORDER CLOSURE TURBULENCE MODEL

2.1 TURBULENCE MODEL POSTULATIONS

2.1.1 Prospect

Complex turbulent flows exist in many circumstances in our natural environment, as well as in human-made, industrial environments. For example, turbulent flow occurs in oceans, rivers, the atmosphere, and even in the lungs of a human being. Turbulent flows also occur when flying an airplane, driving a car, heating or cooling a house, burning gas/coal in furnaces, and for blood flowing past prosthetic heart valves. Although the Navier–Stokes (N-S) equations can properly describe the details of the turbulent motions, it is too costly and often time consuming for engineers and physicists to solve such complex and detailed equations. Instead, ensemble-averaged N-S equations are often sufficient and practical to describe the turbulent motions in engineering and physical problems. However, in taking an average of the N-S equations for turbulent flow that is three-dimensional, unsteady, random, irregular, and rotational, detailed information about fluid motions is lost. In order to recover the information lost during the averaging process, a turbulence model must be introduced. However, no such unified turbulence model is available at the present time, although some prediction capability has been achieved. The objective of this chapter is to discuss the state of the art of turbulence modeling and to present possible ways of improving turbulence modeling in the future. Without a unified turbulence model that is capable of predicting the averaged turbulent flows, our understanding of turbulence phenomena will remain incomplete and will hamper engineering design.

We shall attempt to formulate a general postulation for a turbulence flow model. Before the postulations are stated, let us consider the following questions. If

- in momentum diffusion (N-S), the shearing stress is a function of the gradient of velocity ($\tau_{ij} \sim \nabla U_i$),
- in heat diffusion, the heat flux is a function of the gradient of temperature ($q_i \sim \nabla T$), and
- in mass diffusion, the mass flux is a function of the gradient of the concentration ($M_i \sim \nabla C$),

then is it logical to expect that

- in turbulent momentum diffusion, $\overline{u_i u_j u_l} + \overline{(p/\rho)(\cdot)} \overset{?}{\sim} \nabla \overline{u_i u_j}$, and
- in turbulent heat diffusion, $\overline{u_i u_j \theta} + \overline{p\theta/\rho} \overset{?}{\sim} \nabla \overline{u_i \theta}$?

Also, if one wishes to understand the turbulent phenomenon, it is important to know the proper scale for turbulent length, l, turbulent time, t, and turbulent velocity, u. This is just as if one is planning for a trip, it is important to know the distance of the trip, the time needed, and the car speed on the road. Among relevant variables of significance are turbulent kinetic energy, k, and the rate at which it is consumed, ϵ. This is just as if the money available for the trip is $\$ = k$, and the rate it is spent is $d\$/dt = \epsilon$. Furthermore, one may ask should there be one, two, or multiple scales? Also, do small eddies behave like isotropic turbulence?

Even though we are not entirely clear, we know that to the first approximation, turbulent diffusion of a transport variable should be proportional to its gradient because if there is no gradient, then there will be no net turbulent diffusion. One-scale concept for turbulent scales (k, ϵ) is predominant in the present models. However, a two-scale concept with one scale (k, ϵ) for large eddies and one for small or dissipating eddies (ϵ, ν) also appears very logical. Also, there is some evidence that small eddies are isotropic.

2.1.2 Stokes–Fourier Postulations (Viscous-Conducting Fluid Model)

The historical development of turbulence modeling can be considered a reflection of the historical development of viscous fluid modeling. The mathematical modeling of fluid motion advanced from the continuum, inviscid fluid model when the Euler equation was derived in 1755 [46] to the viscous, conducting fluid model proposed first by Navier in 1827 [108], and later by Stokes in 1845 [169] for viscous fluid motions, and by Fourier in 1822 [47] for heat conduction. It took from 1755 and almost a century of effort by many scientists, mathematicians, physicists, and engineers to develop the viscous-conducting fluid model. The N-S equation and the Fourier energy equation describe viscous-conducting fluid motions so well that it is often forgotten nowadays that these equations are just mathematical models for these motions.

The postulations made by Navier, Stokes, and Fourier for the derivation of the viscous-conducting fluid model have a strong bearing on the modeling of turbulent flows. These postulations, *known as the Stokes–Fourier Postulates*, can be summarized and rephrased as follows:

1. The fluid can be considered to behave as a continuum and in local-thermal equilibrium. Molecular motions are averaged in continuum assumption, and, hence, detailed information on the dynamics of the molecular collisions is lost. A model is required to recover this lost information (modeling requirement).
2. The diffusion of momentum and thermal energy by viscous fluid motions is proportional to the rate of deformation and temperature gradient, respectively (diffusion gradient model).
3. The fluid is assumed to be isotropic (isotropic molecular collision model).
4. The fluid is assumed to be homogeneous in space and time (τ_{ij} and q_i are not explicit functions of (X_i, t)).
5. When the fluid is at rest, the viscous stress is equal to the hydrostatic pressure (consistency and realizability requirement).
6. When the flow is pure dilative, the average viscous stress is equal to the pressure (Stokes's hypothesis). Mathematically, this means $\tau_{\text{avg}} = \bar{\tau} = (\tau_{xx} + \tau_{yy} + \tau_{zz})/3 = -P$.
7. Viscous-conducting fluid model moduli (density, viscosity, specific heat, thermal conductivity, etc.) require experimental calibration and determination (uniqueness of moduli).

For incompressible fluid with constant transport properties (μ, k, C_p, ρ), the N-S and energy equations are

$$\frac{\partial U_i}{\partial X_i} = 0, \tag{2.1}$$

$$\rho \frac{DU_i}{Dt} = -\frac{\partial P}{\partial X_i} + \mu \frac{\partial^2 U_i}{\partial X_j^2} + \rho G_i, \tag{2.2}$$

and

$$\rho C_p \frac{DT}{Dt} = k \frac{\partial^2 T}{\partial X_j^2} + \mu \Phi. \tag{2.3}$$

The postulation that fluid is a continuum eliminates the need for the description of the intermolecular forces and collisions. In order to recover the information lost about molecular effects, a viscous-conducting fluid model must be introduced and the model moduli, such as viscosity and thermal conductivity, must be calibrated by performing experiments. The same is true when an averaging is imposed on the N-S equations for turbulent motions, the details of turbulence motions are lost. In order to recover the information lost during such an averaging process, a turbulence model must be introduced and turbulence model moduli must be determined from experiments. It should be remarked that the N-S equations are not restricted to laminar fluid motions and should be perfectly capable of describing the turbulent fluid motions. Thus, turbulent flow motions can be predicted by directly solving the N-S equations. This approach is known as *direct simulated turbulent flow motions*.

2.1.3 Turbulence Closure Postulations (Incompressible Fluid)

It has been almost over a century since O. Reynolds introduced in 1895 [128] the time-averaged N-S equations. The averaging process produces a set of turbulent stresses, known as *Reynolds stresses*, which are additional unknown variables. Researchers and scientists have attempted to create a turbulence model for Reynolds stresses to recover the information on turbulent motions lost in the averaging process. In the early twentieth century, because of the lack of computing machines, turbulence models were relatively simple and intuitive. These models were very limited in the sense that the model could apply only to a particular problem or problems with similar geometry and were inappropriate for predicting other types of flows with different geometries. An example of such a model is the mixing length model proposed by L. Prandtl in 1925 [124]. These simple models, in general, have a low-prediction capability. However, they possess the advantage of simplicity and are capable of modeling turbulence with similar flow conditions and geometrical shapes.

With the advance of high-speed computing after World War II, turbulence modeling also advanced. From the late 1950s to the present, turbulence modeling in engineering practice has progressed to the second-order turbulence closure model in which the second-order turbulence transport quantities (viz., the Reynolds stresses $\overline{u_i u_j}$, the Reynolds heat fluxes $\overline{u_i \theta}$, the turbulent kinetic energy k, and the rate of dissipation of the turbulent kinetic energy ϵ) are modeled by differential equations. These variables, in general, are related to the turbulent scale such as the turbulent length l, time t, and velocity u_i. Turbulence models with various combinations of the variables have been proposed; for example, k-$k^{1/2}/l$ (Kolmogorov, 1942 [78]), k-k/l^2 (Spalding, 1982 [165], k-kl (Ng and Spalding, 1972 [115]), and the k-ϵ (Launder and Spalding, 1974 [92]). The k-ϵ has emerged as the most popular turbulence model, mainly because ϵ, the rate of dissipation of k, appears naturally in the k equation and can be derived from the averaged N-S equations without introducing additional variables. Because of its popularity, the following discussion is restricted to the k-ϵ type turbulence models.

The second-order turbulence closure model has greatly improved prediction capability over that of the simple turbulence models. We can extract the basic postulations that are adopted by most turbulence models. These postulations can be analogously formulated along the lines of the Stokes–Fourier postulations for the viscous-conducting fluid model. They are summarized as follows:

1. Ensemble averaged N-S equations and the Fourier energy equation can properly describe the turbulent mean motion and turbulent transport properties. Instantaneous fluid motions are averaged, and detailed information about fluid motions is lost. A model is required to recover the information lost (modeling requirement).
2. The diffusion of turbulent transport properties by turbulence is proportional to the gradient of transport properties (diffusion gradient model).
3. Small turbulent eddies are isotropic (isotropic dissipation model).
4. All turbulent transport quantities are local functions of Reynolds stress, turbulent kinetic energy, rate of dissipation of turbulent kinetic energy, mean-flow variables,

and thermodynamics variables ($\overline{u_i u_j}$, k, ϵ, $\overline{u_i \theta}$, U_i, ρ, P, T) (one-point correlation closure statement).

5. All modeled turbulent phenomena must be consistent in symmetry, invariance, permutation, and physical observations (consistency and realizability requirement).
6. Turbulent phenomenon can be characterized by one turbulent scale (k, ϵ) on the basis of turbulent kinetic energy and its rate of dissipation; that is, $u = k^{1/2}$, $l = k^{3/2}/\epsilon$, and $t = k/\epsilon$ (turbulence scale hypothesis).
7. All turbulence model moduli, C_ϵ, $C_{\epsilon 1}$, $C_{\epsilon 2}$, C_k, C_1, C_2, C_T, C_{T1}, and C_{T2}, require experimental calibration and determination (uniqueness of moduli).

It should be remarked here that if a turbulence model is complete, then the turbulence model moduli are unique and valid for any turbulent flow geometry and conditions. However, at the present time, no such turbulence model is available. Therefore, many variations of turbulence models, and hence model moduli, under the above-indicated postulations are in existence, such as a high-*Re* turbulence model, a low-*Re* turbulence model, a near-wall turbulence model, a two-scale turbulence model, and so forth. Some models make ad hoc adjustments to relax the aforementioned postulations. For example, one approach to modifying Postulation 3 would be to consider that the small turbulent eddies are anisotropic. Also, for example, Postulation 6 could be modified by considering that multiple scales are required to characterize the turbulent phenomenon. For example, we introduce the Kolmogorov scale (ν, ϵ) for the smaller eddies as the second scale, which then may give $u = (\nu\epsilon)^{1/4}$, $l = (\nu^3/\epsilon)^{1/4}$, and $t = (\nu/\epsilon)^{1/2}$. Thus, the k-ϵ scale is used to relate the mean-flow kinetic energy to the dissipation of kinetic energy in the larger eddies, and the Kolmogorov scale (ν, ϵ) is used to relate the kinetic energy dissipation in the larger eddies to the viscous dissipation in the smaller eddies.

2.2 MODELING OF $\overline{u_i u_j}$, k, ϵ, AND $\overline{u_i \theta}$ EQUATIONS

2.2.1 Exact Equation for $\overline{u_i u_j}$

The exact equation for the Reynolds stress $\overline{u_i u_j}$ can be derived as

$$\frac{D\overline{u_i u_j}}{Dt} = \frac{\partial}{\partial X_l}\left(-\overline{u_i u_j u_l} - \overline{\frac{p}{\rho}(\delta_{jl} u_i + \delta_{il} u_j)} + \nu \frac{\partial \overline{u_i u_j}}{\partial X_l}\right)$$
$$- \left(\overline{u_i u_l}\frac{\partial U_j}{\partial X_l} + \overline{u_j u_l}\frac{\partial U_i}{\partial X_l}\right) - 2\nu \overline{\frac{\partial u_i}{\partial X_l}\frac{\partial u_j}{\partial X_l}} + \overline{\frac{p}{\rho}\left(\frac{\partial u_i}{\partial X_j} + \frac{\partial u_j}{\partial X_i}\right)}, \qquad (2.4)$$

wherein

- the terms

$$-\overline{u_i u_j u_l} - \overline{\frac{p}{\rho}(\delta_{jl} u_i + \delta_{il} u_j)} : D_{ij}$$

represent the turbulent diffusion,

- the term

$$\nu \frac{\partial \overline{u_i u_j}}{\partial X_l} : D_\nu$$

represents the molecular diffusion,

- the terms

$$\left(\overline{u_i u_l} \frac{\partial U_j}{\partial X_l} + \overline{u_j u_l} \frac{\partial U_i}{\partial X_l} \right) : P_{ij}$$

represent the production of the Reynolds stresses from the mean flow,

- the term

$$\nu \overline{\frac{\partial u_i}{\partial X_l} \frac{\partial u_j}{\partial X_l}} : \epsilon_{ij}$$

represents the dissipation of the Reynolds stresses through the smaller eddies, and

- the term

$$\overline{\frac{p}{\rho} \left(\frac{\partial u_i}{\partial X_j} + \frac{\partial u_j}{\partial X_i} \right)} : \Phi_{ij} \text{ or PS}$$

represent the pressure-strain, that is, the tendency to return to isotropicity by the redistribution of Reynolds stresses.

2.2.2 Modeling of the $\overline{u_i u_j}$ Equations

Modeling of the diffusion term. The diffusion term is modeled according to Postulations 1 and 2. Hence, we take it proportional to the gradient of $\overline{u_i u_j}$, or

$$-\overline{u_i u_j u_l} - \overline{\frac{p}{\rho}(\delta_{jl} u_i + \delta_{il} u_j)} = C_k \left(\frac{l^2}{t} \right) \frac{\partial \overline{u_i u_j}}{\partial X_l} = C_k \frac{k^2}{\epsilon} \frac{\partial \overline{u_i u_j}}{\partial X_l}.$$

In order to keep the dimensions consistent, we need to have (l^2/t) attached to the gradient of $\overline{u_i u_j}$. From the dimensional analysis of (k, ϵ), we have $u = k^{1/2}$, $l = k^{3/2}/\epsilon$, and $t = k/\epsilon$, which implies that $(l^2/t) = k^2/\epsilon$. Because C_k is a dimensionless scalar coefficient, k^2/ϵ is a scalar. $C_k k^2/\epsilon$ simulates isotropic diffusivity and is independent of direction.

For nonisotropic diffusivity, one could consider several alternatives, such as

- $\frac{k}{\epsilon}(\overline{u_i u_l} + \overline{u_j u_l}) \frac{\partial \overline{u_i u_j}}{\partial X_l}$,
- $\frac{k}{\epsilon} \overline{u_i u_j} \frac{\partial \overline{u_i u_j}}{\partial X_l}$,
- $\frac{k}{\epsilon} \overline{u_l^2} \frac{\partial \overline{u_i u_j}}{\partial X_l}$, or
- $\frac{k}{\epsilon} \overline{u_l u_k} \frac{\partial \overline{u_i u_j}}{\partial X_k}$.

It should be noted that all of these models retain symmetry in i and j. However, there is no symmetry in (i, l) and (j, l). If one requires the turbulent diffusivity (i.e., k^2/ϵ, $(\overline{u_i u_l} + \overline{u_j u_l})(k/\epsilon)$, $\overline{u_i u_l} k/\epsilon$, or $\overline{u_l^2} k/\epsilon$) to be positive definite, then the only possible candidates are k^2/ϵ and $\overline{u_l^2}k/\epsilon$. The latter provides a sense of direction and nonisotropic diffusivity. Although nonisotropic diffusivity appears more logical, the isotropic diffusivity model is popular in use because of its simplicity. If one wishes to model only $\overline{u_i u_j u_l}$, then by symmetry, we would have

$$-\overline{u_i u_j u_l} = C_s(t)\left(\overline{u_i u_k}\frac{\partial \overline{u_j u_l}}{\partial X_k} + \overline{u_j u_k}\frac{\partial \overline{u_l u_i}}{\partial X_k} + \overline{u_l u_k}\frac{\partial \overline{u_i u_j}}{\partial X_k}\right),$$

which was suggested by Hanjalic and Launder (1972)[55], with the value of C_s set equal to 0.11, and where (t) denotes a time scale. However, when the fluctuating pressure-velocity term in diffusion is combined with $\overline{u_i u_j u_l}$, the first two terms of the above-indicated expression must be eliminated because no symmetry exists in (i, l) and (j, l). The difference in the modeling of the turbulent diffusion is slight and, as pointed out by Launder (1986)[84], insignificant. Few of the anomalies that one finds in the predictions of inhomogeneous flows can be traced back to the weakness in the diffusion model.

Modeling of the dissipation term. The dissipation term (ϵ_{ij}) is modeled according to Postulation 3 (isotropic dissipation). Thus,

$$2\nu\overline{\frac{\partial u_i}{\partial X_l}\frac{\partial u_j}{\partial X_l}} = \frac{2}{3}\delta_{ij}\epsilon,$$

where

$$\epsilon = \nu\overline{\frac{\partial u_i}{\partial X_l}\frac{\partial u_i}{\partial X_l}} = \nu\overline{\frac{\partial u_n}{\partial X_m}\frac{\partial u_n}{\partial X_m}}.$$

According to Postulation 3, the dissipation rate becomes zero or uncorrelated for the non isotropic terms (i.e., for $i \neq j$). Hence, the Kronecker delta, δ_{ij}, is introduced to account for the dissipation becoming zero for nonisotropic terms.

As the local Reynolds number approaches zero, for instance, in the nearwall region, the energy content of k and the dissipation range of the motions overlap, ϵ and the dissipation rate ϵ_{ij} can then be approximated (Rotta, 1951)[143] as

$$2\nu\overline{\frac{\partial u_i}{\partial X_l}\frac{\partial u_j}{\partial X_l}} = 2\frac{\sqrt{\overline{u_i^2}\,\overline{u_j^2}}}{\frac{2k}{3}}\frac{\epsilon}{3}$$

or,

$$\frac{\sqrt{\overline{u_i^2 u_j^2}}}{k}\epsilon = \frac{\overline{u_i u_j}}{k}\epsilon. \tag{2.5}$$

This is consistent with Postulation 6; that is, when $i = j$ and summing from $i = 1$ to $i = 3$, the model reduces to 2ϵ. The symmetry postulation and the positive definiteness postulation are also obeyed.

Launder has pointed out that the model $\overline{u_i u_j}(\epsilon/k)$ is not totally correct as it underestimates the value of ϵ_{22}, the direction normal to the wall, by a factor of four. Although highly anisotropic, near-the-wall turbulence approaches two-dimensional behavior. Using $i = j = 2$ for the normal directions, Launder (1986)[84] showed the limiting behavior of the dissipation terms ϵ_{ij} for each individual stress equation to be equal to

$$\frac{\epsilon_{11}}{\overline{u^2}} = \frac{\epsilon_{33}}{\overline{w^2}} = \frac{\epsilon_{22}}{4\overline{v^2}} = \frac{\epsilon_{12}}{2\overline{uv}} = \frac{\epsilon}{k}.$$

However, the model $\epsilon_{ij} = \overline{u_i u_j}(\epsilon/k)$ gives

$$\frac{\epsilon_{11}}{\overline{u^2}} = \frac{\epsilon_{33}}{\overline{w^2}} = \frac{\epsilon_{22}}{\overline{v^2}} = \frac{\epsilon_{12}}{\overline{uv}} = \frac{\epsilon}{k}.$$

A more accurate model was put forward by Launder and Reynolds (1983)[89] and is given as follows:

$$\epsilon_{ij} = (\epsilon/k)(\overline{u_i u_j} + \overline{u_i u_k} n_k n_j + \overline{u_j u_k} n_k n_i + \delta_{ij} \overline{u_k u_l} n_k n_l)/(1 + 5\overline{u_p u_q} n_p n_q / 2k), \quad (2.6)$$

where n_i denotes the normal to the surface i.

The asymptotic form of Eq. 2.6 have led a number of workers to propose that, in general, the correlation ϵ_{ij} may be approximated as

$$\epsilon_{ij} = \frac{2}{3} f_\epsilon \delta_{ij} \epsilon + (1 - f_\epsilon)\epsilon_{ij}^w, \quad (2.7)$$

where ϵ_{ij}^w denotes the anisotropic dissipation $\overline{u_i u_j}\epsilon/k$, and f_ϵ is a function of the turbulent Reynolds number R_T $(= k^2/(\nu\epsilon))$. Various functions of $f_\epsilon(R_T)$ have been proposed (Hanjalic and Launder, 1976 [56]), and recently Fu, Launder, and Tselepidakis (1987)[49] have suggested setting $f_\epsilon = (A)^{1/2}$, where A is an anisotropic parameter

$$A = \left(1 - \frac{9}{8}(a_{ij} a_{ij} - a_{ij} a_{jk} a_{ki})\right),$$

with

$$a_{ij} = \left(\frac{\overline{u_i u_j}}{k} - \frac{2}{3}\delta_{ij}\right),$$

which takes the value of zero in two-dimensional turbulence and of one in isotropic turbulence. However, Eqs. 2.6 and 2.7 have not been so widely used, perhaps because of their complex expressions. Alternatively, $\epsilon_{ij} = \overline{u_i u_j}\epsilon/k$ is an improvement over the simple isotropic dissipation $\epsilon_{ij} = 2\delta_{ij}\epsilon/3$ and is more frequently adopted. Although the form $\epsilon_{ij} = \overline{u_i u_j}\epsilon/k$ may underestimate ϵ_{22} by a factor of four and ϵ_{12} by a factor of two, this is not crucial because $\overline{v^2}/k$ and $\overline{uv}/k$ go to zero at the wall so that both ϵ_{22} and ϵ_{12} become zero there.

Modeling of the pressure-strain (PS) term. Let us consider a possible and simple model first.

$$\overline{\frac{p}{\rho}\left(\frac{\partial u_i}{\partial X_j} + \frac{\partial u_j}{\partial X_i}\right)} = -C_1 \frac{\epsilon}{k}\left(\overline{u_i u_j} - \frac{2}{3}\delta_{ij} k\right) - C_2\left(P_{ij} - \frac{2}{3}\delta_{ij} P_k\right),$$

where

$$P_{ij} = -\left(\overline{u_i u_l}\frac{\partial U_j}{\partial X_l} + \overline{u_j u_l}\frac{\partial U_i}{\partial X_l}\right),$$

and

$$P_k = -\overline{u_n u_m}\frac{\partial U_n}{\partial X_m}.$$

The reason for this modeling is as follows.

We have the momentum equations for the turbulent fluctuation u_i from Eq. 1.16 as

$$\frac{Du_i}{Dt} + u_l\frac{\partial U_i}{\partial X_l} + u_l\frac{\partial u_i}{\partial X_l} - \frac{\partial \overline{u_i u_l}}{\partial X_l} = -\frac{1}{\rho}\frac{\partial p}{\partial X_l} + \nu\frac{\partial^2 u_i}{\partial X_l \partial X_l}. \tag{2.8}$$

These can be derived by subtracting the averaged N-S equations from the instantaneous N-S equations. Taking the divergence of Eq. 2.8, we obtain

$$\nabla^2\frac{p}{\rho} = -\left[\frac{\partial^2(u_l u_m - \overline{u_l u_m})}{\partial X_l \partial X_m} + 2\frac{\partial U_l}{\partial X_m}\frac{\partial u_m}{\partial X_l}\right].$$

On integrating over a relatively large volume, and applying Green's theorem, we have

$$\frac{p}{\rho} = \frac{1}{4\pi}\int_{\text{vol}}\left[\frac{\partial^2(u_l u_m - \overline{u_l u_m})}{\partial X_l \partial X_l} + 2\frac{\partial U_l}{\partial X_m}\frac{\partial u_m}{\partial X_l}\right]\frac{\text{dvol}}{\boldsymbol{r}},$$

where $\boldsymbol{r}$ is a vector from the location of Point P to the various points within the volume.

Multiplying the above-indicated equation by

$$\left(\frac{\partial u_i}{\partial X_j} + \frac{\partial u_j}{\partial X_i}\right)$$

and taking the average, we have the PS expression

$$\overline{\frac{p}{\rho}\left(\frac{\partial u_i}{\partial X_j} + \frac{\partial u_j}{\partial X_i}\right)} = \frac{1}{4\pi}\int_{\text{vol}}\overline{\left[\left(\frac{\partial^2 u_l u_m}{\partial X_l \partial X_m}\right)^*\left(\frac{\partial u_i}{\partial X_j} + \frac{\partial u_j}{\partial X_i}\right)\right]}\frac{\text{dvol}}{\boldsymbol{r}^*}$$
$$+ \frac{1}{4\pi}\int_{\text{vol}}\left[2\overline{\frac{\partial U_l^*}{\partial X_m}\frac{\partial u_m^*}{\partial X_l}\left(\frac{\partial u_i}{\partial X_j} + \frac{\partial u_j}{\partial X_i}\right)}\right]\frac{\text{dvol}}{\boldsymbol{r}^*}.$$

In the above-indicated expression, terms without the superscript * are evaluated at the location of Point P, whereas the terms with the superscript are evaluated at $\boldsymbol{r}^*$ away from Point P.

In short, we write the PS term as

$$(\text{PS}) = \Phi_{ij,1} + \Phi_{ij,2}.$$

The first part ($\Phi_{ij,1}$) involves only the fluctuating velocities and is known as slow distortion or slow return to isotropic state. The second part ($\Phi_{ij,2}$) involves the mean velocity gradients and is known as rapid distortion or rapid return to isotropic state.

Recall that when $i = j$, (PS) becomes zero because of the incompressibility requirement,

$$\frac{\partial u_i}{\partial x_i} = 0.$$

Thus, $\Phi_{ii,1} = 0$ and $\Phi_{ii,2} = 0$.

The modeling of $\Phi_{ij,1}$ starts out by shrinking the integral expression down to a small volume, say l^3, on the basis of Postulation 4 to be a one-point correlation. Hence,

$$\Phi_{ij,1} = \frac{1}{4\pi}\int_{\text{vol}} \overline{\left(\frac{\partial^2 u_l u_m}{\partial X_l \partial X_m}\right)^* \left(\frac{\partial u_i}{\partial X_j} + \frac{\partial u_j}{\partial X_i}\right)} \frac{\text{dvol}}{r^*}$$

$$\approx \text{constant}\overline{\left(\frac{\partial^2 u_l u_m}{\partial X_l \partial X_m}\right)\left(\frac{\partial u_i}{\partial X_j} + \frac{\partial u_j}{\partial X_i}\right)} \frac{l^3}{l}$$

Here l may be the size of the small eddies. Note that because $\Phi_{ii,1} = 0$, and that Postulations 5 and 6 require $\Phi_{ij,1} = \mathcal{F}\ (\overline{u_i u_j}, k, \epsilon, \text{etc.})$, this can approximately be modeled as

$$\Phi_{ij,1} = \text{constant}\frac{1}{t}\left(\overline{u_i u_j} - \frac{2}{3}\delta_{ij}k\right).$$

From a dimensional point of view, we can write t as equivalent to (k/ϵ) on the basis of the (k, ϵ) scale. Hence,

$$\Phi_{ij,1} = -C_1\frac{\epsilon}{k}\left(\overline{u_i u_j} - \frac{2}{3}\delta_{ij}k\right) = -C_1 \epsilon a_{ij}, \tag{2.9}$$

wherein

$$a_{ij} = \left(\frac{\overline{u_i u_j}}{k} - \frac{2}{3}\delta_{ij}\right).$$

This model was proposed by Rotta (1951)[143] and is referred to as the *linear model* because $\Phi_{ij,1}$ is linearly proportional to $\overline{u_i u_j}$. The negative sign for the equation is an indication that when the difference $(\overline{u_i u_j} - (2/3)\delta_{ij}k)$ is greater than zero, the $\Phi_{ij,1}$ term promotes isotropy or return to isotropy.

Bradshaw, Ferriss, and Johnson (1964)[7] have commented that the real process of stress redistribution is extremely nonlinear and that any approximation for $\Phi_{ij,1}$ should mirror this nonlinearity. Lumley and Khajeh-Nouri (1973)[100] have added terms up to and including third-order Reynolds stress products, with the resulting formulation expressed as

$$\Phi_{ij,1} = -\left(C_1 + C_1'' a_{ij} a_{ij}\right)\epsilon a_{ij} - C_1'\epsilon\left(a_{im}a_{jm} - \frac{1}{3}\delta_{ij}a_{ij}a_{ij}\right),$$

though later on they concluded that the second-order term should be equal to zero. Launder, Reece, and Rodi (1975)[88] concluded from computations of nine turbulent shear flows that there was no advantage in choosing a nonzero value for the coefficient C_1''.However, their conclusion has been somewhat revised by the discovery that

Table 2.1 Models of slow pressure-strain term $\Phi_{ij,1}$

Authors	Models
Rotta (1951)[143]	$\Phi_{ij,1} = -C_1(\epsilon/k)(\overline{u_i u_j} - 2\delta_{ij}k/3)$
Lumley and Khajeh Nouri (1973)[100]	$\Phi_{ij,1} = -(C_1 + C_1'' a_{ij}a_{ij})\epsilon a_{ij} - C_1'\epsilon(a_{im}a_{jm} - \delta_{ij}a_{ij}a_{ij}/3)$

in the turbulent heat flux equation, it is the counterpart process of $\Phi_{ij,1}$ that could not adequately be represented by a linear expression like Eq. 2.9. Some of the proposals for closing the Reynolds stress equation have assumed that $\Phi_{ij,1}$ is the only significant contributor to the PS term (Rotta, 1951 [143]; Donaldson, 1968 [39]; Daly and Harlow, 1970 [37]). This conclusion is supported when one makes predictions over only a narrow range of turbulent shear flows because the effects of omitting $\Phi_{ij,2}$ can then be absorbed either through the value ascribed to C_1 or through the manner in which the dissipation term is simulated. Reynolds (1970)[129] have shown, however, that prediction of a range of even homogeneous free turbulent flows demands the inclusion of mean-strain rates $\Phi_{ij,2}$ in the PS terms. Moreover, both Crow (1968)[36] and Townsend (1954)[176] have shown that under conditions of rapid distortion, the effect of $\Phi_{ij,2}$ far outweighs that of $\Phi_{ij,1}$. Various models and coefficients of $\Phi_{ij,1}$ are listed in Tables 2.1 and 2.2.

It is found that there are bewildering variations of the C_1 value with no readily apparent trends. However, the two largest values in Table 2.2 are associated with proposals that neglected using $\Phi_{ij,2}$. It is safe to say that although there are many proposals for $\Phi_{ij,1}$, none is particularly persuasive (Launder, 1989 [85]). Some researchers (Lee and Reynolds, 1985 [93]; Weinstock and Burk, 1985 [183]) advocate a different basis for approximating this process, such as linking the $\Phi_{ij,1}$ term to the anisotropic dissipation tensor, ϵ_{ij}, rather than just modeling it as the anisotropy of the stress field.

Usually $\Phi_{ij,2}$ is modeled as follows:

$$\Phi_{ij,2} = \frac{1}{4\pi}\int_{\mathrm{vol}} \overline{2\frac{\partial U_l^*}{\partial X_m}\frac{\partial u_m^*}{\partial X_l}\left(\frac{\partial u_i}{\partial X_j} + \frac{\partial u_j}{\partial X_i}\right)}\frac{\mathrm{dvol}}{\boldsymbol{r}^*}.$$

Table 2.2 Model coefficients of C_1

Authors	C_1	C_1'	C_1''	$\Phi_{ij,2}$ included?
Rotta (1951)[143]	1.4			Yes
Rotta (1962)[144]	3.0			Yes
Daly and Harlow (1970)[37]	$1.5 + P_k/\epsilon$			No
Hanjalic and Launder (1972)[55]	2.8			Yes
Donaldson (1972)[40]	5.0			No
Launder et al. (1975)[88]	1.5			Yes
Lumley and Khajeh Nouri (1973)[100]	4.0			No
Lumley and Khajeh Nouri (1974)[101]	1.6	0	3.0	Yes
Reynolds (1974)[174]	2.5	0		Yes

Shrinking again to a small volume, and using the similar arguments as used for simplifying the expression for $\Phi_{ij,1}$, we obtain

$$\Phi_{ij,2} \approx \text{constant}\overline{\frac{\partial U_l}{\partial X_m}\frac{\partial u_m}{\partial X_l}\left(\frac{\partial u_i}{\partial X_j}+\frac{\partial u_j}{\partial X_i}\right)}\frac{l^3}{l}$$

$$= \text{constant}\left(\frac{\partial U_j}{\partial X_m}\overline{u_m u_i}+\frac{\partial U_i}{\partial X_m}\overline{u_m u_j}-\frac{2}{3}\delta_{ij}\frac{\partial U_n}{\partial X_m}\overline{u_m u_n}\right)$$

$$= -C_2\left(P_{ij}-\frac{2}{3}\delta_{ij}P_k\right),$$

wherein

$$P_{ij} = -\left(\overline{u_i u_m}\frac{\partial U_j}{\partial X_m}+\overline{u_j u_m}\frac{\partial U_i}{\partial X_m}\right),$$

and

$$P_k = -\overline{u_n u_m}\frac{\partial U_n}{\partial X_m}.$$

The model indicated above is one of simple models. A more general model, which is based on consistency in tensorial arguments, is

$$\Phi_{ij,2} = a_{lj}^{mi}\left(\frac{\partial U_m}{\partial X_l}+\frac{\partial U_l}{\partial X_m}\right), \tag{2.10}$$

where a_{lj}^{mi} is a fourth-order tensor. Hanjalic and Launder (1972)[55] used a model that involved linear and quadratic terms in the Reynolds stress and wrote a_{lj}^{mi} as

$$\begin{aligned} a_{lj}^{mi} &= \alpha\overline{u_m u_i}\delta_{lj}+\beta(\overline{u_m u_l}\delta_{ij}+\overline{u_m u_j}\delta_{il}+\overline{u_i u_j}\delta_{ml}+\overline{u_i u_l}\delta_{mj}) \\ &\quad +(\gamma\delta_{mi}\delta_{lj}+\sigma[\delta_{ml}\delta_{ij}+\delta_{mj}\delta_{il}])k+C_2(\overline{u_m u_i u_l u_j})/k \\ &\quad +\nu(\overline{u_m u_j}\,\overline{u_i u_l}+\overline{u_m u_l}\,\overline{u_i u_j})/k, \end{aligned}$$

with all the coefficients (α, β, γ, σ, and ν) written in terms of C_2 as $\alpha = (10-8C_2)/11$, $\beta = -(2-6C_2)/11$, $\gamma = -(4-12C_2)/55$, $\sigma = (6-18C_2)/55$, and $\nu = -C_2$, with $C_2 = 0.45$.

The reason Hanjalic and Launder (1972)[55] included the quadratic Reynolds-stress terms in their model was because $C_1 = 2.8$. Without including the quadratic Reynolds stress terms, the prediction of a homogeneous shear flow with their model led to very small differences between the normal stresses as compared with the experimental data of Champagne, Harris, and Corrsin (1970)[9]. However, as Launder et al. (1975)[88] pointed out, their model failed to satisfy a kinematic constraint, $a_{jj}^{mi} = 2\overline{u_m u_i}$, and was therefore expected to give specious effects in complex strain fields. Hence, Launder et al. (1975)[88] reduced the C_1 value to 1.5 and dropped the quadratic terms and modeled a_{lj}^{mi} as

$$\begin{aligned} a_{lj}^{mi} &= \alpha\overline{u_m u_i}\delta_{lj}+\beta(\overline{u_m u_l}\delta_{ij}+\overline{u_m u_j}\delta_{il}+\overline{u_i u_j}\delta_{ml}+\overline{u_i u_l}\delta_{mj}) \\ &\quad +(\gamma\delta_{mi}\delta_{lj}+\sigma[\delta_{ml}\delta_{ij}+\delta_{mj}\delta_{il}])k+C_2\delta_{mi}\overline{u_l u_j}, \end{aligned} \tag{2.11}$$

wherein $\alpha = (10 + 4C_2)/11$, $\beta = -(2 + 3C_2)/11$, $\gamma = -(4 + 50C_2)/55$, and $\sigma = (6 + 20C_2)/55$, with $C_2 = 0.4$.

On combining Eqs. 2.10 and 2.11, the complete influence of the mean strain on the PS correlation may be expressed in the following compact form:

$$\Phi_{ij.2} = -\frac{C_2 + 8}{11}\left(P_{ij} - \frac{2}{3}\delta_{ij}P_k\right) - \frac{30C_2 - 2}{55}k\left(\frac{\partial U_i}{\partial X_j} + \frac{\partial U_j}{\partial X_i}\right) - \frac{8C_2 - 2}{11}\left(D_{ij} - \frac{2}{3}\delta_{ij}P_k\right), \tag{2.12}$$

where

$$P_{ij} = -\left(\overline{u_i u_k}\frac{\partial U_j}{\partial X_k} + \overline{u_j u_k}\frac{\partial U_i}{\partial X_k}\right),$$

$$D_{ij} = -\left(\overline{u_i u_k}\frac{\partial U_k}{\partial X_j} + \overline{u_j u_k}\frac{\partial U_k}{\partial X_i}\right),$$

and

$$P_k = -\overline{u_i u_j}\frac{\partial U_i}{\partial X_j}.$$

Here, P_k denotes the rate of production of turbulent stress.

Launder et al. (1975)[88] also pointed out that the first group on the right-hand side of Eq. 2.12 turns out to be the dominant one. Moreover, because each of the three groups vanishes under contraction, one may retain simply the first group without causing any loss to the essential redistributive nature of the approximation. Some computations with such a simplified version, $\Phi_{ij.2} = -\chi(P_{ij} - 2\delta_{ij}P_k/3)$, with the value of chi being assigned a magnitude different from the coefficient of the first term to compensate in part for the neglected terms, have been reported by Launder et al. (1975)[88] and Donaldson (1972)[40]. The same model was also later adopted by Gibson and Launder (1978)[51] with chi set equal to 0.6. However, this model gives spectacularly wrong predictions of the effects of swirl on the spreading rate of an axisymmetric jet (Launder and Morse, 1977 [86]).

More complex $\Phi_{ij.2}$ models were proposed recently. Limiting attention to terms strictly quadratic in a_{ij} (Shih and Lumley, 1985 [151]; Fu et al., 1987 [49]) gives

$$\Phi_{ij.2} = -0.6\left(P_{ij} - \frac{2}{3}\delta_{ij}P_k\right) + 0.3\epsilon a_{ij}\left(\frac{P_k}{2\epsilon}\right) - 0.2\left[\frac{\overline{u_k u_j}\,\overline{u_l u_i}}{k}\left(\frac{\partial U_k}{\partial X_l} + \frac{\partial U_l}{\partial X_k}\right) - \frac{\overline{u_l u_k}}{k}\left(\overline{u_i u_k}\frac{\partial U_j}{\partial X_l} + \overline{u_j u_k}\frac{\partial U_i}{\partial X_l}\right)\right]. \tag{2.13}$$

However, in a simple shear flow in local equilibrium, this model gives values of $\overline{u_2^2}$ greater than $\overline{u_3^2}$, the reverse of what the experiments indicate. This weakness was removed by Shih, Lumley, and Chen (1985)[152] by the addition of a corrective term, $\Phi^c_{ij.2}$,

$$\Phi^c_{ij.2} = -0.8\sqrt{A}\left[\frac{1}{15}\left(P_{ij} - \frac{2}{3}\delta_{ij}P_k\right) + \frac{8}{15}\left(D_{ij} - \frac{1}{3}\delta_{ij}D_{kk}\right) + \frac{2}{5}k\left(\frac{\partial U_k}{\partial X_l} + \frac{\partial U_l}{\partial X_k}\right)\right]. \tag{2.14}$$

Fu et al. (1987)[49] set one of the terms to zero to reduce the resultant expression to the relatively simple form

$$
\begin{aligned}
\Phi_{ij,2} = & -0.6\left(P_{ij} - \frac{2}{3}\delta_{ij}P_k\right) + 0.3\epsilon a_{ij}\left(\frac{P_k}{2\epsilon}\right) \\
& - 0.2\left[\frac{\overline{u_k u_j}\,\overline{u_l u_i}}{k}\left(\frac{\partial U_k}{\partial X_l} + \frac{\partial U_l}{\partial X_k}\right) - \frac{\overline{u_l u_k}}{k}\left(\overline{u_i u_k}\frac{\partial U_j}{\partial X_l} + \overline{u_j u_k}\frac{\partial U_i}{\partial X_l}\right)\right] \\
& - 0.6[A_2(P_{ij} - D_{ij}) + 3a_{mi}a_{nj}(P_{mn} - D_{mn})]
\end{aligned}
\tag{2.15}
$$

where A_2 is the second invariant; that is, $A_2 = a_{ij}a_{ji}$. Because $(D_{ij} - P_{ij})$ is proportional to the vorticity tensor, the extra term in the equation indicated above may explicitly be regarded as a correction for mean flow rotation. According to Craft, Fu, Launder, and Tselepidakis (1989)[35], this form brings a good deal more subtlety and variety to the effects of even a simple shear on the stress field.

Various combinations of the models listed above have been used in the literature (Amano and Goel, 1984 [3]; Martinuzzi and Pollard, 1989 [104]; Brankovic and Syed, 1991 [8]). In general, application of these models indicates a much slower return to isotropy than indicated in experiments. However, different components return at decidedly different rates. It can be concluded that current stress models still do not perform very well in handling the return to isotropy. They may however work well in flows dominated by other effects. Most of the works to date in which the differential stress model is used tend to adopt the simpler version of the model for the rapid return to isotropy, that is,

$$
\Phi_{ij,2} = -\chi\left(P_{ij} - \frac{2}{3}\delta_{ij}P_k\right). \tag{2.16}
$$

Thus, the simple modeled $\overline{u_i u_j}$ equation is given as

$$
\begin{aligned}
\frac{D\overline{u_i u_j}}{Dt} = & \frac{\partial}{\partial X_l}\left(C_k\frac{k^2}{\epsilon}\frac{\partial\overline{u_i u_j}}{\partial X_l} + \nu\frac{\partial\overline{u_i u_j}}{\partial X_l}\right) + P_{ij} - \frac{2}{3}\delta_{ij}\epsilon \\
& - C_1\frac{\epsilon}{k}\left(\overline{u_i u_j} - \frac{2}{3}\delta_{ij}k\right) - C_2\left(P_{ij} - \frac{2}{3}\delta_{ij}P_k\right).
\end{aligned}
\tag{2.17}
$$

In this equation, the various coefficients are determined by using experiments and have been found to be $C_k = 0.09 \sim 0.11$, $C_1 = 1.5 \sim 2.2$, and $C_2 = 0.4 \sim 0.5$. The determination of these constants will be discussed in the next section. In short, the modeled $\overline{u_i u_j}$ equation may also be written as

$$
\frac{D\overline{u_i u_j}}{Dt} = D_{ij} + P_{ij} - \frac{2}{3}\delta_{ij}\epsilon + \mathrm{PS},
$$

wherein D_{ij} denotes the diffusion term of the $\overline{u_i u_j}$ equation, P_{ij} is the Reynolds stress production term, ϵ is the rate of dissipation of the turbulent kinetic energy, and PS is the pressure-strain term.

2.2.3 Modeling of k Equation

The k equation requires no additional modeling when the $\overline{u_i u_j}$ equations are modeled. As mentioned in the previous section, we can write the modeled $\overline{u_i u_j}$ equation as

$$\frac{D\overline{u_i u_j}}{Dt} = D_{ij} + P_{ij} - \frac{2}{3}\delta_{ij}\epsilon + \mathrm{PS}. \tag{2.18}$$

Taking $i = j$ and summing from $i = 1$ to 3 and letting $k = \overline{u_i u_i}/2$, we can rewrite this equation as

$$\frac{Dk}{Dt} = \frac{\partial}{\partial X_l}\left(C_k \frac{k^2}{\epsilon}\frac{\partial k}{\partial X_l} + \nu \frac{\partial k}{\partial X_l}\right) - \overline{u_i u_l}\frac{\partial U_i}{\partial X_l} - \epsilon + 0(= \mathrm{PS}), \tag{2.19}$$

where $C_k = 0.09 \sim 0.11$. In short, we may write

$$\frac{Dk}{Dt} = D_k + P_k - \epsilon.$$

Note that in the k equation, the D_k term is the only term that is modeled. Hence, the k equation may be considered to be relatively accurate as compared with the $\overline{u_i u_j}$ or the ϵ equations.

For the case of uniform flow (i e., $U_i =$ constant), we have

$$\frac{Dk}{Dt} = -\epsilon, \qquad \epsilon > 0. \tag{2.20}$$

Thus, turbulence always decays in uniform flow (see Fig. 2.1).

For understanding the relation between k and ϵ, let us consider the analogous example of the money earned versus the money spent by a person. If this person, Mr. Turbulence, earns k amount of money, then ϵ would represent the time rate at which this money gets spent. Hence, if we have a good model to estimate both k and ϵ, then the turbulent phenomena can be approximately simulated. In other words, turbulent

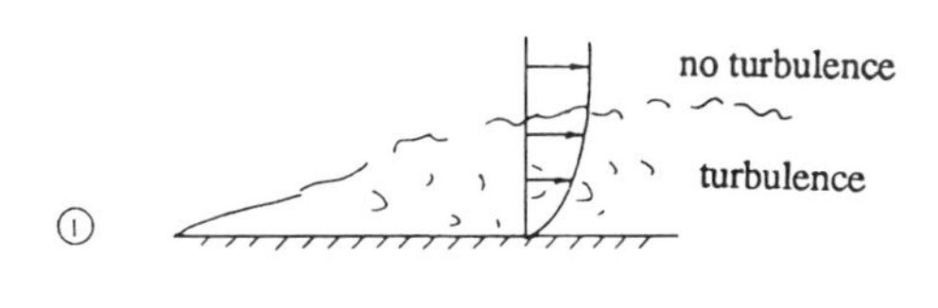

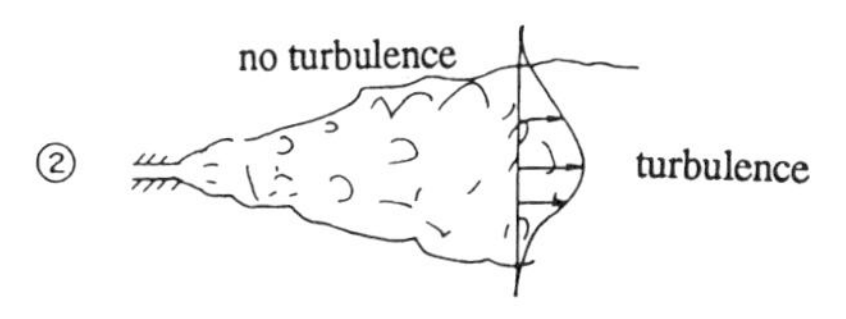

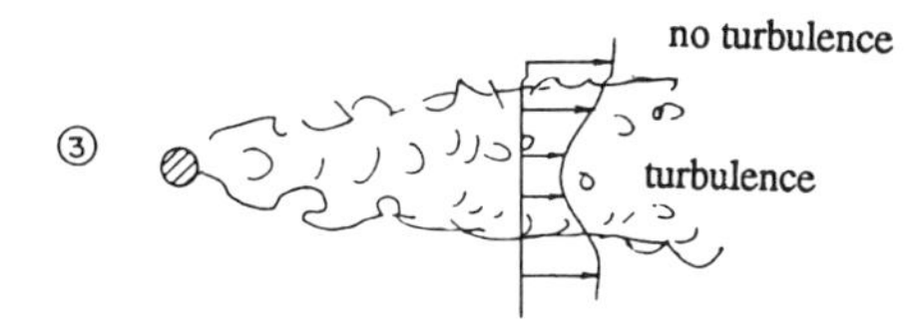

Figure 2.1 Examples of turbulence decay in uniform flow.

kinetic energy and its dissipation rate are the two most important quantities governing the phenomenon of turbulence.

2.2.4 Modeling of ϵ Equation

Remarks and exact ϵ equation. Because the ϵ term arises naturally in the $\overline{u_i u_j}$ and k equations as an additional unknown, it needs to be modeled. If Dk/Dt is the rate of money (k) earned, then ϵ is the rate of money (k) spent (or dissipated). Hence, if k and ϵ are properly modeled and predicted, then the state of turbulence is approximately solved.

The exact ϵ equation is

$$\frac{D\epsilon}{Dt} = \frac{\partial}{\partial X_l}\left(-\overline{\epsilon' u_l} - \frac{2\nu}{\rho}\overline{\frac{\partial u_l}{\partial X_j}\frac{\partial p}{\partial X_j}} + \nu\frac{\partial \epsilon}{\partial X_l}\right) - 2\nu\frac{\partial U_i}{\partial X_j}\left(\overline{\frac{\partial u_l}{\partial X_i}\frac{\partial u_l}{\partial X_j}} + \overline{\frac{\partial u_i}{\partial X_l}\frac{\partial u_j}{\partial X_l}}\right)$$
$$- 2\nu\overline{u_l\frac{\partial u_i}{\partial X_j}}\frac{\partial^2 U_i}{\partial X_l \partial X_j} - 2\nu\overline{\frac{\partial u_i}{\partial X_j}\frac{\partial u_i}{\partial X_l}\frac{\partial u_j}{\partial X_l}} - 2\overline{\left(\nu\frac{\partial^2 u_i}{\partial X_l \partial X_j}\right)^2}, \tag{2.21}$$

wherein

- the term

$$-\overline{\epsilon' u_l} - \frac{2\nu}{\rho}\overline{\frac{\partial u_l}{\partial X_j}\frac{\partial p}{\partial X_j}}$$

  represents the turbulent diffusion of ϵ,
- the term

$$\nu\frac{\partial \epsilon}{\partial X_l}$$

  represents the molecular diffusion of ϵ,
- the terms

$$2\nu\frac{\partial U_i}{\partial X_j}\left(\overline{\frac{\partial u_l}{\partial X_i}\frac{\partial u_l}{\partial X_j}} + \overline{\frac{\partial u_i}{\partial X_l}\frac{\partial u_j}{\partial X_l}}\right) + 2\nu\overline{u_l\frac{\partial u_i}{\partial X_j}}\frac{\partial^2 U_i}{\partial X_l \partial X_j}$$

  represent the production of ϵ, whereas
- the terms

$$2\nu\overline{\frac{\partial u_i}{\partial X_j}\frac{\partial u_i}{\partial X_l}\frac{\partial u_j}{\partial X_l}} + 2\overline{\left(\nu\frac{\partial^2 u_i}{\partial X_l \partial X_j}\right)^2}$$

  represent the destruction of ϵ.

Each term on the right-hand side of Eq. 2.21 must be modeled. This is not good as there will be too much modeling and, hence, too much uncertainty.

A complicated question to ask is what turbulent scale should be used in modeling the ϵ equation or each term in the ϵ equation. Is it the (k, ϵ) scale, some other scale such as the (ν, ϵ) scale developed by Kolmogorov, or a combination of these scales? In the following discussion, we first develop one-scale modeling on the basis of energy-containing scale (k, ϵ).

Modeling of the diffusion term. This is the first term on the right-hand side of Eq. 2.21 and is modeled in the following manner on the basis of Postulation 2:

$$-\overline{\epsilon' u_l} - \overline{\frac{2\nu}{\rho}\frac{\partial u_l}{\partial X_j}\frac{\partial p}{\partial X_j}} = C_\epsilon \left(\frac{l^2}{t}\right)\frac{\partial \epsilon}{\partial X_i}$$
$$= C_\epsilon \frac{k^2}{\epsilon}\frac{\partial \epsilon}{\partial X_i}.$$

From the (k, ϵ) scale, $l = k^{3/2}/\epsilon$, $t = k/\epsilon$, and $u = \sqrt{k}$, we have

$$\left(\frac{l^2}{t}\right) \sim \frac{k^2}{\epsilon}. \tag{2.22}$$

Because the turbulent diffusion is most effective in the range of the large-scale (energy containing) velocity fluctuations, it seems appropriate to use the (k, ϵ) scale. The model indicated above is an isotropic diffusion model because $C_\epsilon k^2/\epsilon$ is scalar (i.e., isotropic) and is independent of flow direction. For nonisotropic diffusion, one may consider

$$C'_\epsilon \frac{k}{\epsilon}\overline{u_l^2}\frac{\partial \epsilon}{\partial X_l}$$

in place of the above-indicated model. Here, $C'_\epsilon \sim 3C_\epsilon/2$.

Modeling of the production term. The symbols l and L are used to denote fluctuation and mean turbulent length scales, respectively. The magnitude of the following two production terms will be

$$2\nu\overline{u_l\frac{\partial u_i}{\partial X_j}}\frac{\partial^2 U_i}{\partial X_l \partial X_j} \quad \text{versus} \quad 2\nu\overline{\frac{\partial u_i}{\partial X_j}\frac{\partial u_i}{\partial X_l}\frac{\partial u_j}{\partial X_l}}$$

$$2\nu\frac{u^2}{l}\frac{U}{L^2} \overset{?}{<} 2\nu\frac{u^3}{l^3}$$

$$\left(\frac{l^2}{L^2}\frac{U}{u}\right) \overset{?}{<} 1.$$

For the present model, we let $[(l/L)^2(U/u)]$ to be smaller than one by arguing that the fluctuation and the gradient of fluctuation are normally not as well correlated as the correlation of the gradient quantities, and, hence,

$$2\nu\overline{u_l\frac{\partial u_i}{\partial X_j}}\frac{\partial^2 U_i}{\partial X_l \partial X_j} \sim 0,$$

although this is questionable. Also, the term

$$\nu\frac{\partial U_i}{\partial X_j}\left(\overline{\frac{\partial u_l}{\partial X_i}\frac{\partial u_l}{\partial X_j}} + \overline{\frac{\partial u_i}{\partial X_l}\frac{\partial u_j}{\partial X_l}}\right) \sim 0 \tag{2.23}$$

because, if $i \neq j$, the term inside the brackets goes to zero due to the isotropic dissipation assumption in Postulation 3, and if $i = j$, then by the incompressibility assumption, $\partial U_i/\partial X_i = 0$.

Alternatively, Hanjalic and Launder (1972)[55] modeled the production term given in Eq. 2.21 by contracting the indices

$$
\begin{aligned}
2\nu \frac{\partial U_i}{\partial X_j}\left(\overline{\frac{\partial u_l}{\partial X_i}\frac{\partial u_l}{\partial X_j}} + \overline{\frac{\partial u_i}{\partial X_l}\frac{\partial u_j}{\partial X_l}}\right) &= \left(C_{\epsilon 1}\frac{\epsilon}{k}\overline{u_i u_j} + C'_{\epsilon 1}\delta_{ij}\epsilon\right)\frac{\partial U_i}{\partial X_j} \\
&= C_{\epsilon 1}\frac{\epsilon}{k}\overline{u_i u_j}\frac{\partial U_i}{\partial X_j},
\end{aligned}
$$

where $C_{\epsilon 1}$ and $C'_{\epsilon 1}$ are constants. For incompressible flows, because $\partial U_i/\partial X_i = 0$, the term containing $C'_{\epsilon 1}$ vanishes when it is multiplied by $\partial U_i/\partial X_j$. Thus, no modeling is required for the $C'_{\epsilon 1}$ term. Note that the modeling of the $C_{\epsilon 1}$ term is consistent with the modeling of anisotropic dissipation rate given in $\epsilon_{ij} = \epsilon(\overline{u_i u_j}/k)$.

Modeling of the destruction term. This term is the most questionable in the present model. Lumley (1970)[97] has reasoned that ϵ must be conserved if the production of k equals the dissipation of k (i.e., local equilibrium). If there is any nonequilibrium effect present, their influences should be proportional to the difference of production and destruction. Hence, the production and destruction terms can be modeled together as

$$
\begin{aligned}
-2\nu\overline{\frac{\partial u_i}{\partial X_j}\frac{\partial u_i}{\partial X_l}\frac{\partial u_j}{\partial X_l}} - 2\overline{\left(\nu\frac{\partial^2 u_i}{\partial X_l \partial X_j}\right)^2} &\sim \text{constant}\left(\frac{\epsilon}{t}\right)\left(\frac{\text{Prod.}k}{\epsilon} - 1\right) \\
&= C_{\epsilon 1}\left(\frac{1}{t}\right)\left(-\overline{u_i u_l}\frac{\partial U_i}{\partial X_l}\right) - C_{\epsilon 2}\left(\frac{\epsilon}{t}\right) \\
&= -C_{\epsilon 1}\left(\frac{\epsilon}{k}\right)\overline{u_i u_l}\frac{\partial U_i}{\partial X_l} - C_{\epsilon 2}\left(\frac{\epsilon^2}{k}\right) \qquad (2.24)
\end{aligned}
$$

where, from the (k, ϵ) scaling, $1/t = \epsilon/k$.

Hanjalic and Launder (1972)[55] modeled the triple correlation and destruction terms as

$$
-2\nu\overline{\frac{\partial u_i}{\partial X_j}\frac{\partial u_i}{\partial X_l}\frac{\partial u_j}{\partial X_l}} - 2\left(\nu\overline{\frac{\partial^2 u_i}{\partial X_l \partial X_j}}\right)^2 = -C_{\epsilon 2}\frac{\epsilon^2}{k} \qquad (2.25)
$$

on the basis of the reasoning (Rodi, 1971 [133]) that at a high turbulent Reynolds number, these two terms may be taken as being controlled by the dynamics of the energy cascade process transporting energy from the lower to the higher wave numbers. Note that in modeling the production and destruction terms, various authors have proposed models that are based on different reasonings but have finally arrived at similar forms. In the subsequent section, Section 4.5, a comparison with the direct numerical simulation (DNS) budgets will clearly indicate which of these models is correct. The question of whether the term $1/t$ should be modeled from the (k, ϵ) scale or the Kolmogorov scale (ν, ϵ) will be examined in Section 3.4.2.

Summarizing the modeled ϵ equation, we have

- The *exact equation*, as indicated in Eq. 2.21:

$$\frac{D\epsilon}{Dt} = \frac{\partial}{\partial X_l}\left(-\overline{\epsilon' u_l} - \frac{2\nu}{\rho}\overline{\frac{\partial u_l}{\partial X_j}\frac{\partial p}{\partial X_j}} + \nu\frac{\partial \epsilon}{\partial X_l}\right)$$
$$-2\nu\frac{\partial U_i}{\partial X_j}\left(\overline{\frac{\partial u_l}{\partial X_j}\frac{\partial u_l}{\partial X_i}} + \overline{\frac{\partial u_i}{\partial X_l}\frac{\partial u_j}{\partial X_l}}\right) - 2\nu\overline{u_l\frac{\partial u_i}{\partial X_j}}\frac{\partial^2 U_i}{\partial X_l \partial X_j}$$
$$-2\nu\overline{\frac{\partial u_i}{\partial X_j}\frac{\partial u_i}{\partial X_l}\frac{\partial u_j}{\partial X_l}} - 2\overline{\left(\nu\frac{\partial^2 u_i}{\partial X_l \partial X_j}\right)^2}.$$

- The *modeled equation*:

$$\frac{D\epsilon}{Dt} = \frac{\partial}{\partial X_l}\left(C_\epsilon\frac{k^2}{\epsilon}\frac{\partial \epsilon}{\partial X_l} + \nu\frac{\partial \epsilon}{\partial X_l}\right) - C_{\epsilon 1}\frac{\epsilon}{k}\overline{u_i u_l}\frac{\partial U_i}{\partial X_l} - C_{\epsilon 2}\left(\frac{\epsilon}{k}\right)\epsilon. \qquad (2.26)$$

If we model the destruction term by the Kolmogorov scale (ν, ϵ), as suggested by Chen and Singh (1990)[22], then $1/t$ will be replaced by $\epsilon/\nu^{1/2}$ in the modeled equation, Eq. 2.24, instead of ϵ/k.

On the basis of experimental data, the various constants are of the order of

$$C_\epsilon = 0.07 \sim 0.09$$
$$C_{\epsilon 1} = 1.41 \sim 1.45$$
$$C_\epsilon = 1.90 \sim 1.92$$

2.2.5 Modeling of the $\overline{u_i\theta}$ Equation

Exact equation:

$$\frac{D\overline{u_i\theta}}{Dt} = \frac{\partial}{\partial X_l}\left(-\overline{u_l u_i \theta} - \delta_{il}\frac{\overline{p\theta}}{\rho} + \alpha\overline{u_i\frac{\partial \theta}{\partial X_l}} + \nu\overline{\theta\frac{\partial u_i}{\partial X_l}}\right)$$
$$-\left(\overline{u_i u_l}\frac{\partial T}{\partial X_l} + \overline{u_l\theta}\frac{\partial U_i}{\partial X_l}\right) - (\alpha+\nu)\overline{\frac{\partial u_i}{\partial X_l}\frac{\partial \theta}{\partial X_l}} + \overline{\frac{p}{\rho}\frac{\partial \theta}{\partial X_i}} + \overline{\phi' u_i}, \qquad (2.27)$$

where α is the thermal diffusivity.

- The terms

$$-\overline{u_l u_i \theta} - \delta_{il}\frac{\overline{p\theta}}{\rho}$$

  represent the turbulent diffusion of the Reynolds heat flux, whereas
- the terms

$$\alpha\overline{u_i\frac{\partial \theta}{\partial X_l}} + \nu\overline{\theta\frac{\partial u_i}{\partial X_l}}$$

  represent the molecular diffusion of the Reynolds heat flux.

- The terms

$$\left(\overline{u_i u_l}\frac{\partial T}{\partial X_l} + \overline{u_l \theta}\frac{\partial U_i}{\partial X_l}\right)$$

represent the production of the Reynolds heat flux from the mean flow quantities, and

- the term

$$(\alpha + \nu)\overline{\frac{\partial u_i}{\partial X_l}\frac{\partial \theta}{\partial X_l}}$$

represents the dissipation of the heat flux.

- The term

$$\overline{\frac{p}{\rho}\frac{\partial \theta}{\partial X_i}}$$

is the pressure-temperature (PT) term, analogous to the PS term, and is responsible for thermal redistribution to reduce anisotropicity, whereas

- the term

$$\overline{\phi' u_i}$$

represents the contribution to the total Reynolds heat flux by the frictional heating.

Modeling of the diffusion term:

$$-\overline{u_l u_i \theta} - \delta_{il}\frac{\overline{p\theta}}{\rho} = C_T\left[\frac{k^2}{\epsilon}\right]\frac{\partial \overline{u_i \theta}}{\partial X_l}$$

wherein we have made use of the relation $l^2/t = k^2/\epsilon$ from the (k, ϵ) scale. Also, if we assume $\alpha = \nu$, (i.e., Prandtl number $= 1$), then the viscous and thermal diffusion terms can be modeled as

$$\alpha \overline{u_i \frac{\partial \theta}{\partial X_l}} + \nu \overline{\theta \frac{\partial u_i}{\partial X_l}} \sim \alpha \frac{\partial \overline{u_i \theta}}{\partial X_l}.$$

Modeling of the dissipation term. Because of the Postulation 4 (i.e., isotropic dissipation), we have

$$-(\alpha + \nu)\overline{\frac{\partial u_i}{\partial X_l}\frac{\partial \theta}{\partial X_l}} \sim 0.$$

Modeling of the pressure-temperature (PT) term. The PT term can be derived as in the PS term of Eq. 1.5:

$$\overline{\frac{p}{\rho}\frac{\partial \theta}{\partial X_i}} = \frac{1}{4\pi}\int_{\text{vol}}\left[\overline{\left(\frac{\partial^2 u_l u_m}{\partial X_l \partial X_m}\right)^* \frac{\partial \theta}{\partial X_i}} + 2\left(\frac{\partial U_l}{\partial X_m}\right)^* \overline{\left(\frac{\partial u_m}{\partial X_l}\right)^* \frac{\partial \theta}{\partial X_i}}\right]\frac{\text{dvol}}{\boldsymbol{r}^*}.$$

By shrinking the integration to a small volume, we can approximate the above integration to

$$= \text{const.}\left(\overline{\frac{\partial^2 u_l u_m}{\partial X_l \partial X_m}\frac{\partial \theta}{\partial X_i}} + 2\frac{\partial U_l}{\partial X_m}\overline{\frac{\partial u_m}{\partial X_l}\frac{\partial \theta}{\partial X_i}}\right) l^2$$

$$= -C_{T1}\left(\frac{1}{t}\right)\overline{u_i\theta} + C_{T2}\frac{\partial U_i}{\partial X_m}\overline{u_m\theta}$$

$$= -C_{T1}\frac{\epsilon}{k}\overline{u_i\theta} + C_{T2}\frac{\partial U_i}{\partial X_m}\overline{u_m\theta},$$

where, again, we make use of the (k, ϵ) scale to obtain $1/t \sim \epsilon/k$.

Modeling of the friction term. Because the order of magnitude of the friction term is small compared with the other terms in the Reynolds heat flux equation,

$$\overline{\phi' u_i} \approx 0,$$

the modeled $\overline{u_i\theta}$ equation, which is based on the modeling of the various terms of the exact equation, is given as

$$\begin{aligned}\frac{D\overline{u_i\theta}}{Dt} &= \frac{\partial}{\partial X_l}\left[C_T\left(\frac{k^2}{\epsilon}\right)\frac{\partial \overline{u_i\theta}}{\partial X_l} + \alpha\frac{\partial \overline{u_i\theta}}{\partial X_l}\right] - \left(\overline{u_i u_l}\frac{\partial T}{\partial X_l} + \overline{u_l\theta}\frac{\partial U_i}{\partial X_l}\right) - 0 \\ &\quad - C_{T1}\frac{\epsilon}{k}\overline{u_i\theta} + C_{T2}\frac{\partial U_i}{\partial X_m}\overline{u_m\theta} + 0,\end{aligned} \tag{2.28}$$

where the various coefficients are of the following order of magnitude:

$$C_T = 0.07$$
$$C_{T1} = 3.2$$
$$C_{T2} = 0.5.$$

In short, we may write

$$\frac{D\overline{u_i\theta}}{Dt} = D_{u\theta} + P_{u\theta} + \text{PT}.$$

It should be noted that this model is based on the Turbulence Postulates 1–7 and holds for $Pr \approx 1$, with isotropic turbulent diffusion.

2.3 SUMMARY OF THE SECOND-ORDER TURBULENCE MODEL

On the basis of the postulations made, one obtains the modeled equations for turbulence. These are summarized and compared with the exact equations, which are shown below.

2.3.1 Reynolds-Stress Equations

- The exact equation is

$$\begin{aligned}\frac{D\overline{u_i u_j}}{Dt} &= \frac{\partial}{\partial X_l}\left(-\overline{u_i u_j u_l} - \overline{\frac{p}{\rho}(\delta_{jl}u_i + \delta_{il}u_j)} + \nu\frac{\partial \overline{u_i u_j}}{\partial X_l}\right) \\ &\quad - \left(\overline{u_i u_l}\frac{\partial U_j}{\partial X_l} + \overline{u_j u_l}\frac{\partial U_i}{\partial X_l}\right) - 2\nu\overline{\frac{\partial u_i}{\partial X_l}\frac{\partial u_j}{\partial X_l}} + \overline{\frac{p}{\rho}\left(\frac{\partial u_i}{\partial X_j} + \frac{\partial u_j}{\partial X_i}\right)}.\end{aligned} \tag{2.4}$$

- The modeled equation is

$$\frac{D\overline{u_i u_j}}{Dt} = \frac{\partial}{\partial X_l}\left(\left(C_k\frac{k^2}{\epsilon}+\nu\right)\frac{\partial\overline{u_i u_j}}{\partial X_l}\right) - \left(\overline{u_i u_l}\frac{\partial U_j}{\partial X_l}+\overline{u_j u_l}\frac{\partial U_i}{\partial X_l}\right)$$
$$-\frac{2}{3}\delta_{ij}\epsilon - C_1\frac{\epsilon}{k}\left(\overline{u_i u_j}-\frac{2}{3}\delta_{ij}k\right)$$
$$+C_2\left(\overline{u_i u_l}\frac{\partial U_j}{\partial X_l}+\overline{u_j u_l}\frac{\partial U_i}{\partial X_l}-\frac{2}{3}\delta_{ij}\overline{u_n u_m}\frac{\partial U_n}{\partial X_m}\right), \tag{2.17}$$

where $C_k = 0.09 \sim 0.11$, $C_1 = 2.30$, and $C_2 = 0.40$.

2.3.2 Turbulent Kinetic Energy Equation

- The exact equation is

$$\frac{Dk}{Dt} = \frac{\partial}{\partial X_l}\left(-\overline{u_l\left(k'+\frac{p}{\rho}\right)}+\nu\frac{\partial k}{\partial X_l}\right) - \overline{u_i u_l}\frac{\partial U_i}{\partial X_l} - \epsilon. \tag{2.18}$$

- The modeled equation is

$$\frac{Dk}{Dt} = \frac{\partial}{\partial X_l}\left(C_k\frac{k^2}{\epsilon}\frac{\partial k}{\partial X_l}+\nu\frac{\partial k}{\partial X_l}\right) - \overline{u_i u_l}\frac{\partial U_i}{\partial X_l} - \epsilon, \tag{2.19}$$

where $C_k = 0.09 \sim 0.11$.

2.3.3 Rate of Dissipation of Kinetic Energy Equation

- The exact equation is

$$\frac{D\epsilon}{Dt} = \frac{\partial}{\partial X_l}\left(-\overline{\epsilon' u_l} - \frac{2\nu}{\rho}\overline{\frac{\partial u_l}{\partial X_j}\frac{\partial p}{\partial X_j}}+\nu\frac{\partial\epsilon}{\partial X_l}\right) - 2\nu\frac{\partial U_i}{\partial X_j}\left(\overline{\frac{\partial u_l}{\partial X_i}\frac{\partial u_l}{\partial X_j}}+\overline{\frac{\partial u_i}{\partial X_l}\frac{\partial u_j}{\partial X_l}}\right)$$
$$-2\nu\overline{u_l\frac{\partial u_i}{\partial X_j}}\frac{\partial^2 U_i}{\partial X_l\partial X_j} - 2\nu\overline{\frac{\partial u_i}{\partial X_j}\frac{\partial u_i}{\partial X_l}\frac{\partial u_j}{\partial X_l}} - 2\overline{\left(\nu\frac{\partial^2 u_i}{\partial X_l\partial X_j}\right)^2}. \tag{2.21}$$

- The modeled equation is

$$\frac{D\epsilon}{Dt} = \frac{\partial}{\partial X_l}\left[\left(C_\epsilon\frac{k^2}{\epsilon}+\nu\right)\frac{\partial\epsilon}{\partial X_l}\right] - C_{\epsilon 1}\frac{\epsilon}{k}\overline{u_i u_l}\frac{\partial U_i}{\partial X_l} - C_{\epsilon 2}\frac{\epsilon^2}{k}, \tag{2.26}$$

where $C_\epsilon = 0.07$, $C_{\epsilon 1} = 1.45$, and $C_{\epsilon 2} = 1.92$.

2.3.4 Reynolds Heat Flux Equations

- The exact equation is given as

$$\frac{D\overline{u_i\theta}}{Dt} = \frac{\partial}{\partial X_l}\left(-\overline{u_l u_i\theta} - \delta_{il}\frac{\overline{p\theta}}{\rho}+\alpha\overline{u_i\frac{\partial\theta}{\partial X_l}}+\nu\overline{\theta\frac{\partial u_i}{\partial X_l}}\right)$$
$$-\left(\overline{u_i u_l}\frac{\partial T}{\partial X_l}+\overline{u_l\theta}\frac{\partial U_i}{\partial X_l}\right) - (\alpha+\nu)\overline{\frac{\partial u_i}{\partial X_l}\frac{\partial\theta}{\partial X_l}} + \overline{\frac{p}{\rho}\frac{\partial\theta}{\partial X_i}} + \overline{\phi' u_i}. \tag{2.27}$$

• The modeled equation is given by

$$\frac{D\overline{u_i\theta}}{Dt} = \frac{\partial}{\partial X_l}\left(\left(C_T\frac{k^2}{\epsilon}+\alpha\right)\frac{\partial\overline{u_i\theta}}{\partial X_l}\right) - \left(\overline{u_iu_l}\frac{\partial T}{\partial X_l}+\overline{u_l\theta}\frac{\partial U_i}{\partial X_l}\right)$$
$$-C_{T1}\frac{\epsilon}{k}\overline{u_i\theta}+C_{T2}\overline{u_m\theta}\frac{\partial U_i}{\partial X_m}, \tag{2.28}$$

where $C_T = 0.07$, $C_{T1} = 3.20$, and $C_{T2} = 0.50$.

2.4 DETERMINATION OF TURBULENCE MODEL CONSTANTS

2.4.1 General Remarks

The N-S and energy equations require experiments to determine μ, (μ_2), C_p, k, and ρ. There are four constants in Eqs. 1.6, 1.7, and 1.8, which are five equations for the five unknowns, U_i, p, and T. The turbulence transport equations require experiments to determine C_k, C_1, C_2; C_ϵ, $C_{\epsilon1}$, $C_{\epsilon2}$; and C_T, C_{T1}, C_{T2}. There are nine constants in the equations for the Reynolds stress $(\overline{u_iu_j})$, kinetic energy (k), rate of dissipation (ϵ), and the Reynolds heat flux $(\overline{u_i\theta})$, which is a total of eleven equations for eleven unknowns. The approximate results for these nine constants (or, more generally, model coefficients) are given in Table 2.3. Although these constants are determined from experiments in air and water, they should be approximately valid for most other fluids. However, for very viscous fluids such as oil and for liquid metal with a high or low Prandtl number, these constants have not yet been tested. More study is required in this area.

We should also remark here that although these model coefficients are called constants, they are nevertheless variable model coefficients. They can depend on fluid properties and nondimensional parameters, such as Prandtl, *Pr*, Reynolds, *Re*, and turbulent Reynolds number Re_T. In the following section, the method of obtaining these constants is outlined.

Table 2.3 Constants for the turbulence transport equations

C_k	C_1	C_2	C_ϵ	$C_{\epsilon1}$	$C_{\epsilon2}$	C_T	C_{T1}	C_{T2}
0.09	2.30	0.40	0.07	1.45	1.92	0.07	3.2	0.5

2.4.2 Determining $C_{\epsilon2}$

This is determined by utilizing the isotropic grid turbulence data (Townsend, 1976 [168]). The experiment for this consists of uniform flow past a grid with $U = U_0 =$ constant and $V = W = 0$, as shown in Fig. 2.2. At large distance from the grid, $X > 100\,M$, where M is the grid spacing, it is found that

$$k = \frac{\overline{u^2}+\overline{v^2}+\overline{w^2}}{2}$$
$$\sim \frac{3\overline{u^2}}{2}.$$

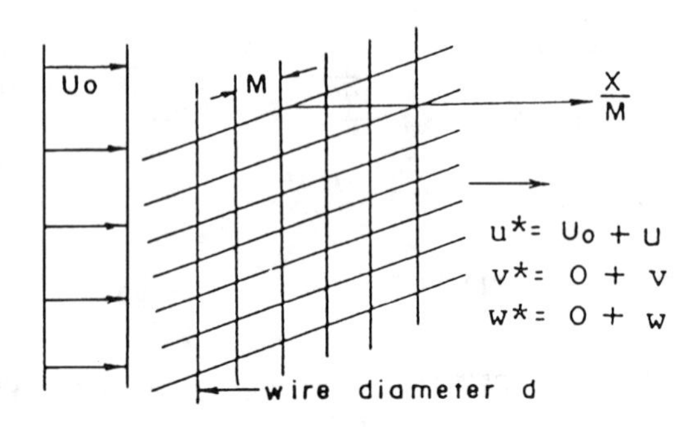

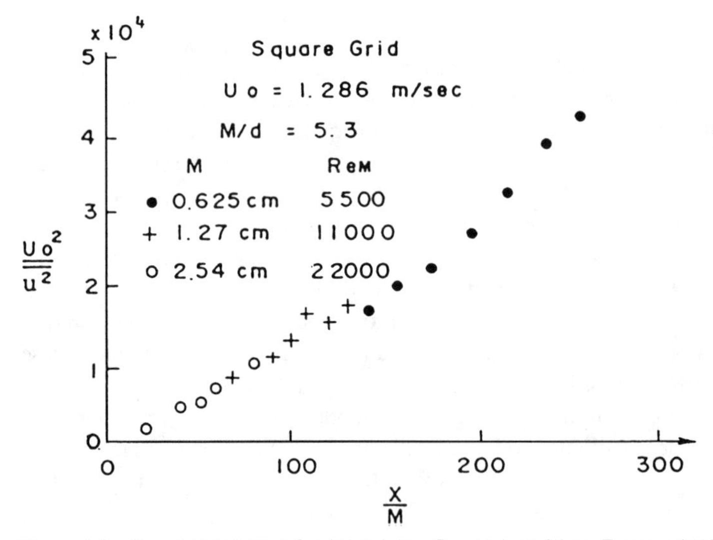

Figure 2.2 Experimental setup for determining $C_{\epsilon 2}$. (Adapted from Townsend (1976).)

It should be noted here that

$$\overline{u^2} > \overline{v^2} = \overline{w^2}$$

is found experimentally, and for small X/M (<100),

$$\overline{u^2} \sim 1.25\overline{v^2}.$$

For uniform flow, the modeled k equation

$$\begin{aligned}\frac{Dk}{Dt} &= \frac{\partial k}{\partial t} + U_l \frac{\partial k}{\partial X_l} \\ &= \frac{\partial}{\partial X_l}\left(C_k \frac{k^2}{\epsilon}\frac{\partial k}{\partial X_l}\right) - \overline{u_i u_l}\frac{\partial U_i}{\partial X_l} - \epsilon\end{aligned}$$

is reduced to

$$U_0 \frac{\partial k}{\partial x} = 0 - 0 - \epsilon, \tag{2.29}$$

and the modeled ϵ equation

$$\frac{D\epsilon}{Dt} = \frac{\partial \epsilon}{\partial t} + U_l \frac{\partial \epsilon}{\partial X_l}$$
$$= \frac{\partial}{\partial X_l}\left(C_\epsilon \frac{k^2}{\epsilon}\frac{\partial \epsilon}{\partial X_l}\right) - C_{\epsilon 1}\frac{\epsilon}{k}\overline{u_i u_l}\frac{\partial U_i}{\partial X_l} - C_{\epsilon 2}\frac{\epsilon^2}{k}$$

is reduced to

$$U_0 \frac{\partial \epsilon}{\partial X} = 0 - 0 - C_{\epsilon 2}\frac{\epsilon^2}{k}. \tag{2.30}$$

Note that the diffusion terms of the k and the ϵ equations in their exact forms are

$$\frac{\partial}{\partial X_l}\left(-\overline{u_l k'} - \frac{\overline{p u_l}}{\rho}\right)$$

and

$$\frac{\partial}{\partial X_l}\left(-\overline{u_l \epsilon'} - \frac{2\nu}{\rho}\overline{\frac{\partial u_l}{\partial X_j}\frac{\partial p}{\partial X_j}}\right),$$

respectively. In isotropic turbulent flows, there should be no directional correlations for triple correlations and for correlations between a vector quantity and a scalar quantity. Hence, the ensemble averages of $\overline{u_l k'}$, $\overline{p u_l}$, $\overline{u_l \epsilon'}$, and $\overline{\partial u_l \partial p / \partial X_j \partial X_j}$ are equal to zero.

Solving for k and ϵ, let $k = AX^n$, where n = constant. From grid turbulence experiments, for the decay of turbulent kinetic energy, we have $n \sim -1.08$ (see Fig. 2.3). Eqs. 2.29 and 2.30 become

$$U_0 A n X^{n-1} = -\epsilon$$
$$-U_0^2 A n(n-1)X^{n-2} = -C_{\epsilon 2}\frac{U_0^2 A^2 n^2 X^{2(n-1)}}{AX^n}$$
$$= -C_{\epsilon 2} U_0^2 A n^2 X^{n-2}.$$

We find that any power of n will satisfy the k and the ϵ equations. Simplifying the above-indicated equation, and rewriting in terms of $C_{\epsilon 2}$ and n, we have

$$C_{\epsilon 2} = \frac{n-1}{n}. \tag{2.31}$$

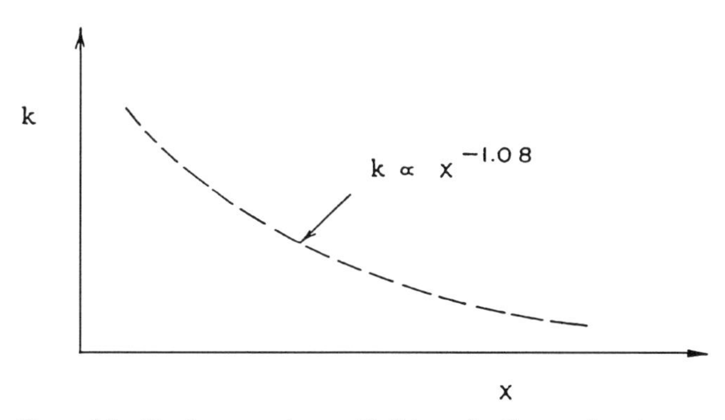

Figure 2.3 Kinetic energy decay with distance for $C_{\epsilon 2}$ experiment.

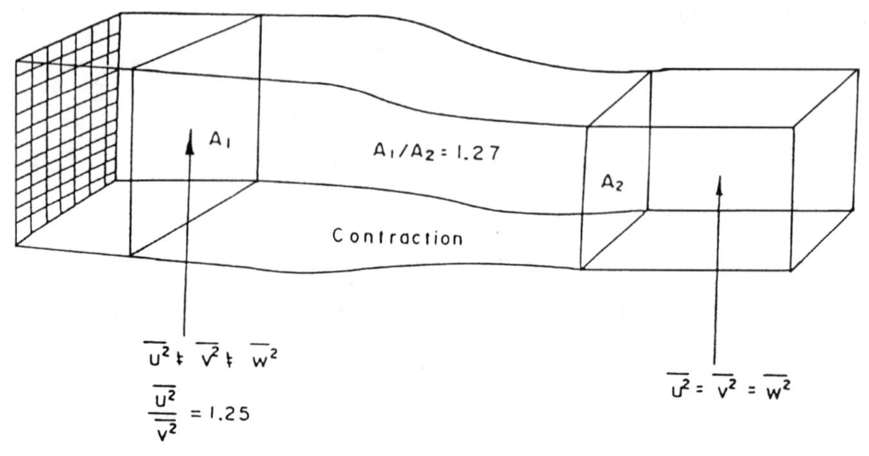

Figure 2.4 Setup improvement for better isotropic flow.

If n equals -1, then $C_{\epsilon 2} = 2.00$; if n equals -1.08, then $C_{\epsilon 2} = 1.92$. It should be remarked here that the flow behind the grid is not truly isotropic turbulence. Sometimes, nonisotropicity can persist farther than $X/M > 100$. To produce better isotropic flow, one may put the grid turbulence through a 1.27:1 contraction, as shown in Fig. 2.4.

2.4.3 Determining C_1

For this, we use the anisotropic grid turbulence. We would still like to use a simple mean flow ($U = U_0 =$ constant and $V = W = 0$). However, this time we make a contraction of flow area to 4:1 ($A_1/A_2 = 4$), as shown in Fig. 2.5, to produce the anisotropic grid turbulence. We shall use the experiment of Uberoi (1957)[178] to determine C_1. The development of the normal Reynolds stresses is indicated in Fig. 2.6

With $U =$ constant and $V = W = 0$, the exact equation for $\overline{u^2}$ becomes

$$U\frac{\partial \overline{u^2}}{\partial X} = -2\nu\overline{\frac{\partial u}{\partial X_l}\frac{\partial u}{\partial X_l}} + \overline{\frac{2p}{\rho}\frac{\partial u}{\partial X}}.$$

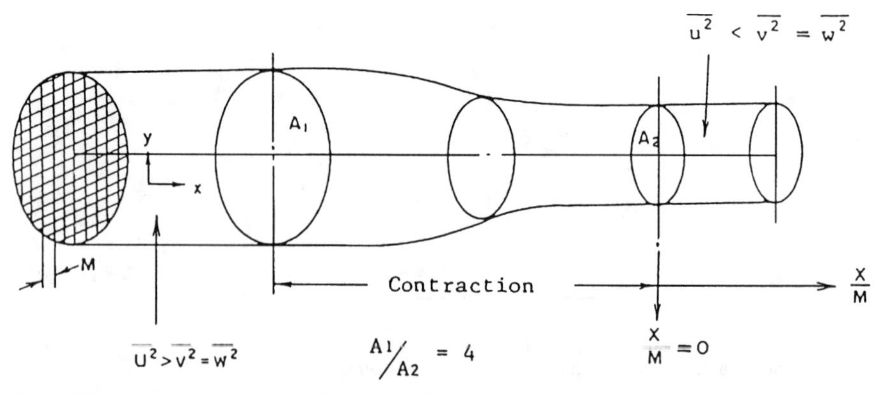

Figure 2.5 Experimental setup for determining C_1.

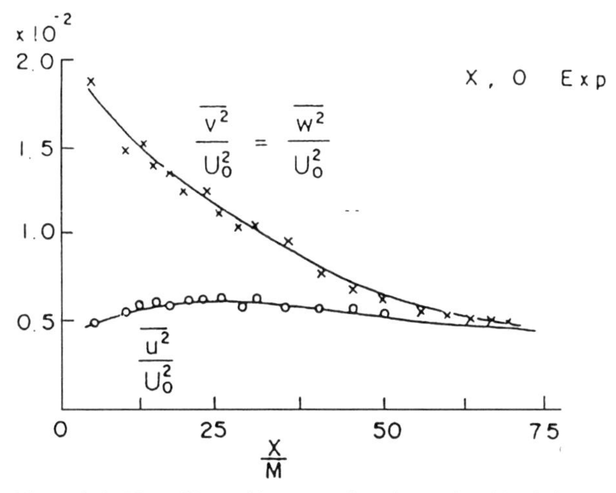

Figure 2.6 Normal Reynolds stresses for anisotropic grid turbulence. (Adapted from Uberoi (1957).)

The modeled equation of $\overline{u^2}$ becomes

$$U\frac{\partial \overline{u^2}}{\partial X} = -\frac{2}{3}\epsilon - C_1\frac{\epsilon}{k}\left(\overline{u^2} - \frac{2}{3}k\right).$$

From the two equations indicated above, we see that

$$\frac{\overline{\frac{2p}{\rho}\frac{\partial u}{\partial X}}}{2\nu\overline{\frac{\partial u}{\partial X_l}\frac{\partial u}{\partial X_l}}} = \frac{-C_1\frac{\epsilon}{k}\left(\overline{u^2} - \frac{2}{3}k\right)}{\frac{2}{3}\epsilon}$$

$$= -C_1\left(\frac{3\overline{u^2}}{2k} - 1\right).$$

Figure 2.7 plots the variation of the numerator and the denominator on the left side of the equation indicated above. From the anisotropic region at $X/M < 75$, the experimental

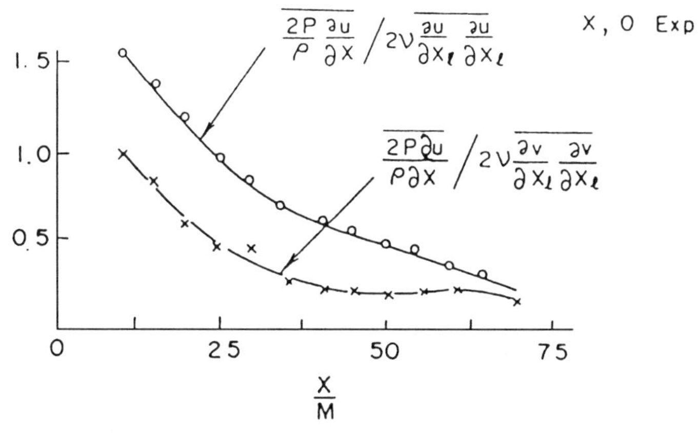

Figure 2.7 Ratio of PS term and isotropic dissipation for two experimental data sets. (Adapted from Uberoi (1957).)

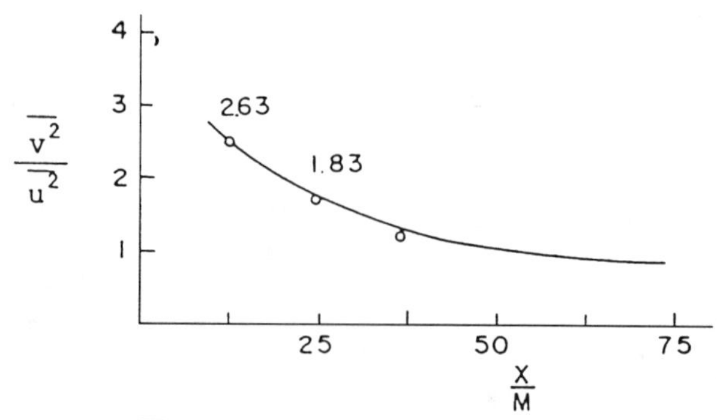

Figure 2.8 $\overline{v^2}/\overline{u^2}$ variation along the axis for C_1.

data are available for the ratio of the PS term to the dissipation term and for the ratio of $\overline{v^2}$ to $\overline{u^2}$, as shown in Fig. 2.8.

At $X/M = 25$, $\overline{v^2} = 1.83\,\overline{u^2}$ and $\overline{v^2} = \overline{w^2}$. Thus,

$$\frac{3}{2}\frac{\overline{u^2}}{k} = \frac{3\overline{v^2}/1.83}{\overline{v^2}(2 + 1/1.83)}$$

$$= 0.644$$

$$\frac{2\overline{\frac{p}{\rho}\frac{\partial u}{\partial X}}}{2\nu\overline{\frac{\partial u}{\partial X_l}\frac{\partial u}{\partial X_l}}} = 1.0$$

$$= -C_1\left(\frac{3}{2}\frac{\overline{u^2}}{k} - 1\right)$$

$$= -C_1(0.644 - 1)$$

$$= 0.356C_1.$$

At $X/M = 25.0$, $C_1 \sim 2.80$; whereas at $X/M = 12.50$, $C_1 \sim 2.88$.

In general, the value of $C_1 = 1.00 \sim 3.00$ is used. Launder et al. (1975)[88] recommended $C_1 = 1.50$. Norris and Reynolds (1975)[119] suggested $C_1 = 1.65$.

2.4.4 Determining C_2 and $C_{\epsilon 1}$

For this we consider homogeneous shear flow behind the grid, which is the next simple flow other than the uniform flow. In other words, $U = U_0(Y/H)$ and $V = W = 0$. The experimental setup of Champagne et al. (1970)[9] for homogeneous shear flow is shown in Fig. 2.9. In Fig. 2.10, the experimental data ($dU/dY = U_0/H =$ constant) are plotted as a function of X/H, illustrating the decay of turbulent intensity with distance.

The modeled equations for $\overline{u_i u_j}$, k, and ϵ in general are

- the Reynolds stress ($\overline{u_i u_j}$) equation:

$$\frac{D\overline{u_i u_j}}{Dt} = \frac{\partial}{\partial X_l}\left(C_k\frac{k^2}{\epsilon}\frac{\partial\overline{u_i u_j}}{\partial X_l} + \nu\frac{\partial\overline{u_i u_j}}{\partial X_l}\right) + P_{ij} - \frac{2}{3}\delta_{ij}\epsilon$$
$$- C_1\frac{\epsilon}{k}\left(\overline{u_i u_j} - \frac{2}{3}\delta_{ij}k\right) - C_2\left(P_{ij} - \frac{2}{3}\delta_{ij}P_k\right). \qquad (2.4)$$

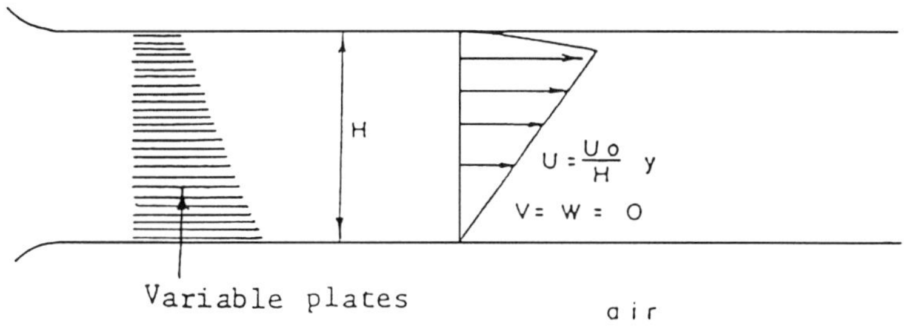

Figure 2.9 Experimental setup for determining C_2. (Adapted from Champagne et al. (1970).)

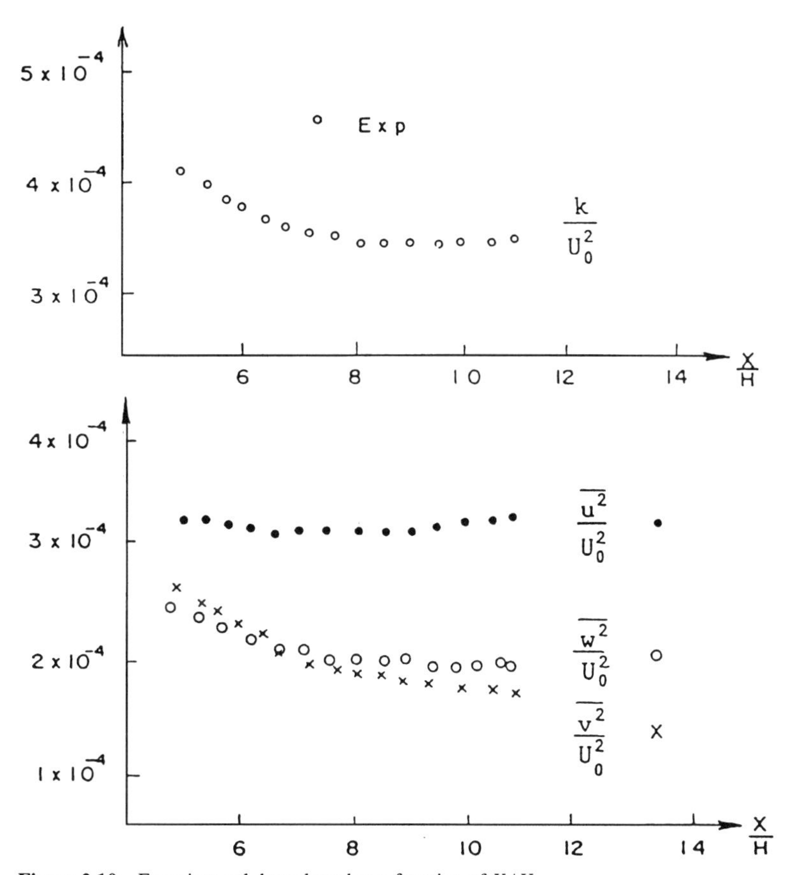

Figure 2.10 Experimental data plotted as a function of X/H.

- the turbulent kinetic energy (k) equation is

$$\frac{Dk}{Dt} = \frac{\partial}{\partial X_l}\left(C_k \frac{k^2}{\epsilon}\frac{\partial k}{\partial X_l} + \nu \frac{\partial k}{\partial X_l}\right) - \overline{u_i u_l}\frac{\partial U_i}{\partial X_l} - \epsilon, \tag{2.19}$$

and

• the rate of dissipation of turbulent kinetic energy (ϵ) equation is

$$\frac{D\epsilon}{Dt} = \frac{\partial}{\partial X_l}\left(C_\epsilon \frac{k^2}{\epsilon}\frac{\partial \epsilon}{\partial X_l} + \nu \frac{\partial \epsilon}{\partial X_l}\right) - C_{\epsilon 1}\frac{\epsilon}{k}\overline{u_i u_l}\frac{\partial U_i}{\partial X_l} - C_{\epsilon 2}\frac{\epsilon^2}{k}. \tag{2.26}$$

For the flow under consideration, $U = U_0 Y/H$, $V = W = 0$ and $\partial/\partial X = \partial/\partial Y = 0$.

The modeled equations for k becomes

$$-\overline{uv}\frac{\partial U}{\partial Y} - \epsilon = 0,$$

or

$$\epsilon = -\overline{uv}\frac{\partial U}{\partial Y}, \tag{2.32}$$

and for ϵ, the modeled equation becomes

$$-C_{\epsilon 1}\frac{\epsilon}{k}\overline{uv}\frac{\partial U}{\partial y} - C_{\epsilon 2}\frac{\epsilon^2}{k} \approx 0. \tag{2.33}$$

Subtracting Eq. 2.33 from Eq. 2.32, we have

$$C_{\epsilon 1} \approx C_{\epsilon 2} = 1.92,$$

where the value for $C_{\epsilon 2}$ had been obtained earlier.

However, in most calculations, $C_{\epsilon 1}$ is usually taken to be 1.45. This shows some inadequacies of the modeling process. We will discuss this issue in more detail later when we consider the two-scale (k, ϵ) and (ϵ, ν) turbulence models (see Section 3.4.2).

More recent data of Harris, Graham, and Corrsin (1977)[58], Tavoularis and Corrsin (1981)[174], and Rohr et al. (1988 [140]; see Fig. 2.11) showed that in a longer span of

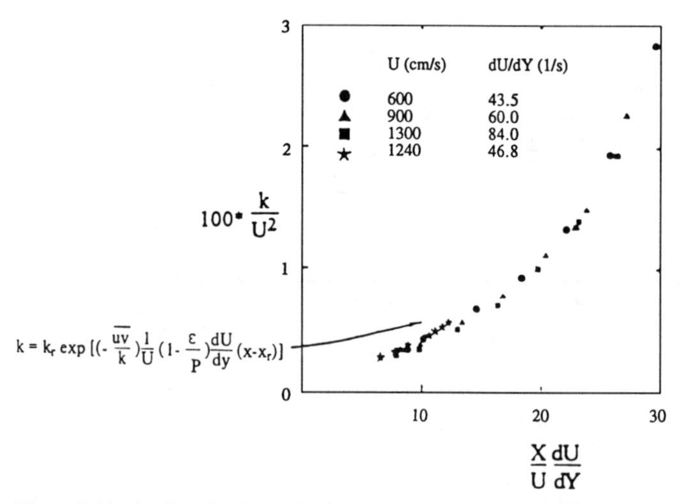

Figure 2.11 Profile of turbulent kinetic energy with distance for homogeneous shear flow. (Adapted from Rohr et al. (1988).)

data recording, the turbulent kinetic energy k increases exponentially with respect to X, or

$$k = k_r \exp\left[\left(-\frac{\overline{uv}}{k}\right)\frac{1}{U}\left(1-\frac{\epsilon}{P_k}\right)\frac{dU}{dY}(X-X_r)\right].$$

This equation is obtained by integrating the equation

$$\begin{aligned} U\frac{dk}{dX} &= -\overline{uv}\frac{\partial U}{\partial Y} - \epsilon \\ &= P_k - \epsilon \\ &= \epsilon\left(\frac{P_k}{\epsilon} - 1\right) \end{aligned}$$

with ϵ/P_k being constant. Here k_r is the reference k at $X = X_r$. Typically the diffusion effects in the homogeneous shear flow are less than 3 percent of the mean convection and are negligible compared with the production and dissipation of k and ϵ. From experimental measurements performed in air, Tavoularis and Corrsin (1981)[174] showed that ϵ/P_k, dU/dY, and $\overline{uv}/k$ remain constant with X in the asymptotic range. From these relations, it can be derived that ϵ/k is also a constant. Hence, we may write the ϵ equation as

$$\begin{aligned} U\frac{d\epsilon}{dX} &= -C_{\epsilon 1}\frac{\epsilon}{k}\overline{uv}\frac{\partial U}{\partial Y} - C_{\epsilon 2}\frac{\epsilon}{k}\epsilon \\ &= C_{\epsilon 1}\frac{\epsilon}{k}P_k - C_{\epsilon 2}\frac{\epsilon}{k}\epsilon \end{aligned}$$

as

$$\begin{aligned} \left(\frac{\epsilon}{k}\right)U\frac{dk}{dX} &= -C_{\epsilon 1}\frac{\epsilon}{k}\overline{uv}\frac{\partial U}{\partial Y} - C_{\epsilon 2}\frac{\epsilon}{k}\epsilon \\ &= C_{\epsilon 1}\frac{\epsilon}{k}P_k - C_{\epsilon 2}\frac{\epsilon}{k}\epsilon. \end{aligned}$$

Replacing Udk/dX in the ϵ equation by the k equation, we have

$$P_k - \epsilon = C_{\epsilon 1}P_k - C_{\epsilon 2}\epsilon,$$

or

$$C_{\epsilon 1} = 1 + \frac{\epsilon}{P_k}(C_{\epsilon 2} - 1).$$

If $\epsilon/P_k = 1$, then the same result, $C_{\epsilon 1} = C_{\epsilon 2}$, mentioned in Section 2.4.4 is obtained. The value of $C_{\epsilon 1} = 1.45$ popularly in use corresponds to $\epsilon/P_k \sim 0.5$.

In order to determine C_2, from Fig. 2.10 we recall the mathematical conditions, that is, $U = U_0 Y/H$, $V = W = 0$ and $\partial/\partial X = \partial/\partial Z = 0$. Hence, for $i = j$, the Eq. 2.17 can be written for $\overline{u^2}$, $\overline{v^2}$, and $\overline{w^2}$, for $X/H = 10$, as

- The $\overline{u^2}$ equation:

$$0 = 0 - 2\overline{uv}\frac{\partial U}{\partial y} - \frac{2}{3}\epsilon - C_1\frac{\epsilon}{k}\left(\overline{u^2} - \frac{2}{3}k\right) + C_2\left(2 - \frac{2}{3}\right)\overline{uv}\frac{\partial U}{\partial Y}$$

Simplifying,

$$\left(\overline{u^2}-\frac{2}{3}k\right)\Big/k\sim 0.22=4(1-C_2)/3C_1.$$

- Similarly, the $\overline{v^2}$ equation:

$$\left(\overline{v^2}-\frac{2}{3}k\right)\Big/k\sim -0.17=-2(1-2C_2)/3C_1.$$

- Similarly, the $\overline{w^2}$ equation:

$$\left(\overline{w^2}-\frac{2}{3}k\right)\Big/k\sim -0.18=-2(1-2C_2)/3C_1.$$

With $C_1=2.8$, we have $C_2\approx 0.3\sim 0.6$. Generally, the values of $C_1=2.3$ and $C_2=0.4$ are used. Some recent models used $C_1=1.5$.

The more recent experimental data of Tavoularis and Corrsin (1981)[174] used to determine C_2, the modeled $\overline{u^2}$, $\overline{v^2}$, and $\overline{w^2}$ equations are first simplified as

$$U\frac{d\overline{u^2}}{dX}=-2\overline{uv}\frac{\partial U}{\partial Y}-\frac{\epsilon}{k}\overline{u^2}-C_1\frac{\epsilon}{k}\left(\overline{u^2}-\frac{2}{3}k\right)+C_2\left(2-\frac{2}{3}\right)\overline{uv}\frac{\partial U}{\partial Y} \tag{2.34}$$

$$U\frac{d\overline{v^2}}{dX}=-\frac{\epsilon}{k}\overline{v^2}-C_1\frac{\epsilon}{k}\left(\overline{v^2}-\frac{2}{3}k\right)+C_2\left(2-\frac{2}{3}\right)\overline{uv}\frac{\partial U}{\partial Y} \tag{2.35}$$

$$U\frac{d\overline{w^2}}{dX}=-\frac{\epsilon}{k}\overline{w^2}-C_1\frac{\epsilon}{k}\left(\overline{w^2}-\frac{2}{3}k\right)+C_2\left(2-\frac{2}{3}\right)\overline{uv}\frac{\partial U}{\partial Y}. \tag{2.36}$$

The k equation is written as before,

$$U\frac{dk}{dX}=-\overline{uv}\frac{\partial U}{\partial Y}-\epsilon.$$

Note that anisotropic dissipation, $\epsilon_{ij}=(\epsilon/k)\overline{u_iu_j}$, is adopted in this case. In principle, C_2 can be determined from the above-indicated equations, if $\partial U/\partial Y$, $P\ (=-\overline{uv}\partial U/\partial Y)$, $\epsilon, k, \overline{u^2}, \overline{v^2}, \overline{w^2}$, and C_1 are known. According to Tavoularis and Corrsin (1981), the ratio of Reynolds stress to turbulent kinetic energy, $\overline{u_iu_j}/k$, does not vary appreciably among the various available experiments of Harris et al. (1977)[58], or Tavoularis and Corrsin (1981)[174], for homogeneous shear flows. For example, the ratio of $\overline{u^2}/k$ calculated from the experiments of Harris et al. [58] is about 1.02 and from the experiments of Tavoularis and Corrsin [174] is about 1.06. An expression for C_2 may be derived from the $\overline{u^2}$ and k equations as

$$C_2=0.75\left[2-\frac{\overline{u^2}}{k}-C_1\left(\frac{\overline{u^2}}{k}-\frac{2}{3}\right)\frac{\epsilon}{P_k}\right].$$

Again, C_2 is also a function of ϵ/P_k. When $\overline{u^2}/k$ is taken to be 1.04, an average of the values 1.02 and 1.06 is obtained in the experiments; with $C_1=2.3$, C_2 is found:

$$C_2=0.72-0.64\frac{\epsilon}{P_k}.$$

For $\epsilon/P_k = 0.5$, C_2 is 0.4, the same as the value $C_2 = 0.4$, which is generally adopted. Similar expressions of C_2 can be obtained from the $\overline{v^2}$, $\overline{w^2}$, and k equations as

$$C_2 = 1.5\left[\frac{\overline{v^2}}{k} + C_1\left(\frac{\overline{v^2}}{k} - \frac{2}{3}\right)\frac{\epsilon}{P_k}\right]$$

$$C_2 = 1.5\left[\frac{\overline{w^2}}{k} + C_1\left(\frac{\overline{w^2}}{k} - \frac{2}{3}\right)\frac{\epsilon}{P_k}\right].$$

If the ratios of $\overline{v^2}/k$ and $\overline{w^2}/k$ are both approximated to $k - \overline{u^2}/2$ $(=0.48)$, which is necessary to satisfy the relation $\overline{u^2} + \overline{v^2} + \overline{w^2} = 2k$, C_2 derived from the above-indicated two equations becomes

$$C_2 = 0.72 - 0.64\frac{\epsilon}{P_k},$$

which is consistent with the C_2 relation derived from the modeled $\overline{u^2}$ equation. However, the measurements of Harris et al. (1977)[58] and Tavoularis and Corrsin (1981)[174] showed that $\overline{v^2}/k$ is smaller than $\overline{w^2}/k$ for a homogeneous shear flow, with $\overline{v^2}/k = 0.38$ and $\overline{w^2}/k = 0.58$, respectively.

As Launder et al. (1975)[88] pointed out, the inconsistency of the C_2 relations derived from the modeled $\overline{v^2}$ and the $\overline{w^2}$ equations for the homogeneous shear flow with $\overline{v^2}/k \neq \overline{w^2}/k$ is due to the simplicity of the PS model being

$$\overline{\frac{p}{\rho}\left(\frac{\partial u_i}{\partial X_j} + \frac{\partial u_j}{\partial X_i}\right)} = -C_1\frac{\epsilon}{k}\left(\overline{u_iu_j} - \frac{2}{3}\delta_{ij}k\right) + C_2\left(\overline{u_iu_m}\frac{\partial U_j}{\partial X_m} + \overline{u_ju_m}\frac{\partial U_i}{\partial X_m} - \frac{2}{3}\delta_{ij}\overline{u_nu_m}\frac{\partial U_n}{\partial X_m}\right).$$

To obtain a more consistent coefficient for C_2, we must use a more sophisticated PS model.

2.4.5 Determining C_{T1} and C_{T2}

For these coefficients, we consider a homogeneous shear flow with a linear temperature gradient. Consider the same setup as in Section 2.4.4, but add a temperature gradient to the experiment, as shown in Fig. 2.12. The modeled $\overline{u_i\theta}$ equation, in general, is

$$\frac{D\overline{u_i\theta}}{Dt} = \frac{\partial}{\partial X_l}\left[C_T\left(\frac{k^2}{\epsilon}\right)\frac{\partial\overline{u_i\theta}}{\partial X_l} + \alpha\frac{\partial\overline{u_i\theta}}{\partial X_l}\right] - \left(\overline{u_iu_l}\frac{\partial T}{\partial X_l} + \overline{u_l\theta}\frac{\partial U_i}{\partial X_l}\right) - C_{T1}\frac{\epsilon}{k}\overline{u_i\theta} + C_{T2}\frac{\partial U_i}{\partial X_m}\overline{u_m\theta}. \tag{2.28}$$

In order to determine C_{T1}, we set $i = 2$ (i.e., solve for $\overline{v\theta}$), then with $v = 0$

$$0 = 0 - \left(\overline{v^2}\frac{\partial T}{\partial Y} + 0\right) - C_{T1}\frac{\epsilon}{k}\overline{v\theta} + C_{T2}\cdot 0.$$

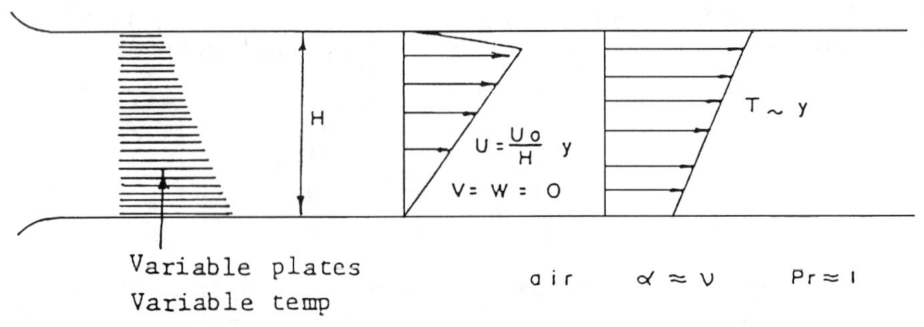

Figure 2.12 Experimental setup for determining C_{T1} and C_{T2}.

Hence,

$$C_{T1} = -\frac{\overline{v^2}k}{\epsilon}\frac{\frac{\partial T}{\partial Y}}{\overline{v\theta}} = \frac{\overline{v^2}k}{\overline{uv}\,\overline{v\theta}}\frac{\frac{\partial T}{\partial Y}}{\frac{\partial U}{\partial Y}}$$

because $\epsilon = -\overline{uv}\partial U/\partial Y$. From the experimental data for $\overline{uv}$, $\overline{v^2}$, $\overline{v\theta}$, T and U, we find $C_{T1} \sim 3.2$ on the basis of Webster's (1964)[182] measurements.

For determining C_{T2}, we consider $i = 1$ (i.e., solve for $\overline{u\theta}$), hence,

$$0 = 0 - \left(\overline{uv}\frac{\partial T}{\partial Y} + \overline{v\theta}\frac{\partial U}{\partial Y}\right) - C_{T1}\frac{\epsilon}{k}\overline{u\theta} + C_{T2}\overline{v\theta}\frac{\partial U}{\partial Y},$$

which gives

$$\frac{\overline{u\theta}}{\overline{v\theta}} = \frac{\overline{uv}}{\overline{v^2}} + \frac{(1 - C_{T2})}{C_{T1}}\frac{k}{\overline{uv}}.$$

From the experimental data for $\overline{uv}$, $\overline{u\theta}$, $\overline{v\theta}$, $\overline{v^2}$, and k, we find $C_{T2} \sim 0.5$.

2.4.6 Determining the Diffusion Constant, C_ϵ

Consider the near-wall turbulence (not too near), where the flow is not in a laminar sublayer. Then the flow is nearly parallel, the convective terms are negligible, and

$$\frac{D\epsilon}{Dt} = \frac{\partial}{\partial X_l}\left(C_\epsilon \frac{k^2}{\epsilon}\frac{\partial \epsilon}{\partial X_l}\right) - C_{\epsilon 1}\frac{\epsilon}{k}\overline{u_i u_l}\frac{\partial U_i}{\partial X_l} - C_{\epsilon 2}\frac{\epsilon^2}{k}. \tag{2.26}$$

Near the wall, if parallel flow and zero-pressure gradient are assumed, we can then set $\partial/\partial X = 0$ and $\partial/\partial Z = 0$. The mean U equation gives

$$-\frac{d\overline{uv}}{dY} + \frac{d}{dY}\nu\frac{dU}{dY} = 0.$$

With

$$\frac{\tau_w}{\rho} = \nu\left.\frac{dU}{dY}\right|_0,$$

we have

$$-\overline{uv} = u^{*2} = \frac{\tau_w}{\rho} = \text{constant}$$

Defining $u^+ = U/u^*$ and $y^+ = yu^*/\nu$, we have from experimental measurements of Nikuradse (1932)[116]

$$u^+ = \frac{1}{\kappa} \ln y^+ + \text{const.}$$

Assuming that the turbulence is nearly at equilibrium near the wall (i.e., the dissipation rate, ϵ, is equal to production rate), we have from k equation (2.19)

$$\epsilon = -\overline{uv}\frac{\partial U}{\partial Y} = u^{*3}\frac{1}{\kappa Y},$$

where κ, known as the von Kármán constant, is equal to 0.4. With these approximations, the ϵ equation becomes

$$0 = C_\epsilon \frac{\partial}{\partial Y}\left(\frac{k^2}{\epsilon}\frac{\partial \epsilon}{\partial Y}\right) + (C_{\epsilon 1} - C_{\epsilon 2})\frac{\epsilon^2}{k}$$

or

$$C_\epsilon = (C_{\epsilon 2} - C_{\epsilon 1})\frac{u^{*6}}{\kappa^2 k^3}.$$

From the near-wall data, $u^{*2}/k \approx 0.30$, $C_{\epsilon 2} = 1.92$, and $C_{\epsilon 1} = 1.45$. Then, $C_\epsilon \approx 0.07$. Generally, $C_\epsilon = 0.07$ is used.

2.4.7 Determining the Diffusion Constant, C_k

Consider the oscillating grid turbulence where the mean flow has a zero velocity. In this case, both the convective and the production terms are zero. For the turbulent kinetic energy, k, the exact equation becomes

$$\frac{d}{dz}\left(-\frac{1}{\rho}\overline{wp} - \overline{wk'}\right) - \epsilon = 0.$$

Here z is the direction of oscillation. The modeled k equation becomes

$$\frac{d}{dz}\left(C_k \frac{k^2}{\epsilon}\frac{\partial k}{\partial z}\right) - \epsilon = 0.$$

Thus, by direct comparison, we have, approximately,

$$-\frac{1}{\rho}\overline{wp} - \overline{wk'} = C_k \frac{k^2}{\epsilon}\frac{\partial k}{\partial z}.$$

To determine the constant C_k, we consider the experiment of Komatsu et al. (1986)[79]. The experimental setup is shown in Fig. 2.13. It consists of an oscillating grid device and a tank with 1.0-m length, 0.254-m width and 0.4-m depth. Turbulence is generated by a vertically oscillated grid in the fresh water. The grid is square with mesh size, $M = 5.0$ cm. The width of the square bar is 1.0 cm. The strokes (S_0) of the grid oscillation are 1.0, 4.0, and 8.0 cm, and the frequency (f_0) ranges from 1.5 to 6.0 Hz. Scotch–Crank

Table 2.4 Experimental data for the Komatsu et al. (1986)[167] experiment

Experiment	Viscosity ν (cm^2/s)	Stroke S_0 (cm)	Frequency f_0 (Hz)
	• symbol		
1	0.15	4.0	4.0
2	0.16	4.0	2.0
3	0.17	4.0	6.0
4	0.17	8.0	2.0
	▲ symbol		
5	0.10	8.0	2.0
6	0.10	4.0	2.0
7	0.10	4.0	4.0
8	0.11	4.0	6.0
	○ symbol		
9	0.073	4.0	2.0
10	0.073	4.0	4.0
11	0.068	4.0	6.0
12	0.075	8.0	2.0
	△ symbol		
13	0.010	8.0	2.0
14	0.010	4.0	2.0
15	0.010	4.0	4.0
16	0.010	4.0	6.0

systems are used to generate the sinusoidal oscillating motion. Two components of the turbulence velocity are measured with a V-shaped hot-film velocity probe attached to a motor-driven carriage. The probe is calibrated beforehand by driving the carriage at a constant speed in the range of 2.28 to 15.40 cm/s. The variables x and z represent the horizontal and vertical coordinates, respectively. The center of the grid oscillations is taken at $z = 0$. The x and z components of the turbulence velocity u and w are recorded. The signal is digitized at a sampling frequency of 500 Hz, and 4,096 data points are taken. The experimental data are indicated in Table 2.4, and the experimental plots are shown in Fig. 2.14. The turbulent kinetic energy, k, and the triple correlation, $\overline{wk'}$, can be measured with $k = 1/2(\overline{u^2} + \overline{v^2} + \overline{w^2}) \sim 1/2(2\overline{u^2} + \overline{w^2})$ and $\overline{wk'} = 1/2(2\overline{wu^2} + \overline{w^3})$. Komatsu et al. (1986)[79] neglected the pressure fluctuation term, $\overline{wp}/\rho$, and determined the dissipation rate of turbulent kinetic energy, ϵ, from the exact k equation by taking the derivative of the data for $\overline{wk'}$:

$$\epsilon = \frac{d}{dz}(-\overline{wk'}) = -\frac{d}{dz}\left[\frac{1}{2}(2\overline{wu^2} + \overline{w^3})\right].$$

Equating the above exact k equation with the modeled k equation, we have

$$\frac{1}{2}(2\overline{wu^2} + \overline{w^3}) = -C_k \frac{k^2}{\epsilon}\frac{\partial k}{\partial z}. \tag{2.37}$$

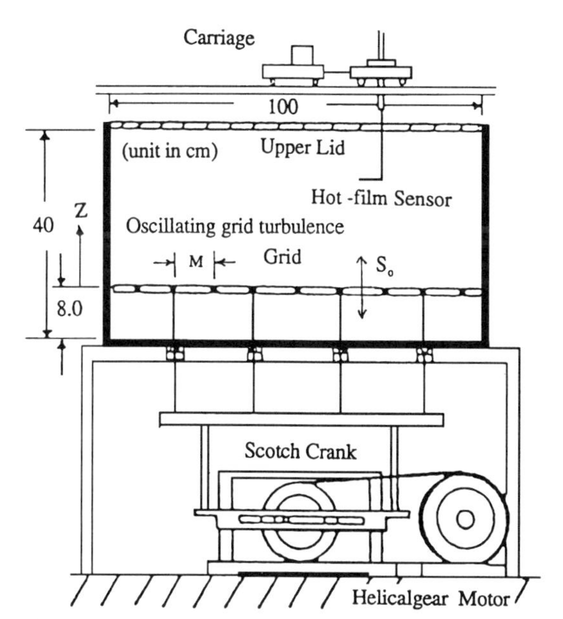

Figure 2.13 Experimental setup for determining C_k.

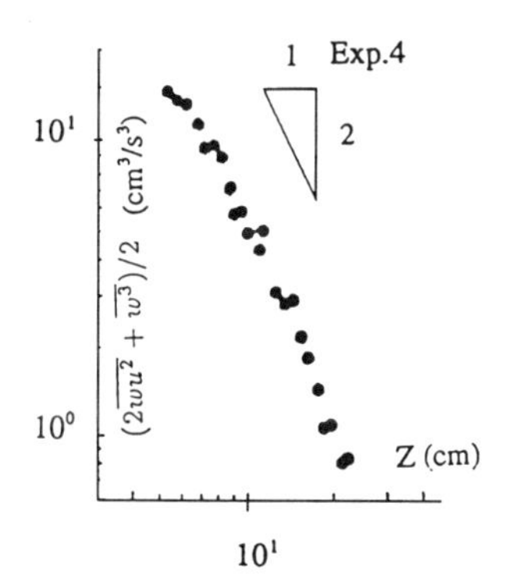

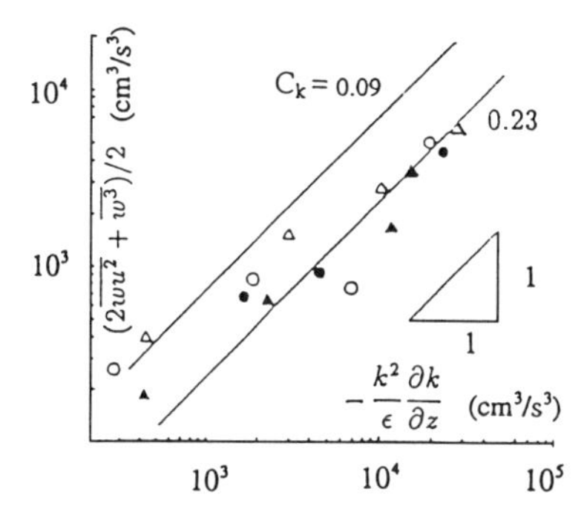

Figure 2.14 Experimental data plots for determining C_k. (Adapted from Komatsu et al. (1986).)

Plotting $(\overline{2wu^2}+\overline{w^3})/2$ versus $-k^2\partial k/\epsilon\partial z$ on a log-log coordinate, C_k can be determined from the intersection with the x-axis and is found to be about 0.23. The C_k value usually used, $C_k = 0.09$, is also plotted for comparison in Fig. 2.15. The discrepancy between the experimentally determined value of 0.23 and the computer optimization value of 0.09 may be due to the omission of the pressure velocity correlation in Komatsu et al.'s data [79].

2.4.8 Determining the Thermal Eddy Diffusion Constant, C_T

C_T can be determined from similar experimental data with oscillating grid turbulence and heating. C_T is also often taken to be of similar value to C_k. A commonly used value is $C_T = 0.07$. It should be noted that in high-shear flows, turbulent diffusion is not effective. Thus, some variations in the values of turbulent diffusion coefficients for C_k, C_ϵ, and C_T have small effects on the prediction results.

2.5 SUMMARY AND CONCLUSIONS

2.5.1 Remarks

In this chapter, we attempted to develop a second-order closure turbulence model, which is based on a logical set of postulations, for an incompressible, viscous, thermally conducting fluid. Similarities between the Stokes postulations and the turbulence model postulations were pointed out. Turbulence constants were calibrated with selected experiments. The model must now be tested for its validity.

2.5.2 Summary of Modeled Equations

- The Reynolds stress equation is given by

$$\begin{aligned}\frac{D\overline{u_iu_j}}{Dt} &= \frac{\partial}{\partial X_l}\left[\left(C_k\frac{k^2}{\epsilon}+\nu\right)\frac{\partial\overline{u_iu_j}}{\partial X_l}\right] - \left(\overline{u_iu_l}\frac{\partial U_j}{\partial X_l}+\overline{u_ju_l}\frac{\partial U_i}{\partial X_l}\right)\\ &\quad -\frac{2}{3}\delta_{ij}\epsilon - C_1\frac{\epsilon}{k}\left(\overline{u_iu_j}-\frac{2}{3}\delta_{ij}k\right)\\ &\quad + C_2\left(\overline{u_iu_l}\frac{\partial U_j}{\partial X_l}+\overline{u_ju_l}\frac{\partial U_i}{\partial X_l}-\frac{2}{3}\delta_{ij}\overline{u_nu_m}\frac{\partial U_n}{\partial X_m}\right)\\ &= D_{ij}+P_{ij}-\frac{2}{3}\delta_{ij}\epsilon+\text{PS},\end{aligned} \tag{2.17}$$

  where $C_k = 0.09$, $C_1 = 2.30$, and $C_2 = 0.40$.
- The kinetic energy, k, equation is given by

$$\begin{aligned}\frac{Dk}{Dt} &= \frac{\partial}{\partial X_l}\left[\left(C_k\frac{k^2}{\epsilon}+\nu\right)\frac{\partial k}{\partial X_l}\right] - \overline{u_iu_l}\frac{\partial U_i}{\partial X_l} - \epsilon\\ &= D_k + P_k - \epsilon,\end{aligned} \tag{2.19}$$

  where $C_k = 0.09 \sim 0.11$.

- The rate of dissipation, ϵ, equation is given by

$$\begin{aligned}\frac{D\epsilon}{Dt} &= \frac{\partial}{\partial X_l}\left[\left(C_\epsilon \frac{k^2}{\epsilon} + \nu\right)\frac{\partial \epsilon}{\partial X_l}\right] - C_{\epsilon 1}\frac{\epsilon}{k}\overline{u_i u_l}\frac{\partial U_i}{\partial X_l} - C_{\epsilon 2}\frac{\epsilon^2}{k} \\ &= D_\epsilon + P_\epsilon - \Phi_\epsilon, \end{aligned} \quad (2.26)$$

  where $C_\epsilon = 0.07$, $C_{\epsilon 1} = 1.45$, and $C_{\epsilon 2} = 1.92$.
- The Reynolds heat flux, $\overline{u_i\theta}$, equation is given by

$$\begin{aligned}\frac{D\overline{u_i\theta}}{Dt} &= \frac{\partial}{\partial X_l}\left[\left(C_T \frac{k^2}{\epsilon} + \alpha\right)\frac{\partial \overline{u_i\theta}}{\partial X_l}\right] - \left(\overline{u_i u_l}\frac{\partial T}{\partial X_l} + \overline{u_l\theta}\frac{\partial U_i}{\partial X_l}\right) \\ &\quad - C_{T1}\frac{\epsilon}{k}\overline{u_i\theta} + C_{T2}\overline{u_m\theta}\frac{\partial U_i}{\partial X_m} \\ &= D_{u\theta} + P_{u\theta} - \mathrm{PT}, \end{aligned} \quad (2.28)$$

  where $C_T = 0.07$, $C_{T1} = 3.20$, and $C_{T2} = 0.50$.

2.5.3 Closure Problem

With the averaged flow equations and the turbulent equations, we have sixteen equations and sixteen unknowns. This closes the turbulent flow problem. There are nine constants generated by the modeling processes. These values are determined from several turbulent flow experiments just as the molecular viscosity, thermal conductivity, and specific heats are determined from laminar flow experiments.

CHAPTER

THREE

DISCUSSIONS OF TURBULENCE MODELS

3.1 VARIATION OF SECOND-ORDER TURBULENCE MODELS

3.1.1 General Remarks

The most comprehensive second-order averaged governing equations for turbulent flow problems are, of course, the complete set of modeled equations, the $\overline{u_i u_j}$, k, ϵ, and $\overline{u_i \theta}$ equations presented in chapter 2. However, to economize the computational effort, several simplified versions of the turbulence model may be devised from the modeled governing equations.

It can also be shown that the various first-order closure models, such as the Boussinesq eddy viscosity model or the Prandtl mixing length model, can be all embedded in the second-order turbulence model. The predictability of the second-order turbulence model has vastly improved over that of the first-order models.

There are some improvements yet to be made in modeling the ϵ equation, the pressure-strain (PS) term, nonisotropic turbulent diffusion, nonisotropic dissipation, near-wall conditions, near-free-surface conditions, and secondary flows.

3.1.2 The Differential Model

This is sometimes called *the Reynolds stress model (RSM)*. That is, the differential forms of $\overline{u_i u_j}$ and $\overline{u_i \theta}$ are used, or

$$\frac{D\overline{u_i u_j}}{Dt} = D_{ij} + P_{ij} - \frac{2}{3}\delta_{ij}\epsilon + \text{PS} \tag{3.1}$$

$$\frac{Dk}{Dt} = D_k + P_k - \epsilon \tag{3.2}$$

$$\frac{D\epsilon}{Dt} = D_\epsilon + P_\epsilon - \Phi_\epsilon \tag{3.3}$$

$$\frac{D\overline{u_i\theta}}{Dt} = D_{u\theta} + P_{u\theta} + \text{PT}, \tag{3.4}$$

where

$$\epsilon = \nu \overline{\frac{\partial u_i \partial u_i}{\partial X_l \partial X_l}}.$$

The above-indicated equations are solved along with the mean equations,

- the mass conservation equation,

$$\frac{\partial U_i}{\partial X_i} = 0, \tag{3.5}$$

- the averaged momentum equation,

$$\rho \frac{DU_i}{Dt} = \rho G_i - \frac{\partial P}{\partial X_i} + \frac{\partial \tau_{ij}}{\partial X_j} + \frac{\partial \tau^t_{ij}}{\partial X_j}, \tag{3.6}$$

where

$$\tau_{ij} = \mu\left(\frac{\partial U_i}{\partial X_j} + \frac{\partial U_j}{\partial X_i}\right),$$

$$\tau^t_{ij} = -\rho\overline{u_i u_j},$$

and

- the averaged energy equation,

$$\rho C_p \frac{DT}{Dt} = \tau_{ij}\frac{\partial U_i}{\partial X_j} - \frac{\partial q_i}{\partial X_i} + \frac{\partial q^t_i}{\partial X_i} + \Phi^t, \tag{3.7}$$

where

$$q_i = -K\frac{\partial T}{\partial X_i}$$

$$q^t_i = -\rho C_p \overline{u_i\theta},$$

and

$$\Phi^t = \overline{\tau'_{ij}\frac{\partial u_i}{\partial X_j}}$$

$$= \mu\overline{\left(\frac{\partial u_i}{\partial X_j} + \frac{\partial u_j}{\partial X_i}\right)\frac{\partial u_i}{\partial X_j}}$$

$$= \epsilon\rho.$$

Note that normally $\Phi^t \ll \tau_{ij}\partial U_i/\partial X_j$. Hence, Φ^t is neglected. However, the Φ^t term shows that the turbulent kinetic energy that disappears in the k equation reappears in the averaged energy equation in the form of frictional heating.

3.1.3 The Algebraic Stress Model (ASM)

This is also known as the k-ϵ-A model (ASM). To solve a turbulent flow with the differential turbulence model (Reynolds stress model; RSM), we need to solve sixteen equations: five equations in U_i, P, and T and eleven turbulent transport equations (six in $\overline{u_i u_j}$, two in k and ϵ, and three in $\overline{u_i \theta}$). It is too costly to solve eleven turbulent transport equations. As an engineering solution, we would like to make a proper approximation.

For flows when turbulent convection and diffusion are small (high-shear flows) or convection and diffusion are approximately equal (local equilibrium), then the $\overline{u_i u_j}$ equation may be approximated by dropping the convection and diffusion terms. Hence, the modeled $\overline{u_i u_j}$ equation, Eq. 2.17, becomes

$$0 = P_{ij} - \frac{2}{3}\delta_{ij}\epsilon - C_1 \frac{\epsilon}{k}\left(\overline{u_i u_j} - \frac{2}{3}\delta_{ij} k\right) - C_2\left(P_{ij} - \frac{2}{3}\delta_{ij} P_k\right), \tag{3.8}$$

where

$$P_{ij} = -\left(\overline{u_i u_l}\frac{\partial U_j}{\partial X_l} + \overline{u_j u_l}\frac{\partial U_i}{\partial X_l}\right)$$

and

$$P_k = -\overline{u_n u_m}\frac{\partial U_n}{\partial X_m}.$$

The resulting equation is an algebraic equation because there is no derivative of $\overline{u_i u_j}$ in the equation. The gain is that we need now to solve six algebraic $\overline{u_i u_j}$ equations instead of six differential $\overline{u_i u_j}$ equations.

Similarly, if the flows are high-shear flows with a high-temperature gradient or are in local turbulent equilibrium, then the $\overline{u_i \theta}$ equation can be approximated by dropping the convection and diffusion terms. Hence, the modeled $\overline{u_i \theta}$ equation, Eq. 2.28, becomes

$$0 = -\left(\overline{u_i u_l}\frac{\partial T}{\partial X_l} + \overline{u_l \theta}\frac{\partial U_i}{\partial X_l}\right) - C_{T1}\frac{\epsilon}{k}\overline{u_i \theta} + C_{T2}\overline{u_m \theta}\frac{\partial U_i}{\partial X_m}. \tag{3.9}$$

Again, the approximation leads to an algebraic $\overline{u_i \theta}$ equation. The gain is that we need to solve three algebraic equations as compared with three differential equations. The turbulence model is summarized in Table 3.1.

If the flow is not high shear with a large temperature gradient and local equilibrium, then we cannot neglect the convection and diffusion effects. However, Rodi (1972)[134] suggested that because $\overline{u_i u_j} \sim k$, one may approximate the left-hand side of the $\overline{u_i u_j}$ equation,

$$\frac{D\overline{u_i u_j}}{Dt} - D_{ij} = P_{ij} - \frac{2}{3}\delta_{ij}\epsilon - C_1 \frac{\epsilon}{k}\left(\overline{u_i u_j} - \frac{2}{3}\delta_{ij} k\right) - C_2\left(P_{ij} - \frac{2}{3}\delta_{ij} P_k\right), \tag{3.10}$$

as

$$\begin{aligned}\frac{D\overline{u_i u_j}}{Dt} - D_{ij} &= \frac{D\left(\frac{\overline{u_i u_j}}{k}\right)k}{Dt} - \left(\frac{\overline{u_i u_j}}{k}\right)D_{ij} \\ &= \frac{\overline{u_i u_j}}{k}\left(\frac{Dk}{Dt} - D_k\right).\end{aligned} \tag{3.11}$$

Table 3.1 Summary of turbulence model (High-shear–high-temperature gradient and local equilibrium flows) (k-ϵ-A)

Variable	Equation
$\overline{u_i u_j}$	$(1-C_2)P_{ij} - C_1\frac{\epsilon}{k}\left(\overline{u_i u_j} - \frac{2}{3}\delta_{ij}k\right) - \frac{2}{3}\delta_{ij}(\epsilon - C_2 P_k) = 0$
$\overline{u_i \theta}$	$\overline{u_l u_i}\frac{\partial T}{\partial X_l} + C_{T1}\frac{\epsilon}{k}\overline{u_i\theta} + (1-C_{T2})\overline{u_m\theta}\frac{\partial U_i}{\partial X_m} = 0$
k	$\frac{Dk}{Dt} = D_k + P_k - \epsilon$
	$P_{ij} = -\left(\overline{u_i u_l}\frac{\partial U_j}{\partial X_l} + \overline{u_j u_l}\frac{\partial U_i}{\partial X_l}\right)$
	$P_k = -\overline{u_i u_l}\frac{\partial U_i}{\partial X_l}$
ϵ	$\frac{D\epsilon}{Dt} = D_\epsilon - C_{\epsilon 1}\frac{\epsilon}{k}\overline{u_i u_l}\frac{\partial U_i}{\partial X_l} - C_{\epsilon 2}\frac{\epsilon^2}{k}$

Hence using the modeled k equation, and Eq. 3.11, in Eq. 3.10, we have

$$\frac{\overline{u_i u_j}}{k}(P_k - \epsilon) = P_{ij} - \frac{2}{3}\delta_{ij}\epsilon - C_1\frac{\epsilon}{k}\left(\overline{u_i u_j} - \frac{2}{3}\delta_{ij}k\right) - C_2\left(P_{ij} - \frac{2}{3}\delta_{ij}P_k\right). \quad (3.12)$$

This is an algebraic equation that retains some effects of convection–diffusion.

Similarly, we can assume that $\overline{u_i\theta} \sim k$ and replace the left-hand side of the $\overline{u_i\theta}$ equation,

$$\frac{D\overline{u_i\theta}}{Dt} - D_{u\theta} = -\left(\overline{u_i u_l}\frac{\partial T}{\partial X_l} + \overline{u_l\theta}\frac{\partial U_i}{\partial X_l}\right) - C_{T1}\frac{\epsilon}{k}\overline{u_i\theta} + C_{T2}\overline{u_m\theta}\frac{\partial U_i}{\partial X_m}, \quad (3.13)$$

by

$$\frac{D\overline{u_i\theta}}{Dt} - D_{u\theta} = \frac{\overline{u_i\theta}}{k}\left(\frac{Dk}{Dt} - D_k\right)$$

$$= \frac{\overline{u_i\theta}}{k}(P_k - \epsilon). \quad (3.14)$$

Hence, by applying Eq. 3.14 in Eq. 3.13, we have

$$\frac{\overline{u_i\theta}}{k}(P_k - \epsilon) = -\left(\overline{u_i u_l}\frac{\partial T}{\partial X_l} + \overline{u_l\theta}\frac{\partial U_i}{\partial X_l}\right) - C_{T1}\frac{\epsilon}{k}\overline{u_i\theta} + C_{T2}\overline{u_m\theta}\frac{\partial U_i}{\partial X_m}. \quad (3.15)$$

The performance of Rodi's assumption in general improves over the model that neglects the convection–diffusion terms. However, the assumpuon is not appropriate near the central-line region of a jet, wake, or annular flow where diffusion and convection effects are dominant.

3.1.4 Eddy Viscosity Model

This is also known as the k-ϵ-E model. In the eddy viscosity model, one gives up the modeled $\overline{u_i u_j}$ and $\overline{u_i\theta}$ equations and adopts the generalized Boussinesq eddy viscosity

Table 3.2 The k-ϵ-E model

Variable	Equation
$\overline{u_i u_j}$	$-\overline{u_i u_j} = C_\mu \frac{k^2}{\epsilon}\left(\frac{\partial U_i}{\partial X_j} + \frac{\partial U_j}{\partial X_i}\right) - \frac{2}{3}\delta_{ij}k$
$\overline{u_i\theta}$	$-\overline{u_i\theta} = \frac{C_\mu}{Pr_t}\frac{k^2}{\epsilon}\frac{\partial T}{\partial X_i}$
	$C_\mu \approx 0.09$
k	$\frac{Dk}{Dt} = D_k + P_k - \epsilon$
	$P_k = -\overline{u_i u_l}\frac{\partial U_i}{\partial X_l}$
ϵ	$\frac{D\epsilon}{Dt} = D_\epsilon - C_{\epsilon 1}\frac{\epsilon}{k}\overline{u_i u_j}\frac{\partial U_i}{\partial X_j} - C_{\epsilon 2}\frac{\epsilon^2}{k}$

model. Thus, the Reynolds stress equation is given by

$$-\overline{u_i u_j} = \nu_t\left(\frac{\partial U_i}{\partial X_j} + \frac{\partial U_j}{\partial X_i}\right) - \frac{2}{3}\delta_{ij}k, \tag{3.16}$$

whereas the Reynolds heat flux equation is given by

$$-\overline{u_i\theta} = \alpha_t\left(\frac{\partial T}{\partial X_i}\right), \tag{3.17}$$

where by dimensional analysis of k and ϵ, we find

$$\nu_t = C_\mu\frac{k^2}{\epsilon}$$

$$\alpha_t = C_\alpha\frac{k^2}{\epsilon} = \frac{C_\mu}{Pr_t}\frac{k^2}{\epsilon}.$$

Here k and ϵ are solved from the k and the ϵ equations, and $C_\mu = 0.09$ and $Pr_t = 0.8 \sim 1.3$, where Pr_t is the turbulent Prandtl number. The k-ϵ-E model is summarized in Table 3.2. In general, the performance of the k-ϵ-E model is not as good as that of the k-ϵ-A model in complex flows, but it is about the same as that of the k-ϵ-A model in boundary layer like flows.

Modified k-ϵ-E models or models with different coefficients have been proposed. In summary, these can be expressed as

$$\frac{Dk}{Dt} = \frac{\partial}{\partial X_l}\left(C_k\frac{k^2}{\epsilon}\frac{\partial k}{\partial X_l} + \nu\frac{\partial k}{\partial X_l}\right) - \overline{u_i u_l}\frac{\partial U_i}{\partial X_l} - \epsilon \tag{3.18}$$

$$\frac{D\epsilon}{Dt} = \frac{\partial}{\partial X_l}\left(C_\epsilon\frac{k^2}{\epsilon}\frac{\partial \epsilon}{\partial X_l} + \nu\frac{\partial \epsilon}{\partial X_l}\right) - C_{\epsilon 1}\frac{\epsilon}{k}\overline{u_i u_l}\frac{\partial U_i}{\partial X_l} - C_{\epsilon 2}\frac{\epsilon^2}{k} - C_{\epsilon 3}\frac{\epsilon}{k}\frac{\partial}{\partial X_l}\left(C_k\frac{k^2}{\epsilon}\frac{\partial k}{\partial X_l}\right), \tag{3.19}$$

with model coefficients listed in Table 3.3. Jones and Launder (1972)[70] have pointed out that the difference between the predictions based on Set 1 and Set 2 coefficients is

Table 3.3 Coefficients used in various models

Set	Authors	C_μ	C_k	C_ϵ	$C_{\epsilon 1}$	$C_{\epsilon 2}$	$C_{\epsilon 3}$
1	Hanjalic and Launder (1972)[55]	0.07	0.07	0.064	1.45	2.00	0
2	Jones and Launder (1972)[72]	0.09	0.09	0.069	1.55	2.00	0
3	Launder et al. (1972)[87]	0.09	0.09	0.069	1.44	1.92	0
4	Yakhot and Orszag (1986)[189]	0.084	0.117	0.117	1.063	1.722	0
5	Jaw and Chen (1991)[69]	0.09	0.103	0.110	1.23	1.92	1.67

barely discernable. Launder et al. (1972)[87] has recommended the coefficients of Set 3 for practical applications after extensive examinations of turbulent free-shear flows. This set of model coefficients is used by most researchers at present. The k-ϵ model with Set 3 model coefficients has been applied to a large number of different flows, (e.g., plane jets, mixing layers, boundary layer flows, and so forth) so that it is now one of the best tested turbulence models (Rodi, 1980[135]). Compilations of further examples of the application of the k-ϵ model can be found in Launder and Spalding (1974)[92], Lumley (1978)[98], Rodi (1980, 1981)[135, 136], Chen (1983)[12], and Markatos (1986)[103]. Some references (Chen and Nikitopoulos, 1979[18]; Chen and Rodi, 1980[20]) also discuss a few exceptional flow situations that cannot be predicted satisfactorily with the given coefficients, notably axisymmetric jets and weak free-shear layers (e.g., far wakes), where overall turbulence production is small compared with dissipation. Launder and Spalding (1974)[92], Rodi (1980)[135], and Michelassi and Shih (1991)[107] also give empirical functions to replace some of the coefficients such as C_μ and $C_{\epsilon 1}$ so that more satisfactory predictions can be obtained from the k-ϵ model. However, little universality can be claimed for these empirical functions.

Special attention needs to be drawn to the model proposed by Yakhot and Orszag (1986)[189] who used the renormalization group (RNG) theory. Their model was derived on the basis of the assumption that at the small eddies, the length scale is approximated by the Kolmogorov energy spectrum. The model coefficients given in Set 4 can then be determined directly from renormalization analysis. When using the RNG theory, turbulence is considered to be created by a random force specified by a two-point correlation regardless of initial and boundary conditions. The analysis starts with fluctuations from small wavelengths and renormalizes those at a long wavelength. Yakhot and Orszag [189] have shown that the renormalization procedure that leads to the recurrence relations and fixed points for the turbulence viscosity can be applied to any correlation of fluctuating quantities in the flow field. The results of the theoretical developments are turbulence models for Reynolds-averaged and large-eddy simulation, (LES), with all model coefficients established directly from theory, as well as differential recursion relations for a low-Reynolds-number regime. Set 4 is the model coefficient for a high-Reynolds-number k-ϵ model.

It must be mentioned, however, that some problems associated with the specific numerical values of the constants in the RNG k-ϵ model have been reported by Smith and Reynolds (1992)[156]. For example, Speziale et al. (1989)[167] showed that the value of $C_{\epsilon 1} = 1.063$ yields excessively large growth rates for the turbulent kinetic energy in homogeneous shear flow in comparison to both physical and numerical experiments.

This led Yakhot et al. (1992)[190] to reformulate the derivation of the ϵ transport equation. In contrast to the earlier development of RNG models, the only new feature is the infrared cutoff of the random force in the forced Navier–Stokes (N-S) equations when wave number is smaller than a certain amount. This property, which is usually unimportant, is needed to derive an equation for the mean rate of energy dissipation ϵ. The ϵ transport equation thus derived and modeled has the form

$$\frac{D\epsilon}{Dt} = \frac{\partial}{\partial X_l}\left(C_\epsilon \frac{k^2}{\epsilon}\frac{\partial \epsilon}{\partial X_l} + \nu \frac{\partial \epsilon}{\partial X_l}\right) - C_{\epsilon 1}^* \frac{\epsilon}{k}\overline{u_i u_j}\frac{\partial U_i}{\partial X_l} - C_{\epsilon 2}\frac{\epsilon^2}{k}, \tag{3.20}$$

where coefficient $C_{\epsilon 1}^*$ is given by

$$C_{\epsilon 1}^* = 1.42 - \frac{\eta(1 - \eta/\eta_o)}{1 + \beta\eta^3}. \tag{3.21}$$

Here η is the ratio of the turbulent to mean-strain time scale $\eta = SK/\epsilon$ and $S = (2S_{ij}S_{ij})^{1/2}$, and $S_{ij} = \frac{1}{2}(\frac{\partial U_i}{\partial X_j} + \frac{\partial U_j}{\partial X_i})$ is the mean strain. η_o is the single fixed point obtained by setting $\frac{d\eta}{dt} = 0$ and is given as

$$\eta_o = \sqrt{\frac{C_{\epsilon 2} - 1}{C_\mu(C_{\epsilon 1} - 1)}}.$$

The constant β is related to the von Kármán constant and is determined as $\beta = 0.012$. With this modification, $C_{\epsilon 2}$ is reevaluated to be 1.68, and all of the rest constants are kept the same. Yakhot et al. (1992)[190] successfully applied this model to predict a backward facing step flow.

Set 5, the model proposed by Jaw and Chen (1991)[69], differs from the other models by including an additional term, cross-diffusion of k, in the ϵ equation. They reasoned that small turbulent eddies can be anisotropic and hence rederived the turbulence model based on this more realistic assumption. The two-equation model they produced is the same as the model shown above, but with $C_{\epsilon 3} \neq 0$. The inclusion of the cross-diffusion term is found to improve prediction. It is particularly helpful in flows in which the production is not equal to dissipation (i.e., nonequilibrium turbulent flows, such as round jet or far wake flows). The coefficients listed in Set 5 were also determined from some specific experiments and computer optimizations by predicting several turbulent free-shear flows (Jaw, 1991 [67]). For turbulent wall or high-shear flows, it is found that the effect of cross-diffusion is quite small and can be neglected. However, Set 5 requires further verification in wall-shear flows.

It is interesting to point out that the k-ω model proposed by Wilcox (1988)[185] can be transformed to a k-ϵ type model by using the relation $\omega = \epsilon/0.09k$. The equivalent ϵ equation is

$$\frac{D\epsilon}{Dt} = \frac{\partial}{\partial X_l}\left(0.045\frac{k^2}{\epsilon}\frac{\partial \epsilon}{\partial X_l} + \nu\frac{\partial \epsilon}{\partial X_l}\right) - 1.56\frac{\epsilon}{k}\overline{u_i u_l}\frac{\partial U_i}{\partial X_l} - 1.83\frac{\epsilon^2}{k}$$
$$- 0.09\frac{k^2}{\epsilon}\frac{\partial k}{\partial X_l}\frac{\partial(\epsilon/k)}{\partial X_l}, \tag{3.22}$$

which is quite similar to the anisotropic model proposed by Jaw and Chen (1991)[69] except that the model coefficients are different.

Several turbulence models including cross-diffusion terms have also been proposed by different researchers. A more recent one is the model proposed by Yoshizawa (1987)[191], who showed that a turbulent characteristic length-scale l and eddy viscosity ν_t can be expressed as

$$l = 1.84\frac{k^{3/2}}{\epsilon} + 4.95\frac{k^{3/2}}{\epsilon^2}\frac{Dk}{Dt} - 2.91\frac{k^{5/3}}{\epsilon^3}\frac{D\epsilon}{Dt} \tag{3.23}$$

$$\nu_t = 0.0349 l^{4/3}\epsilon^{1/3}, \tag{3.24}$$

respectively. Retaining the first term of the l equation and substituting it into the ν_t expression, the familiar expression for ν_t is recovered as $\nu_t = 0.0785k^2/\epsilon$. However, to impose the condition for the vanishing effect from the last two terms of the l equation, it is required that

$$\begin{aligned}\frac{D\epsilon}{Dt} &= \frac{4.95}{2.91}\frac{\epsilon}{k}\frac{Dk}{Dt}\\ &= 1.7\frac{\epsilon}{k}\frac{Dk}{Dt}.\end{aligned}$$

From this relation, Yoshizawa (1987)[191] derived his version of the k-ϵ model as

$$\frac{Dk}{Dt} = \frac{\partial}{\partial X_l}\left(C_k\frac{k^2}{\epsilon}\frac{\partial k}{\partial X_l} + \nu\frac{\partial k}{\partial X_l}\right) - \overline{u_i u_l}\frac{\partial U_i}{\partial X_l} - \epsilon + \frac{\partial}{\partial X_l}\left(C_{k\epsilon}\frac{k^3}{\epsilon^2}\frac{\partial \epsilon}{\partial X_l}\right) \tag{3.25}$$

$$\begin{aligned}\frac{D\epsilon}{Dt} &= \frac{\partial}{\partial X_l}\left(C_\epsilon\frac{k^2}{\epsilon}\frac{\partial \epsilon}{\partial X_l} + \nu\frac{\partial \epsilon}{\partial X_l}\right) - C_{\epsilon 1}\frac{\epsilon}{k}\overline{u_i u_l}\frac{\partial U_i}{\partial X_l}\\ &\quad - C_{\epsilon 2}\frac{\epsilon^2}{k} + \frac{\partial}{\partial X_l}\left(C_{\epsilon k}k\frac{\partial k}{\partial X_l}\right) + C_{\epsilon k1}\left(\frac{\partial k}{\partial X_l}\right)^2\\ &\quad + C_{\epsilon k2}\left(\frac{k}{\epsilon}\frac{\partial k}{\partial X_l}\frac{\partial \epsilon}{\partial X_l}\right) + C_{\epsilon k3}\left(\frac{\partial \epsilon}{\partial X_l}\right)^2\end{aligned} \tag{3.26}$$

with $C_k = 1.27$, $C_{k\epsilon} = -0.72$, $C_\epsilon = -1.22$, $C_{\epsilon 1} = C_{\epsilon 2} = 1.70$, $C_{\epsilon k} = 2.16$, $C_{\epsilon k1} = 2.16$, $C_{\epsilon k2} = -3.38$, and $C_{\epsilon k3} = 1.22$. One of the notable features of the Yoshizawa (1987)[191] model is the appearance of cross-diffusion terms in both the k and the ϵ equations. The validity of the model requires more study and applications in different turbulent flows. Takemitsu (1986, 1990)[172, 173] simplified the ϵ equation by neglecting the last three terms and successfully applied it to predict two-dimensional channel flow.

The statistical mechanics approach is an area worthy of further experimentation within the structure of two-equation models.

3.1.5 The k-ϵ-Nonlinear RSM

The k-ϵ-nonlinear Reynolds stress turbulence model is essentially a two-equation model that solves for k and ϵ from differential models, whereas the Reynolds stress is determined from a nonlinear algebraic equation by generalizing the eddy viscosity model (Barton, Rubinstein, and Kirtly, 1991 [4]). In other words, additional nonlinear terms for the mean-strain rate are added to the Boussinesq eddy viscosity model.

The rationale for this is that the k-ϵ-eddy viscosity models presented so far have adopted the notion of a scalar turbulence viscosity. This assumption forces the principal axes of $\overline{u_i u_j}$ and the mean-strain rate, $S_{ij} = (\partial U_i/\partial X_j) + (\partial U_j/\partial X_i)$, to be aligned. This is true in pure strain, but is not true in any flow with mean vorticity. In practice, this assumption has proved adequate in two-dimensional flows without swirl, in which only one stress component exerts much influence on the flow development. In flows with swirl, however, and indeed in three-dimensional flows generally, the measured flow distribution can be predicted in detail only by choosing a different level of viscosity for each active stress component. It is thus clear that a mechanism must be provided in the turbulence model that permits inequality of the normal stresses (i.e., that allows for anisotropy). There are two approaches available: (1) development of prognostic equations for the individual Reynolds stresses and (2) development of a nonlinear Reynolds stress model by using a tensor expression for viscosity to allow for directional dependence of the transport coefficients. The former approach has been presented; the latter is introduced in the following discussion.

Several nonlinear RSMs in algebraic form have been proposed to date; for example Crow (1968)[36], Lumley (1970)[97], Pope (1975)[123], and Speziale (1987)[166]. In general, the nonlinear RSM can be written in the form of

$$-\overline{u_i u_j} = -\frac{2}{3}\delta_{ij}k + \nu_t\left[\frac{\partial U_i}{\partial X_j} + \frac{\partial U_j}{\partial X_i}\right] + \frac{\delta_{ij}}{3}\sum_{m=1}^{3} C_{\tau m}\frac{k^3}{\epsilon^2}S_{mll} - \sum_{m=1}^{3} C_{\tau m}\frac{k^3}{\epsilon^2}S_{mij}. \tag{3.27}$$

Here, S_{mij} is defined as

$$S_{1ij} = \frac{\partial U_i}{\partial X_l}\frac{\partial U_j}{\partial X_l},$$

$$S_{2ij} = \frac{1}{3}\left(\frac{\partial U_i}{\partial X_l}\frac{\partial U_l}{\partial X_j} + \frac{\partial U_j}{\partial X_l}\frac{\partial U_l}{\partial X_i}\right),$$

and

$$S_{3ij} = 7\frac{\partial U_l}{\partial X_i}\frac{\partial U_l}{\partial X_j},$$

respectively, and ν_t is defined as before, $\nu_t = C_\mu k^2/\epsilon$, with $C_\mu = 0.09$. Coefficients $C_{\tau 1}, C_{\tau 2}$, and $C_{\tau 3}$ of the three most recent models by Speziale (1987)[166], Nisizima and Yoshizawa (1987)[117], and Rubinstein and Barton (1990)[147] are listed in Table 3.4 for comparison.

Speziale (1987)[166] derived the nonlinear constitutive model for anisotropic turbulence model in a way analogous to the modeling of the laminar flow of a non-Newtonian fluid. Rivlin (1957)[132] pointed out that in the non-Newtonian case, flow in a noncircular duct was accomplished by secondary flow in the plane normal to the flow direction, thus terming its origin *a normal stress effect*. Constitutive modeling of non-Newtonian flow requires a nonlinear stress–strain rate relation, in which the stresses depend quadratically on the mean velocity gradients. The same philosophy was recommended later by

Table 3.4 Coefficients of the nonlinear Reynolds stress model

Authors	$C_{\tau 1}$	$C_{\tau 2}$	$C_{\tau 3}$
Speziale (1987)[166]	0.041	0.014	−0.014
Nisizima and Yoshizawa (1987)[117]	0.057	−0.167	−0.0067
Rubinstein and Barton (1990)[147]	0.034	0.104	−0.014

Liepmann (1964)[96] for turbulent flow of a Newtonian fluid, which was then followed by the recommendations of Crow (1968)[36] and Lumley (1970)[97]. Speziale's model was successfully applied to plane channel flow.

The model of Nisizima and Yoshizawa (1987)[117] was derived from statistical approaches, with turbulence regarded as a phenomenon comprising both universal and nonuniversal behavior. Universal behavior can conceivably be simulated with a model having a wide range of applications. Nonuniversal behavior is brought about by large-scale inhomogeneities, such as in geometry or in boundary conditions. Such inhomogeneities must be resolved by the governing equations and cannot be modeled in any general manner. A statistical approach is thus looking for ways to model the universal behavior or at least to model the gross effects of the universal behavior on the nonuniversal part of the flow. Nisizima and Yoshizawa (1987)[117] applied the nonlinear Reynolds stress model given in Eq. 3.27 to square duct flow, along with their version of a k-ϵ model:

$$\frac{Dk}{Dt} = \frac{\partial}{\partial X_l}\left(C_k \frac{k^2}{\epsilon}\frac{\partial k}{\partial X_l} + \nu \frac{\partial k}{\partial X_l}\right) - \overline{u_i u_l}\frac{\partial U_i}{\partial X_l} - \epsilon \tag{3.28}$$

$$\frac{D\epsilon}{Dt} = \frac{\partial}{\partial X_l}\left(C_\epsilon \frac{k^2}{\epsilon}\frac{\partial \epsilon}{\partial X_l} + \nu \frac{\partial \epsilon}{\partial X_l}\right) - C_{\epsilon 1} k \left(\frac{\partial U_i}{\partial X_j} + \frac{\partial U_j}{\partial X_i}\right)^2 - C_{\epsilon 2}\frac{\epsilon^2}{k}. \tag{3.29}$$

Here the ϵ equation is slightly different from that which is generally adopted. The production term in ϵ has the additional mean-strain term. With $C_k = 0.09$, $C_\epsilon = 0.069$, $C_{\epsilon 1} = 0.13$, and $C_{\epsilon 2} = 1.90$, Nisizima and Yoshizawa (1987)[117] produced reasonable results.

Rubinstein and Barton (1990)[147] derived a nonlinear generalization of the form proposed by Yoshizawa by using the Yakhot–Orszag RNG theory. Application of RNG theory to the Reynolds stresses leads to a series expansion for the stresses in powers of a parameter. The terms of orders zero and one produce the simple linear model; the terms of order two produce a quadratically nonlinear model, as the last two terms listed in Eq. 3.27. The model of Rubinstein and Barton (1990)[147] was shown to be able to predict anisotropic turbulent flows for both high and low Reynolds numbers. All model coefficients are derived from RNG theory and are not subject to empirical adjustments other than the basic assumptions mentioned earlier. Listed in Table 3.4 are the model coefficients of the high-Reynolds-number k-ϵ model with the nonlinear RSM. The model was tested against measurements for equilibrium homogeneous shear flow and fully developed turbulent flow in a square duct. Satisfactory results were obtained for both applications.

The nonlinear Reynolds stress models mentioned above consist of algebraic equations alone, and, therefore, no transport effects in convection and diffusion are considered.

Hence, these models are strictly applicable only to flows in which production and dissipation of turbulence energy are in balance.

3.1.6 Multiscale Models

All of the turbulence transport models discussed so far employed only a single time scale and a single length scale to characterize turbulent motion. However, it is recognized (Hinze, 1975)[60] that turbulence comprises fluctuating motions with a wide range of eddy sizes and time scales. Because different turbulent interactions are known to be associated with different parts of the eddy spectrum, the idea that one can mimic the response of the Reynolds stresses with a mathematical model containing just one-time or one-length scale seems highly simplistic. Hanjalic, Launder, and Schiestel (1980)[57] have pointed out that although numerous successful applications of the single-time-scale model have been reported in the literature, the success may be due to the flow considered having been fairly close to a spectral equilibrium in which the single-time-scale hypothesis is adequate. On the other hand, there are a few applications in which the present single-scale turbulence model fails, such as for round jet, wake, buoyant flow, or flow with separation. Therefore, several researchers proposed multiscale models by using more than one time scale in the modeled transport equations.

Models of the multiple-time-scale zone method are presented by the following equations, where the k and the ϵ equations are split and remodeled into two zones. In general, researchers have adopted the idea of dividing the whole energy spectrum into two parts: the energy-containing eddies and the energy-dissipating eddies. Separate transport equations are solved for the turbulence energy across the spectrum. The model is mathematically expressed as

$$\begin{aligned}\frac{Dk_p}{Dt} &= \frac{\partial}{\partial X_l}\left[\left(\frac{\nu_t}{\sigma_{kp}}+\nu\right)\frac{\partial k_p}{\partial X_l}\right] - \overline{u_i u_l}\frac{\partial U_i}{\partial X_l} - \epsilon_p \\ &= \frac{\partial}{\partial X_l}\left[(\nu_t+\nu)\frac{\partial k_p}{\partial X_l}\right] + P_k - \epsilon_p\end{aligned} \tag{3.30}$$

$$\frac{Dk_T}{Dt} = \frac{\partial}{\partial X_l}\left[\left(\frac{\nu_t}{\sigma_{kT}}+\nu\right)\frac{\partial k_T}{\partial X_l}\right] + \epsilon_p - \epsilon_T \tag{3.31}$$

$$\frac{D\epsilon_p}{Dt} = \frac{\partial}{\partial X_l}\left[\left(\frac{\nu_t}{\sigma_{\epsilon p}}+\nu\right)\frac{\partial \epsilon_p}{\partial X_l}\right] + C_{p1}\frac{\epsilon_p}{k_p}P_k - C_{p2}\frac{\epsilon_p^2}{k_p} + C_{p3}\frac{P_k^2}{k_p} \tag{3.32}$$

$$\frac{D\epsilon_T}{Dt} = \frac{\partial}{\partial X_l}\left[\left(\frac{\nu_t}{\sigma_{\epsilon T}}+\nu\right)\frac{\partial \epsilon_T}{\partial X_l}\right] + C_{T1}\frac{\epsilon_p\epsilon_T}{k_p} - C_{T2}\frac{\epsilon_T^2}{k_T} + C_{T3}\frac{\epsilon_p^2}{k_T}. \tag{3.33}$$

Here k_p and k_T are the turbulent kinetic energy in the production and dissipation ranges, P_k is the rate at which turbulent energy is produced (or extracted) from the mean motion, ϵ_p is the rate at which energy is transferred out of the production range, and ϵ_T is the rate at which energy is transferred into the dissipation range from the inertial range. For simplicity, ϵ_T is assumed equal to ϵ, the rate at which turbulence energy is dissipated. The model coefficients used are listed in Table 3.5. Generalization of the model for a multiple split-spectrum case and extension of the multiple-time-scale concept to convection–diffusion of scalar variables was given by Schiestel (1987)[148].

Table 3.5 Coefficients of multiple-time-scale turbulence model

	Authors	
Coefficient	Hanjalic and Launder (1980)	Kim and Chen (1989)
ν_t	$0.1\frac{kk_p}{\epsilon_p}$	$0.09\frac{k^2}{\epsilon_p}$
σ_{kp}	1.11	0.75
σ_{kT}	1.11	0.75
$\sigma_{\epsilon p}$	1.11	1.15
$\sigma_{\epsilon T}$	1.11	1.15
C_{p1}	2.20	1.24
C_{p2}	$1.8 - 0.3\left(\frac{\frac{k_p}{k_T}-1}{\frac{k_p}{k_T}+1}\right)$	1.84
C_{p3}	0.00	0.21
C_{T1}	$1.08\frac{\epsilon_p}{\epsilon_T}$	1.28
C_{T2}	1.15	1.66
C_{T3}	0.00	0.29

The multiple-time-scale turbulence model of Hanjalic et al. (1980)[57] has been applied to predict turbulent free-shear flows, underexpanded supersonic jets (Abdol-Hamid and Wilmoth, 1989 [1]), and confined swirling jets (Chen, 1985)[26], and improved results have been obtained. In the model, a fixed ratio of the turbulent kinetic energy of large eddies k_p to that of the fine scale eddies k_T was used to partition the turbulent kinetic energy spectrum. Because the turbulent kinetic energy in complex turbulent flows is characterized by an inhomogeneous turbulent kinetic energy spectrum distributed over a wide range of wave numbers, the following difficulties may arise in the model. If the partition is located in a too-high wave number region, then the multiple-scale turbulence model will reduce to a single-scale model. On the other hand, if the partition is located in a too-low wave number region, then production of turbulent kinetic energy will be contained in the dissipation region.

This difficulty has been eliminated by use of a variable partitioning, as proposed by Kim and Chen (1989)[75], of the local turbulent kinetic energy spectrum in such a way that the partition is moved toward the high wave number region when the production is high and the partition is moved toward the low wave number region when the production vanishes. That is, the location of the partition, or the ratio k_p/k_T, is determined as part of the solution. This is accomplished by introducing the $C_{p3}P_k^2/k_p$ and $C_{T3}\epsilon_p^2/k_T$ terms in the ϵ_p and ϵ_T transport equations, respectively. The former term increases the energy transfer rate when the production is high, and the latter term increases the dissipation rate when the energy transfer rate is high. The variable partitioning method causes the effective eddy viscosity coefficient to increase when the production is high and to decrease when the production vanishes. The same effect of the production rate on the eddy viscosity coefficient can also be found in the generalized algebraic stress turbulence model (Launder, 1982)[83]. Turbulent flows to which this multiple-time-scale turbulence model has been applied include a class of turbulent boundary layer flows, a class of separated flows, and a confined coaxial swirling jet. A multiscale turbulence model, which is based on the k-ω turbulence model, has also been proposed by Wilcox (1988)[184]. In his model, the turbulent shear stress tensor is computed in addition to the turbulent kinetic energy and the specific dissipation rate, k and ω. The details of the multiscale turbulence model of Wilcox are significantly different from those of the multiple-time-scale model of Hanjalic and Launder, or Kim and Chen, even though the underlying physics are similar. Interested readers are referred to Wilcox's (1988)[184] paper.

Instead of using the idea of dividing the energy spectrum, Chen and Singh (1986, 1990)[21, 22] and Jaw and Chen (1990)[68] proposed two-scale models by adopting both the large-eddy, or energy-containing, scale (k, ϵ) and the small-eddy, or energy-dissipating, scale (ν, ϵ) in the modeling of the ϵ transport equation. Detailed descriptions of these methods will be given in Section 3.4.

3.1.7 One-Equation Model

There were attempts to simplify the turbulence model from the two-equation k-ϵ model. This led to the development of the k equation, or one-equation, model. However, in general, the performance is not as satisfactory, and, in particular, the predictability decreases when the flows are complex. As an example, we consider a one-equation model for k,

$$\frac{Dk}{Dt} = \frac{\partial}{\partial X_l}\left[\left(C_k\frac{k^2}{\epsilon} + \nu\right)\frac{\partial k}{\partial X_l}\right] - \overline{u_i u_l}\frac{\partial U_i}{\partial X_l} - \epsilon. \tag{3.34}$$

From the (k, ϵ) scale, $l = k^{3/2}/\epsilon$, or $\epsilon = k^{3/2}/l$, and thus we have

$$\frac{Dk}{Dt} = \frac{\partial}{\partial X_l}\left[(C_k\sqrt{k}l + \nu)\frac{\partial k}{\partial X_l}\right] - \overline{u_i u_l}\frac{\partial U_i}{\partial X_l} - \frac{k^{3/2}}{l}. \tag{3.35}$$

The Reynolds stress and heat flux are

$$-\overline{u_i u_j} = \nu_t\left(\frac{\partial U_i}{\partial X_j} + \frac{\partial U_j}{\partial X_i}\right) - \frac{2}{3}\delta_{ij}k. \tag{3.36}$$

$$-\overline{u_i \theta} = \frac{\nu_t}{Pr_t}\frac{\partial T}{\partial X_i}, \tag{3.37}$$

where $\nu_t = C_\mu k^2/\epsilon = C_\mu\sqrt{k}l$ and $C_\mu = C_k \approx 0.09$. Here l is a mixing length and must be provided. In general, l depends on the problem. Thus, a universal one-equation model is difficult to create because the mixing length cannot be specified universally. The one-equation model is summarized in Table 3.6.

Table 3.6 The one-equation model (l-problem function and Pr_t-turbulent Prandtl number)

Variable	Equation
$\overline{u_i u_j}$	$-\overline{u_i u_j} = C_\mu \sqrt{k} l \left(\frac{\partial U_i}{\partial X_j} + \frac{\partial U_j}{\partial X_i} \right) - \frac{2}{3} \delta_{ij} k$
k	$\frac{Dk}{Dt} = \frac{\partial}{\partial X_l} \left[(C_k \sqrt{k} l + \nu) \frac{\partial k}{\partial X_l} \right] - \overline{u_i u_l} \frac{\partial U_i}{\partial X_l} - \frac{k^{3/2}}{l}$
$\overline{u_i \theta}$	$-\overline{u_i \theta} = \frac{C_\mu \sqrt{k} l}{Pr_t} \frac{\partial T}{\partial X_i}$
	$Pr_t = 0.8 \sim 1.3$

3.2 TURBULENT FLOW PREDICTIONS: ONE (FREE-SHEAR FLOWS)

3.2.1 Examples of Free-Shear Flows

Free-shear flows are flows in which the shear is confined to a thin layer but is free from interaction with the solid wall. The flows confined to a thin layer near the wall are called boundary layer flows. Although there are similarities between free-shear flows and boundary layer flows, there are more problems in turbulence modeling near the wall. Therefore, we shall examine the validity of turbulence models in free-shear flows first (see Fig. 3.1).

Turbulent free-shear flows are often selected to verify the predictability of turbulence models because the pressure gradient and viscous diffusion in free-shear flows are negligible and will not play a major role in determining the flow field. Therefore, prediction of free-shear flows is most sensitive to the modeling of the Reynolds stress, turbulent kinetic energy, and dissipation rate. Also, the complexity of the near-wall turbulence is absent in free-shear flows so that the accuracy of the turbulence model in predicting the general flow field can carefully be scrutinized without the interference of the wall turbulence. Besides, there are sufficient data available for comparison with both mean velocity and turbulent transport properties. It should also be pointed out that some of the turbulence model coefficients recommended in the existing k-ϵ models were determined after extensive comparison of predictions with experimental data in turbulent free-shear flows (Launder et al., 1972 [87]).

3.2.2 Differential Model (Two-Dimensional RSM)

For two-dimensional and axisymmetric incompressible turbulent free-shear flows, the transport equations can be simplified considerably. The assumptions made in obtaining turbulent free-shear flow equations are as follows.

1. Diffusion in the direction normal (y coordinate) to the flow is much larger than the diffusion in the direction parallel (x coordinate) to the flow,

$$\frac{\partial^2}{\partial X^2} \ll \frac{\partial^2}{\partial Y^2},$$

   and U dominates over V and W.

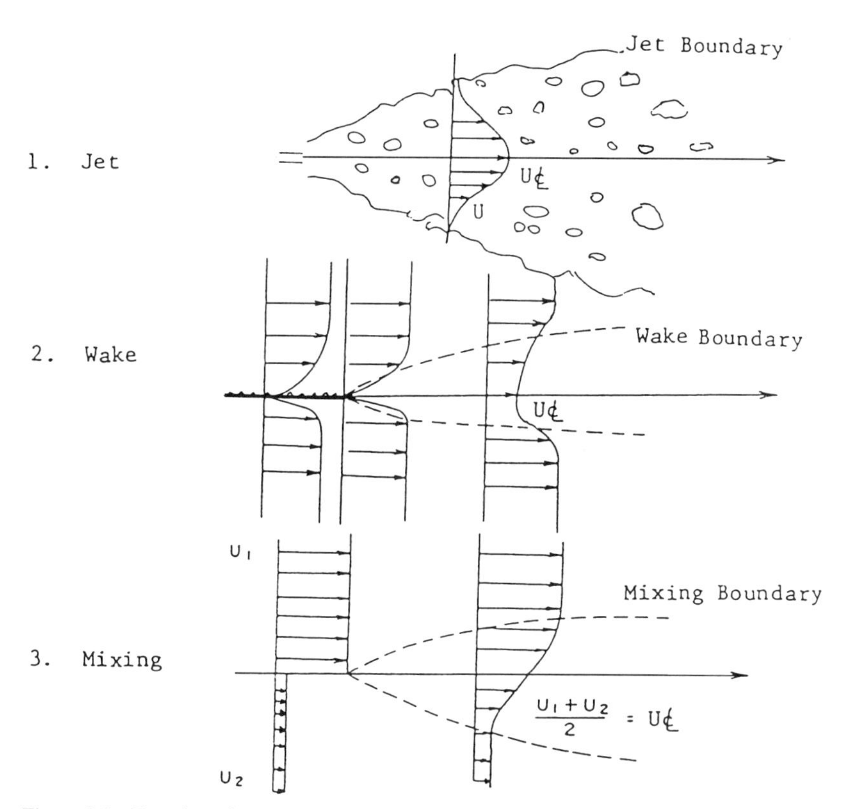

Figure 3.1 Free-shear flows.

2. The ambient pressure is uniform such that the pressure gradient is small in the flow and can be neglected,

$$P \neq f(X).$$

3. Viscous shear stress is much smaller than the turbulent shear stress and can be neglected,

$$\tau_{ij} < \tau_{ij}^t.$$

The mean equations are

- the continuity equation,

$$\frac{\partial U}{\partial X} + \frac{\partial V}{\partial Y} = 0, \tag{3.38}$$

  and
- the X-direction momentum equation,

$$U\frac{\partial U}{\partial X} + V\frac{\partial U}{\partial Y} = -\frac{\partial \overline{uv}}{\partial Y}. \tag{3.39}$$

The turbulent transport equations for the complete RSM are

$$\frac{D\overline{uv}}{Dt} = C_k \frac{\partial}{\partial Y}\left(\frac{k^2}{\epsilon}\frac{\partial \overline{uv}}{\partial Y}\right) - \overline{v^2}\frac{\partial U}{\partial Y} - C_1 \frac{\epsilon}{k}\overline{uv} + C_2 \overline{v^2}\frac{\partial U}{\partial Y} \tag{3.40}$$

$$\frac{D\overline{v^2}}{Dt} = C_k \frac{\partial}{\partial Y}\left(\frac{k^2}{\epsilon}\frac{\partial \overline{v^2}}{\partial Y}\right) - \frac{2}{3}\epsilon - C_1 \frac{\epsilon}{k}\left(\overline{v^2} - \frac{2}{3}k\right) - C_2 \frac{2}{3}\overline{uv}\frac{\partial U}{\partial Y} \tag{3.41}$$

$$\frac{Dk}{Dt} = C_k \frac{\partial}{\partial Y}\left(\frac{k^2}{\epsilon}\frac{\partial k}{\partial Y}\right) - \overline{uv}\frac{\partial U}{\partial Y} - \epsilon \tag{3.42}$$

$$\frac{D\epsilon}{Dt} = C_\epsilon \frac{\partial}{\partial Y}\left(\frac{k^2}{\epsilon}\frac{\partial \epsilon}{\partial Y}\right) - C_{\epsilon 1}\frac{\epsilon}{k}\overline{uv}\frac{\partial U}{\partial Y} - C_{\epsilon 2}\frac{\epsilon^2}{k} \tag{3.43}$$

Hanjalic and Launder (1972)[55] proposed that the Eq. 3.41 could be replaced by the expression

$$\overline{v^2} \approx 0.5k, \tag{3.44}$$

on the basis of experimental evidence, and used $C_1 = 2.8$, $C_2 = 0.6$, $C_k = 0.064$, $C_\epsilon = 0.065$, $C_{\epsilon 1} = 1.45$, and $C_{\epsilon 2} = 2.0$.

Equations 3.38 through 3.44 are solved in the subsequent sections.

3.2.3 Plane Jet Flows (RSM)

For plane jet flows, the Reynolds stress model is used (Hanjalic and Launder, 1972)[55]. The U, $\overline{uv}$, and k predictions are as shown in Fig. 3.2. The solution for $\overline{uv}$ shows 15 percent under prediction.

We can also examine the turbulent energy balance from the k equation.

- The exact k equation is

$$U\frac{\partial k}{\partial X} + V\frac{\partial k}{\partial Y} = -\frac{\partial}{\partial Y}\left(\frac{\overline{uuv}}{2}\right) - \overline{uv}\frac{\partial U}{\partial Y} - \epsilon, \tag{3.45}$$

whereas

- the modeled k equation is

$$U\frac{\partial k}{\partial X} + V\frac{\partial k}{\partial Y} = C_k \frac{\partial}{\partial y}\left(\frac{k^2}{\epsilon}\frac{\partial k}{\partial Y}\right) - \overline{uv}\frac{\partial U}{\partial Y} - \epsilon. \tag{3.46}$$

From Fig. 3.3, it is quite clear that (a) the sum of the convection and the diffusion terms is approximately zero, and (b) the production and dissipation of turbulent kinetic energy are much greater than the convection and diffusion of the turbulent kinetic energy. Hence, we can conclude that except near the center line

$$\left(U\frac{\partial k}{\partial X} + V\frac{\partial k}{\partial Y}\right) + \frac{1}{2}\frac{\partial \overline{uuv}}{\partial Y} \approx 0$$

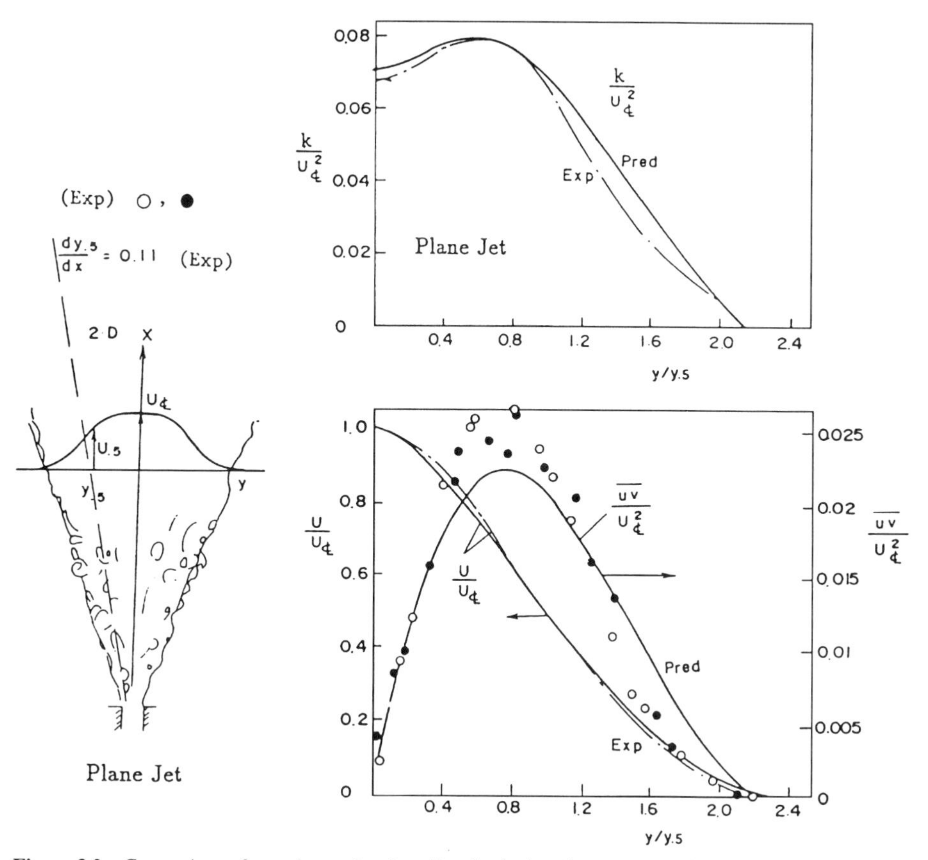

Figure 3.2 Comparison of experimental and predicted velocity, shear stress, and energy profiles for a plane jet. (Adapted from Hanjalic and Launder (1992).)

and

$$\epsilon \approx -\overline{uv}\frac{\partial U}{\partial Y}.$$

Unfortunately, the same set of turbulent transport equations in cylindrical coordinates cannot predict satisfactorily the spread of a round jet, unless some modifications are made to $C_{\epsilon 1}$ and $C_{\epsilon 2}$. This aspect will be discussed in more detail later.

3.2.4 Mixing Layer (RSM)

Using the same differential equations given in Section 3.2.2, Hanjalic and Launder (1972)[55] made the predictions shown in Fig. 3.4. From experimental analysis, we find that the rate of spread of the mixing layer varies. For example, experimental measurements show that $dY_\delta/dX = 0.20$ (Wygnanski and Fielder, 1970 [188]) and that $dY_\delta/dX = 0.15$ (Liefmann and Laufer, 1957 [95]). The rate of spread predicted by Hanjalic and Launder is $dY_\delta/dX = 0.15$. We thus conclude that the prediction may be considered satisfactory.

Table 3.7 Model coefficients for wake, mixing layer, and jet flow

Model	C_k	C_1	C_2	C_ϵ	$C_{\epsilon 1}$	$C_{\epsilon 2}$
1	0.11	1.50	0.40	0.075	1.44	1.90
2	0.11	1.50	0.60	0.075	1.44	1.90

3.2.5 Wake, Mixing Layer, and Jet Flow (RSM)

For the prediction of wake, mixing layer, and jet flow, Launder et al. (1975)[88] used the same equations as in Section 3.2.2, except that different $\Phi_{ij,2}$ model coefficients are adopted. The model coefficients are given in Table 3.7. Figures 3.5–3.7 compare the predicted profiles with the experimental values for a jet, mixing layer and wake, respectively.

3.2.6 *k*-ϵ-*A* or *k*-ϵ-*E* Model

For thin-shear flow, the governing equations for the differential model are given by Eqs. 3.38–3.43. If we assume the algebraic stress model (k-ϵ-A), we then omit convection

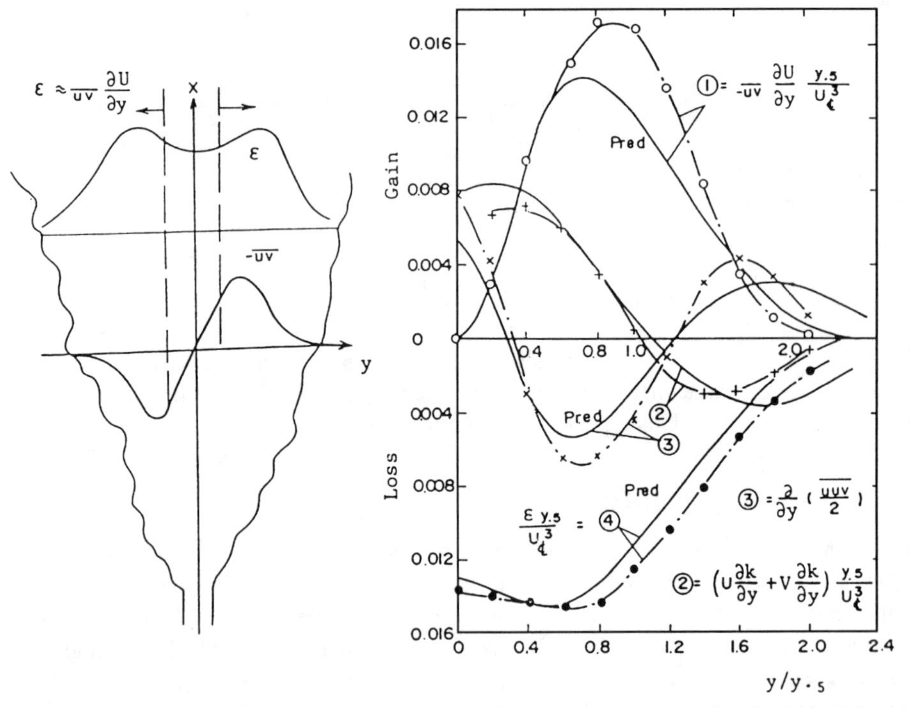

Figure 3.3 Comparison of the various terms of the kinetic energy equation for a plane jet flow. (Adapted from Hanjalic and Launder (1972).)

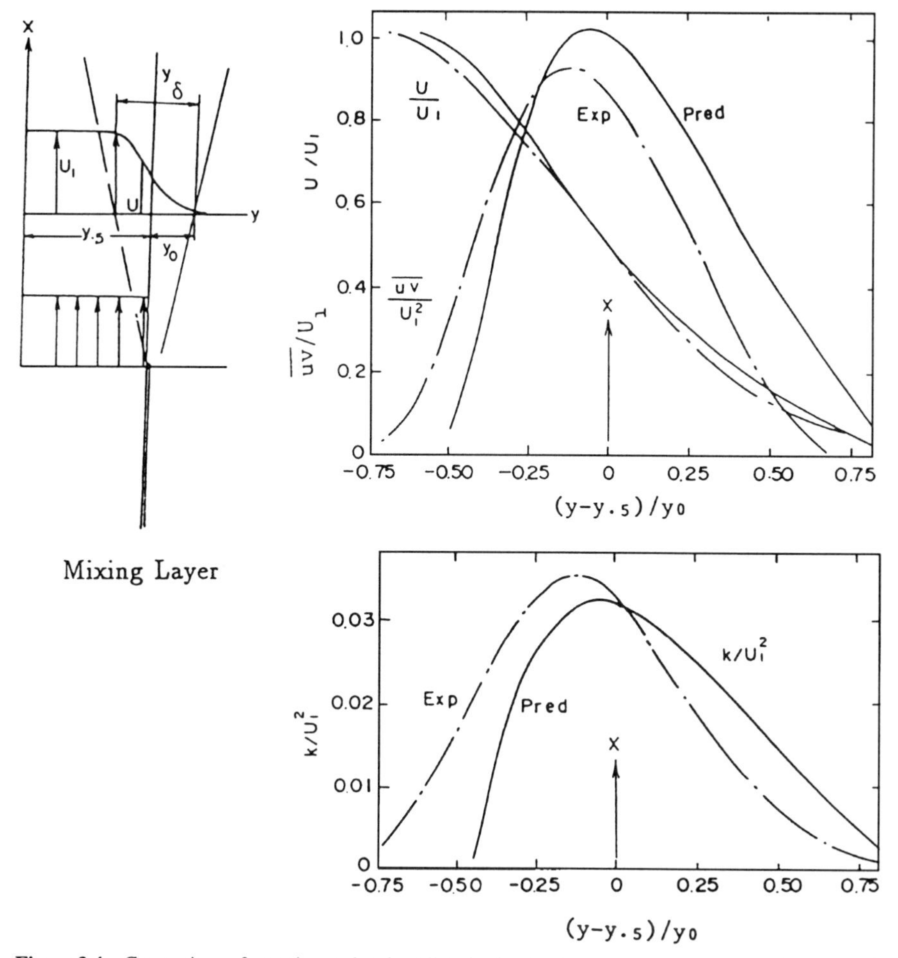

Figure 3.4 Comparison of experimental and predicted velocity, shear stress, and energy profiles for a mixing layer. (Adapted from Hanjalic and Launder (1974).)

and diffusion terms in the $\overline{uv}$ and $\overline{v^2}$ equations (i.e., Eqs. 3.40 and 3.41, respectively). Thus, we have k-ϵ-A model.

- The $\overline{v^2}$ equation becomes

$$0 \approx 0 - \frac{2}{3}\epsilon - C_1\frac{\epsilon}{k}\left(\overline{v^2} - \frac{2}{3}k\right) - C_2\frac{2}{3}\overline{uv}\frac{\partial U}{\partial Y} \tag{3.47}$$

$$\overline{v^2} = \frac{2}{3}\frac{(C_1 - 1 + C_2)}{C_1}k, \tag{3.48}$$

where we have made use of the relation, $\overline{uv}(\partial U/\partial y) = -\epsilon$. If $C_1 = 2.3$ and $C_2 = 0.4$, then $\overline{v^2} \approx 0.49k$, and if $C_1 = 2.8$ and $C_2 = 0.4$, then $\overline{v^2} \approx 0.52k$.

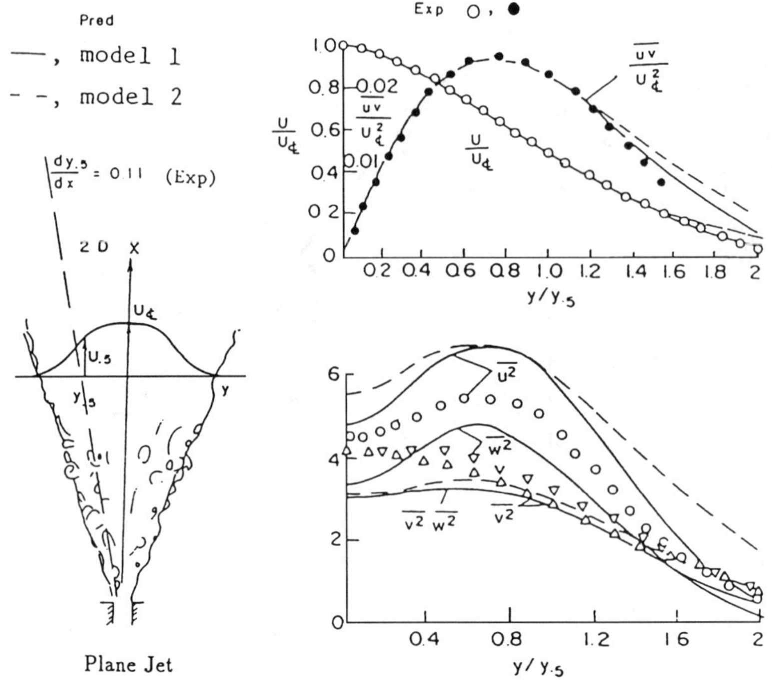

Figure 3.5 Comparison of predicted and experimental profiles for a jet. (Adapted from Launder et al. (1975).)

- The $\overline{uv}$ equation becomes

$$-\overline{uv} = \frac{k\overline{v^2}}{\epsilon}\frac{(1-C_2)}{C_1}\frac{\partial U}{\partial Y} \tag{3.49}$$

$$= 0.128\frac{k^2}{\epsilon}\frac{\partial U}{\partial Y}, \tag{3.50}$$

where $C_1 = 2.3$, $C_2 = 0.4$, and $C_\mu = 0.128$. Substituting $\overline{v^2}$ and $\overline{uv}$ relations into the k, ϵ, and mean flow equations (i.e., Eqs. 3.42, 3.43, 3.38, and 3.39), we have the following:

- the k equation is given by

$$\frac{Dk}{Dt} = C_k\frac{\partial}{\partial Y}\left(\frac{k^2}{\epsilon}\frac{\partial k}{\partial Y}\right) + C_\mu\frac{k^2}{\epsilon}\left(\frac{\partial U}{\partial Y}\right)^2 - \epsilon, \tag{3.51}$$

- the ϵ equation is given by

$$\frac{D\epsilon}{Dt} = C_\epsilon\frac{\partial}{\partial Y}\left(\frac{k^2}{\epsilon}\frac{\partial \epsilon}{\partial Y}\right) + C_{\epsilon 1}C_\mu k\left(\frac{\partial U}{\partial Y}\right)^2 - C_{\epsilon 2}\frac{\epsilon^2}{k}, \tag{3.52}$$

- the X-direction momentum equation is given by

$$U\frac{\partial U}{\partial X} + V\frac{\partial U}{\partial Y} = C_\mu\frac{\partial}{\partial Y}\left(\frac{k^2}{\epsilon}\frac{\partial U}{\partial y}\right), \tag{3.53}$$

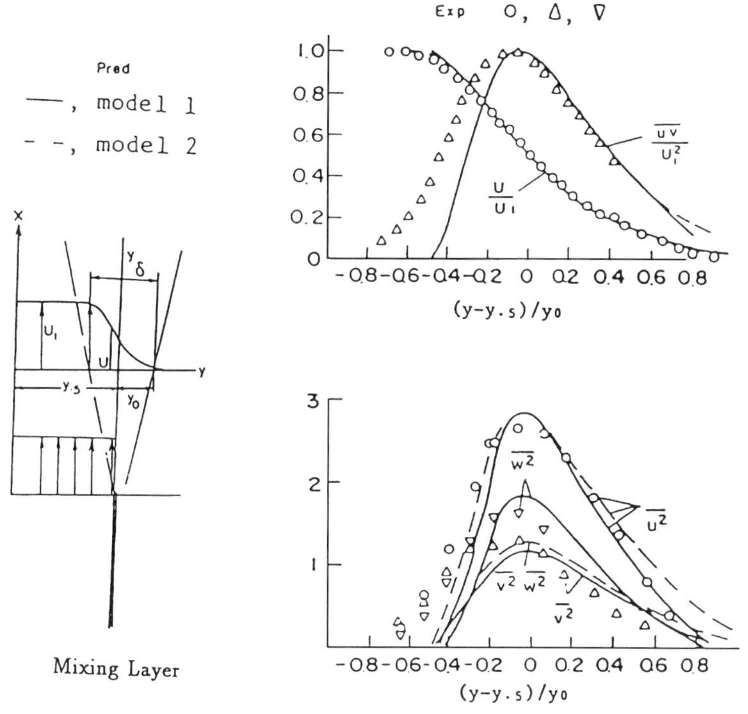

Figure 3.6 Comparison of predicted and experimental profiles for a two-dimensional mixing layer. (Adapted from Launder et al. (1975).)

whereas

- the continuity equation is given by

$$\frac{\partial U}{\partial X} + \frac{\partial V}{\partial Y} = 0. \tag{3.54}$$

Here $C_\mu \approx 0.11$, $C_k = 0.09$, $C_\epsilon = 0.07$, $C_{\epsilon 1} = 1.44$, and $C_{\epsilon 2} = 1.92$.

Consider the Boussinesq eddy viscosity model (k-ϵ-E). The general Reynolds stress equation is

$$-\overline{u_i u_j} = \nu_t \left(\frac{\partial U_i}{\partial X_j} + \frac{\partial U_j}{\partial X_i} \right) - \frac{2}{3}\delta_{ij} k. \tag{3.55}$$

For thin-shear flows, this becomes

$$-\overline{uv} \approx \nu_t \frac{\partial U}{\partial Y}, \tag{3.56}$$

where $\nu_t = C_\mu k^2/\epsilon$. Substituting Eq. 3.56 into the k, ϵ, and the mean-flow equations, we have a set of equations for the k-ϵ-E model similar to the set of equations for the k-ϵ-A model. Thus,

- the k equation is given by

$$\frac{Dk}{Dt} = C_k \frac{\partial}{\partial Y}\left(\frac{k^2}{\epsilon} \frac{\partial k}{\partial Y} \right) + C_\mu \frac{k^2}{\epsilon} \left(\frac{\partial U}{\partial Y} \right)^2 - \epsilon, \tag{3.57}$$

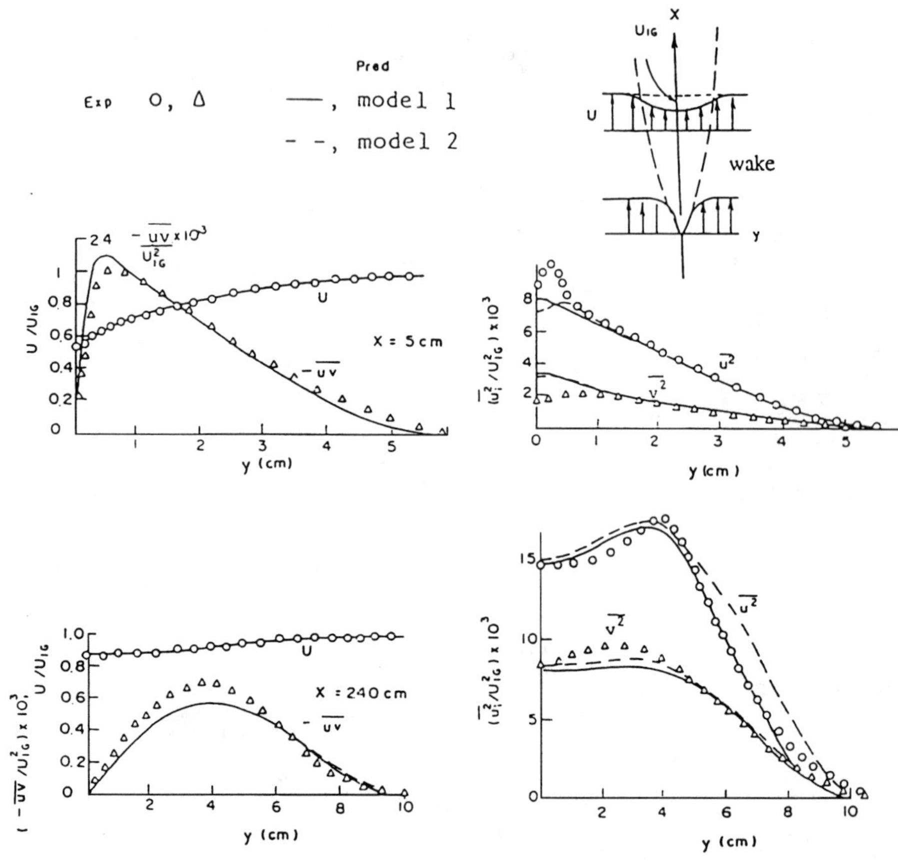

Figure 3.7 Comparison of predicted and experimental profiles for a wake behind a flat plate. (Adapted from Launder et al. (1975).)

- the ϵ equation is given by

$$\frac{D\epsilon}{Dt} = C_\epsilon \frac{\partial}{\partial Y}\left(\frac{k^2}{\epsilon}\frac{\partial \epsilon}{\partial Y}\right) + C_{\epsilon 1} C_\mu k \left(\frac{\partial U}{\partial Y}\right)^2 - C_{\epsilon 2}\frac{\epsilon^2}{k}, \tag{3.58}$$

- the X-direction momentum equation is given by

$$U\frac{\partial U}{\partial X} + V\frac{\partial U}{\partial Y} = C_\mu \frac{\partial}{\partial Y}\left(\frac{k^2}{\epsilon}\frac{\partial U}{\partial Y}\right), \tag{3.59}$$

whereas

- the continuity equation is given by

$$\frac{\partial U}{\partial X} + \frac{\partial V}{\partial Y} = 0. \tag{3.60}$$

Here $C_\mu = 0.09$, $C_k = 0.09$, $C_\epsilon = 0.07$, $C_{\epsilon 1} = 1.44$, and $C_{\epsilon 2} = 1.92$. The k-ϵ-A and the k-ϵ-E models are identical except $C_\mu = 0.11$ (k-ϵ-A) and $C_\mu = 0.09$ (k-ϵ-E). Thus one expects that both the models will have similar performance in thin-shear flow (and boundary layer flow).

For heat transfer problems in thin-shear flows, we have, in general,

$$\frac{DT}{Dt} = -\frac{\partial \overline{v\theta}}{\partial Y} \tag{3.61}$$

$$\frac{D\overline{v\theta}}{Dt} = C_T \frac{\partial}{\partial Y}\left(\frac{k^2}{\epsilon}\frac{\partial \overline{v\theta}}{\partial Y}\right) - \overline{v^2}\frac{\partial T}{\partial Y} - C_{T1}\frac{\epsilon}{k}\overline{v\theta}. \tag{3.62}$$

With the algebraic heat flux approximation (k-ϵ-A), we use

$$0 = 0 - \overline{v^2}\frac{\partial T}{\partial Y} - C_{T1}\frac{\epsilon}{k}\overline{v\theta}$$

$$-\overline{v\theta} = \frac{\overline{v^2}k}{\epsilon C_{T1}}\frac{\partial T}{\partial Y}$$

$$-\overline{v\theta} = C_\alpha \frac{k^2}{\epsilon}\frac{\partial T}{\partial Y} \tag{3.63}$$

with $\overline{v^2} = 0.5k$, $C_{T1} = 3.2$, $C_\alpha = 0.156$, $C_\mu = 0.11$, and $Pr_t = C_\mu/C_\alpha = 0.7$.

3.2.7 Two-Dimensional Jet Flow

For the prediction of two-dimensional jet flow based on the k-ϵ-A model, Chen and Rodi (1975)[19] obtained the slope of the half width, $dy_{.5}/dx = 0.112$. This compares to the experimental value of $dy_{.5}/dx = 0.11$. Figure 3.8 shows the comparison of the

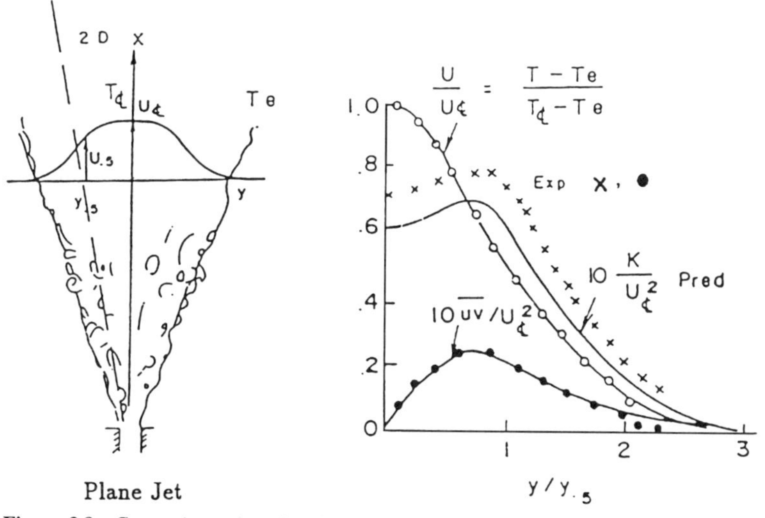

Figure 3.8 Comparison of predicted and experimental profiles for a two-dimensional jet. (Adapted from Chen and Rodi (1975).)

predicted and experimental jet profiles. Here U_c is the centerline velocity and T_e is the ambient temperature.

In general, the k-ϵ-A and k-ϵ-E models perform about the same as the differential model (RSM) for free-shear flows. This is because in free-shear flows, production and dissipation dominate over most of the flow region and also because the flow velocity in one direction dominates over that in the other direction, which makes the eddy viscosity model a reasonably accurate model to relate the mean flow and turbulent transport of $\overline{uv}$.

3.2.8 Prediction of Round Jet Flow by RSM and k-ϵ-A Models

Chen and Jaw (1990)[16] studied the prediction of the RSM for the round jet and found that, as shown in Fig. 3.9, the RSM (Section 2.5.2), and the k-ϵ-A models without modification of the turbulent constants substantially overpredicted the spreading rate and Reynolds stress of the round jet. In particular, the RSM, as it is, still has difficulty in predicting simple flows. Therefore, any attempt to improve the turbulence model must first demonstrate that the improved model is capable of accurately predicting two-dimensional flows as well as axisymmetric jet and wake flows. No RSM turbulence model at the present time has clearly been shown to predict satisfactorily for the flows mentioned.

Examining the postulations made in deriving the turbulence model, we find that Postulations 3 and 6 (chapter 2, Section 2.1.3) seem to be too restrictive and too simplistic. The postulation of isotropic dissipation leads to the omission of $P_{\epsilon 1}$, the production term due to interaction between mean flow and dissipating eddies in the ϵ equation (chapter 2, Section 2.2.4, Eq. 2.23). The Postulation 6 assumes that the turbulent flow can be characterized by one turbulence scale based on (k, ϵ) from the energy-containing eddies. That has led us to believe that the turbulence time scale is $t \sim k/\epsilon$. This time scale was employed by Lumley (1974) to model the last two terms of the ϵ equation (Eq. 2.24) by setting

$$-2\nu\overline{\frac{\partial u_i}{\partial X_j}\frac{\partial u_i}{\partial X_l}\frac{\partial u_j}{\partial X_l}} - 2\overline{\left(\nu\frac{\partial^2 u_i}{\partial X_l \partial X_j}\right)^2} \approx \left(\frac{1}{t}\right)(P_k - \epsilon)$$

$$\approx \frac{\epsilon}{k}(C_{\epsilon 1}P_k - C_{\epsilon 2}\epsilon).$$

All of these may contribute to the necessity of adjusting the $C_{\epsilon 1}$ value in the ϵ equation from a value close to $C_{\epsilon 2} = 1.92$ to $C_{\epsilon 2} = 1.42 \approx 1.45$, and many studies show that both $C_{\epsilon 1}$ and $C_{\epsilon 2}$ must be modified with a correction function. It should also be remarked here that although most turbulent predictions and calculations are not as sensitive as they should be to small variations of moduli C_k, C_1, C_2, and C_ϵ, many computations are quite sensitive to small variations of $C_{\epsilon 1}$ and $C_{\epsilon 2}$. Three significant digits for $C_{\epsilon 1}$ and $C_{\epsilon 2}$ must generally be specified if a stable result for a computation is to be expected. This symptom has led to the current thinking that a multiple scale is necessary to make fundamental progress in turbulence modeling.

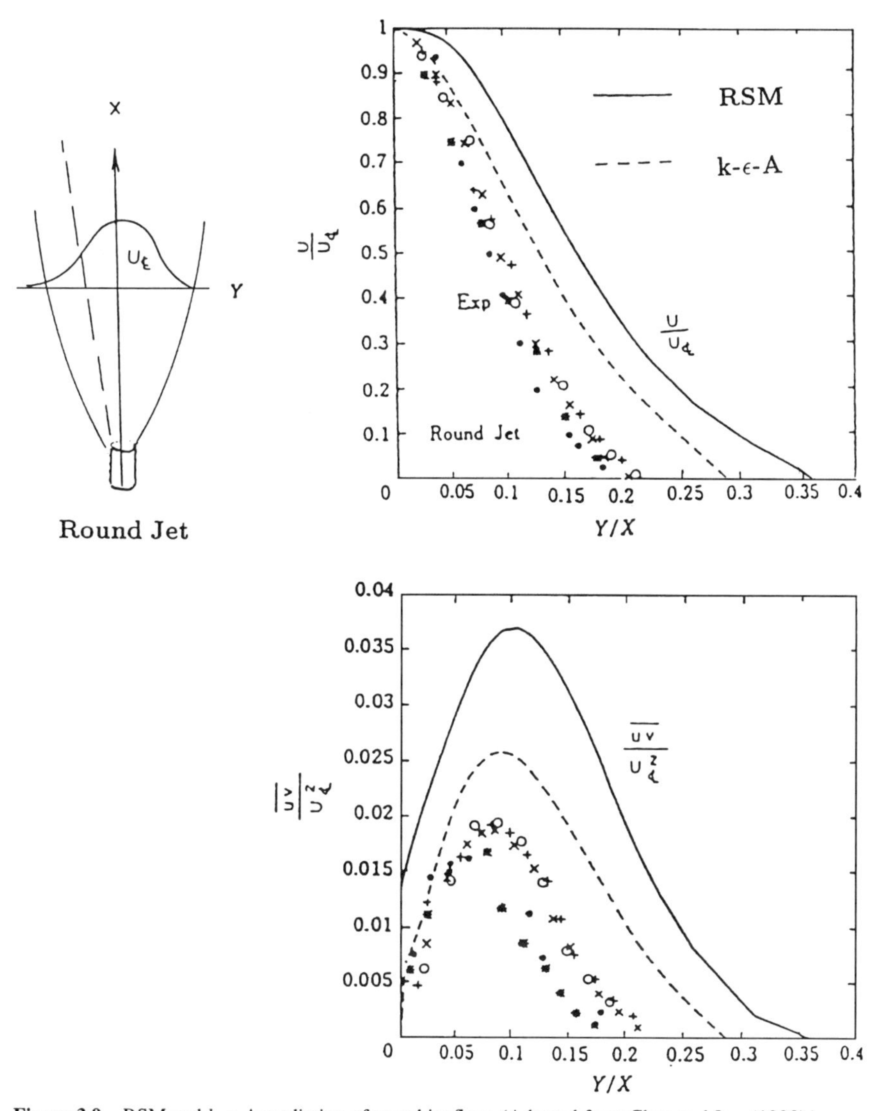

Figure 3.9 RSM and k-ϵ-A prediction of round jet flow. (Adapted from Chen and Jaw (1990).)

3.2.9 Prediction of Turbulent Free-Shear Flows with Anisotropic Turbulence Models

The mean equations of continuity and momentum for turbulent free-shear flows are

$$\frac{\partial U}{\partial X} + \frac{\partial V}{\partial Y} = 0 \tag{3.64}$$

$$U\frac{\partial U}{\partial X} + V\frac{\partial U}{\partial Y} = -\frac{1}{Y^j}\frac{\partial}{\partial Y}(Y^j\overline{uv}) \tag{3.65}$$

with $j = 0$ for plane flows and $j = 1$ for axisymmetric flows. From Jaw and Chen (1991)[69] the generalized equations for the turbulent transport properties can be reduced to

$$U\frac{\partial \overline{uv}}{\partial X} + V\frac{\partial \overline{uv}}{\partial Y} = \frac{1}{Y^j}\frac{\partial}{\partial Y}\left(Y^j C_k \frac{k^2}{\epsilon}\frac{\partial \overline{uv}}{\partial Y}\right) - \frac{\epsilon}{k}\overline{uv} - \overline{v^2}\frac{\partial U}{\partial Y} - C_1\frac{\epsilon}{k}\overline{uv} + C_2\overline{v^2}\frac{\partial U}{\partial Y} \tag{3.66}$$

$$U\frac{\partial \overline{v^2}}{\partial X} + V\frac{\partial \overline{v^2}}{\partial Y} = \frac{1}{Y^j}\frac{\partial}{\partial y}\left(Y^j C_k \frac{k^2}{\epsilon}\frac{\partial \overline{v^2}}{\partial Y}\right) - \frac{\epsilon}{k}\overline{v^2} - C_1\frac{\epsilon}{k}\left(\overline{v^2} - \frac{2}{3}k\right) - C_2\frac{2}{3}\overline{uv}\frac{\partial U}{\partial Y} \tag{3.67}$$

$$U\frac{\partial k}{\partial X} + V\frac{\partial k}{\partial Y} = \frac{1}{Y^j}\frac{\partial}{\partial Y}\left(Y^j C_k \frac{k^2}{\epsilon}\frac{\partial k}{\partial Y}\right) - \overline{uv}\frac{\partial U}{\partial Y} - \epsilon \tag{3.68}$$

$$U\frac{\partial \epsilon}{\partial X} + V\frac{\partial \epsilon}{\partial Y} = \frac{1}{Y^j}\frac{\partial}{\partial Y}\left(Y^j C_\epsilon \frac{k^2}{\epsilon}\frac{\partial \epsilon}{\partial Y}\right) - C_{\epsilon 1}\frac{\epsilon}{k}\overline{uv}\frac{\partial U}{\partial Y} - C_{\epsilon 2}\frac{\epsilon^2}{k}$$
$$+ C_{\epsilon 3}\frac{\epsilon}{k}\frac{1}{Y^j}\frac{\partial}{\partial Y}\left(Y^j C_k \frac{k^2}{\epsilon}\frac{\partial k}{\partial Y}\right). \tag{3.69}$$

For k-ϵ-EX (introduced in Table 3.3, Set 5, or the k-ϵ-eddy viscosity model with a cross diffusion term in the ϵ equation) and the k-ϵ-E model, the Reynolds stress $\overline{uv}$ is solved from the Boussinesq eddy viscosity model,

$$-\overline{uv} = \nu_t \frac{\partial U}{\partial Y}$$

with $\nu_t = C_\mu k^2/\epsilon$. The model coefficients of the anisotropic RSM and the k-ϵ-EX models are determined in a similar way as those presented in chapter 2. The values adopted in them are listed in Table 3.8.

One criterion that can immediately manifest the predictability of the adopted models is to verify the predicted gross parameter, the rate of spread of free-shear flows. If a model is unable to predict accurately the rate of spread of the free-shear flows, it is not very useful in further examining the detailed profiles of the flow. Table 3.9 lists the definitions of the rate of spread for various free-shear flows and compares the predicted results with the experimental data.

In Table 3.9, $Y_{1/2}$, known as the half width, is the normal distance from the center or symmetry line to the location where the velocity in the stream direction, U_i, is one half of the centerline velocity, U_c. Similar definitions apply to $Y_{0.1}$ and $Y_{0.9}$, except that U_c is replaced by U_E, the free-stream velocity. From this table, it is found that the models with

Table 3.8 Turbulence model coefficients for free–shear flows

Model	C_μ	C_k	C_1	C_2	C_ϵ	$C_{\epsilon 1}$	$C_{\epsilon 2}$	$C_{\epsilon 3}$
RSM		0.09	2.14	0.51	0.11	1.23	1.92	1.67
k-ϵ-EX	0.09	0.09			0.11	1.23	1.92	1.67
k-ϵ-E	0.09	0.09			0.07	1.44	1.92	0.00

Table 3.9 Comparison of predicted spread rate

Flow type	Spread rate	Experimental data	k-ϵ-E	k-ϵ-EX	RSM
Round jet	$S=\frac{dY_{1/2}}{dX}$	0.0713–0.086	0.123	0.096	0.101
Plane jet	$S=\frac{dY_{1/2}}{dX}$	0.102–0.11	0.112	0.107	0.106
Mixing layer	$S=\frac{d(Y_{0.1}-Y_{0.9})}{dX}$	0.16–0.165	0.156	0.165	0.173
Plane wake	$S=\frac{U_E}{U_E-U_C}\frac{dY_{1/2}}{dX}$	0.09–0.11	0.098	0.098	0.098

a cross-diffusion term, both the RSM and the k-ϵ-EX model, predict more accurately the rates of spread, whereas the k-ϵ-E model overpredicts this value for the round jet flow.

Note that predictions of wake flow are sensitive to the initial conditions specified. In order to make a fair comparison, the results presented are obtained by adjusting the initial conditions so that the predicted rates of spread are the same for different models. Predictions of plane jet, mixing-layer, and plane wake flows are presented in Figs. 3.10–3.12.

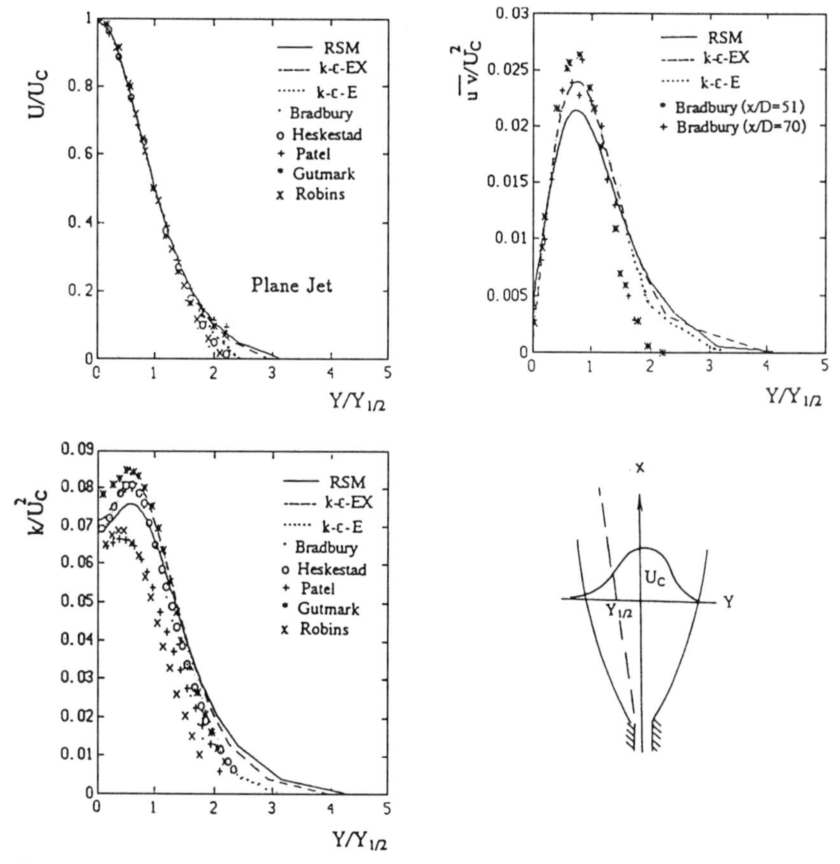

Figure 3.10 Prediction of plane jet flow. (Adapted from Chen and Jaw (1990).)

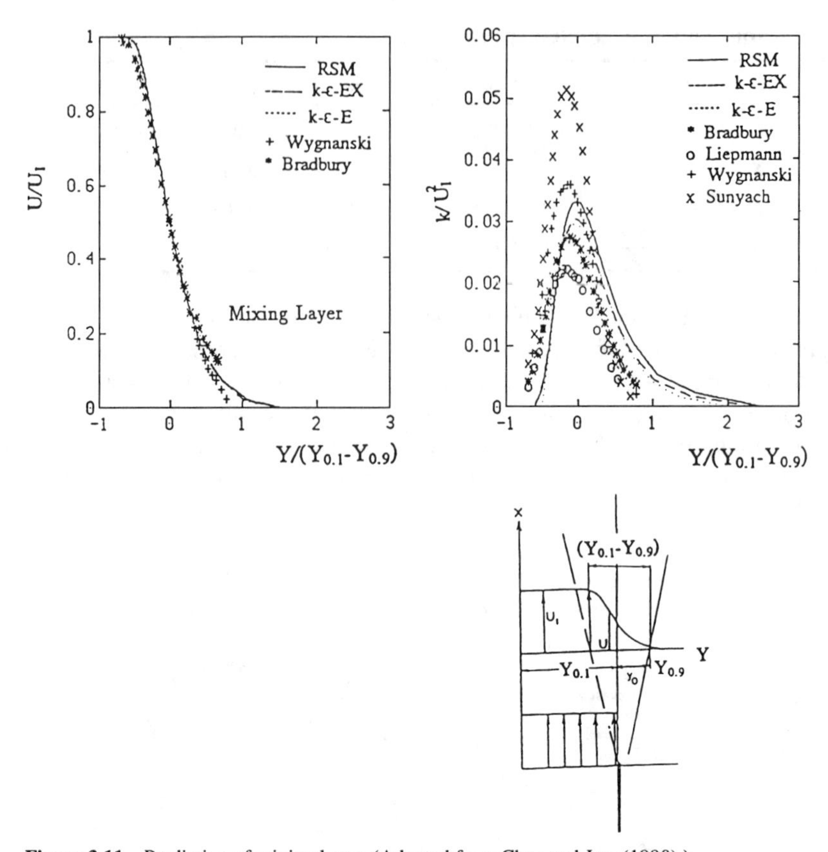

Figure 3.11 Prediction of mixing layer. (Adapted from Chen and Jaw (1990).)

3.2.10 Prediction of Round Jet Flows by Anisotropic Turbulence Models

Because the prediction of the spreading rate of a round jet flow is different among the three models, the transverse profiles are plotted in the similarity variable Y/X to exhibit more clearly the differences of these profiles, as shown in Figs. 3.13 and 3.14. It is obvious that the predicted results of the k-ϵ-E model deviate appreciably from the measured data in both mean and turbulent quantities. The mean-velocity profiles predicted from the k-ϵ-EX and RSM models fit the experimental data fairly well except in the tail region in which larger diffusion profiles are predicted as compared with the experimental profiles. For the kinetic energy and Reynolds-stress $\overline{uv}$ distributions, the RSM model predicts a little higher peak values whereas the k-ϵ-EX predictions match well with the experimental data. For turbulent free-shear flows, the mean strain and stress are approximately parallel; hence, the Boussinesq eddy viscosity approximation is adequate and can be used in prediction of such flows. Figure 3.13(d) shows the distribution of turbulent eddy viscosity, $\nu_t = C_\mu k^2/\epsilon$. In general, the k-ϵ-EX and RSM models predict smaller peak values and wider cross-stream distribution than the k-ϵ-E model. It should be pointed out that Hanjalic et al. (1980)[57] proposed a multiple-time-scale turbulence model that

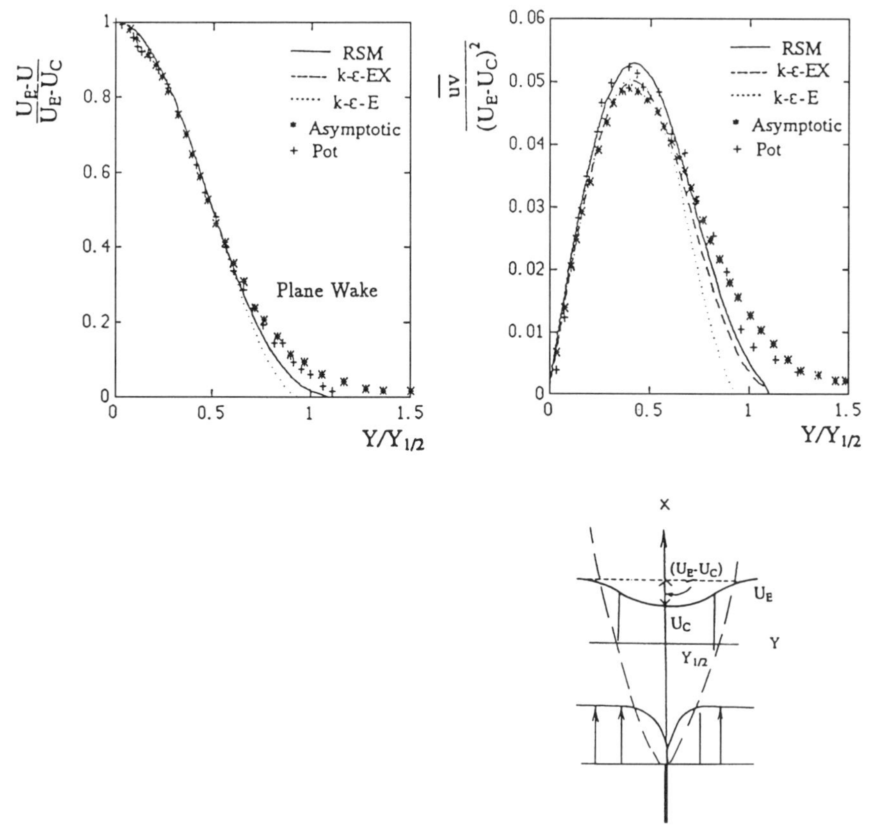

Figure 3.12 Prediction of plane wake. (Adapted from Chen and Jaw (1990).)

also improves the prediction of round jet flow. The multiple-time-scale model predicts a higher ϵ level and thus leads to lower eddy viscosity. According to Hanjalic et al. [57], it is the smaller eddy viscosities that, in turn, cause the somewhat lower spread rate.

Figure 3.14 manifests how influential the cross-diffusion term is in predicting the round jet flow. In these figures, the dotted line and the solid line represent the predicted profiles of the k-ϵ-EX and RSM models, respectively, with $C_{\epsilon 3}$ equal to zero, and all of the other model coefficients unchanged, whereas the dashed line represents the predicted profile of the k-ϵ-EX model with a cross-diffusion effect (i.e., $C_{\epsilon 3} = 1.67$). It is found that without the cross-diffusion term in the ϵ equation, the predicted results deviate appreciably from the experimental data.

3.3 PROBLEM FUNCTION

3.3.1 Use of a Problem Function

A problem function is the model function that one must introduce for one individual or one class of problem. It reduces the predictability of a turbulence model because it

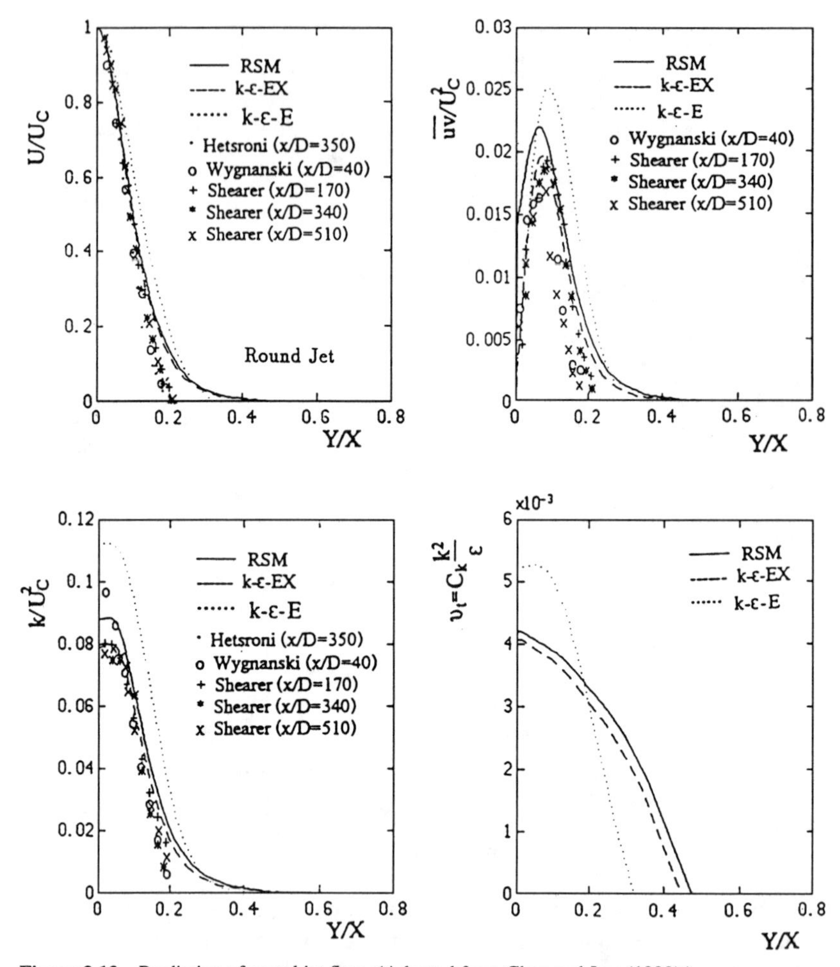

Figure 3.13 Prediction of round jet flow. (Adapted from Chen and Jaw (1990).)

depends on the problem. This is fundamentally undesirable. However, to make a model predict reasonable results when no alternative model is available, it is often used.

We have observed that the RSM and the k-ϵ models perform reasonably well in some shear flows presented in Section 3.2 without modification of the turbulent constants C_k, C_1, C_2, C_ϵ, $C_{\epsilon 1}$, and $C_{\epsilon 2}$ (the constant C_μ is derived from the modeled equations). However, in computing round jet and far wake problems, the second-order turbulence model encounters some difficulty in predicting some of the general characteristics of jets and wakes, namely, the rate of spread and the delay of the decay of the centerline velocity.

Although the second-order turbulence model in some flows requires a problem function in the prediction, in general, the second-order (RSM or k-ϵ) turbulence model is still a much superior model to the first-order model (zero-equation model) or the second-order, one-equation model.

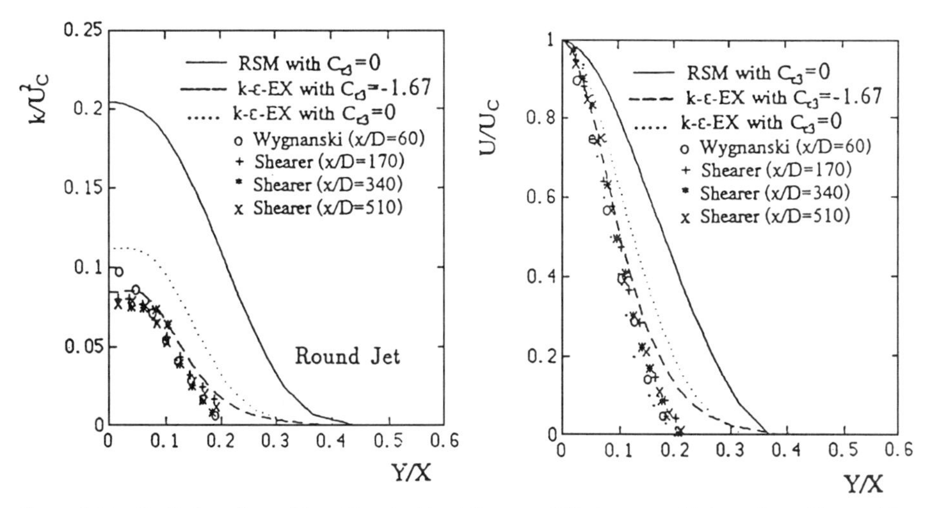

Figure 3.14 Prediction of round jet with and without the cross-diffusion term. (Adapted from Chen and Jaw (1990).)

3.3.2 Zero-Equation Model (Mixing-Length Function)

For thin-shear and boundary layer flows, the continuity equation is given by

$$\frac{\partial U}{\partial X} + \frac{\partial V}{\partial Y} = 0, \tag{3.70}$$

and the mean-flow momentum equation is given by

$$U\frac{\partial U}{\partial X} + V\frac{\partial U}{\partial Y} = -\frac{1}{\rho}\frac{\partial P}{\partial X} + \frac{\partial}{\partial Y}\left[(\nu + \nu_t)\frac{\partial U}{\partial Y}\right], \tag{3.71}$$

where we have

$$-\overline{uv} = \nu_t \frac{\partial U}{\partial Y},$$

and the eddy viscosity is defined in terms of the Prandtl mixing-length theory as

$$\nu_t = l_m^2 \left|\frac{\partial U}{\partial Y}\right|.$$

Some problem functions (or mixing-length functions) for various problems are based on the ratio of the mixing length function, l_m, to the shear layer thickness, $\delta(X)$:

- for two-dimensional mixing, $l_m/\delta(X) = 0.07$;
- for two-dimensional jet, $l_m/\delta(X) = 0.09$;
- for round jet, $l_m/\delta(X) = 0.075$;
- for two-dimensional wake, $l_m/\delta(X) = 0.16$; and
- for two-dimensional boundary layer,

$$l_m = \kappa y \longrightarrow 0 < \frac{y}{\delta} < \frac{\lambda}{\kappa}$$

$$l_m = \lambda\delta \longrightarrow \frac{y}{\delta} > \frac{\lambda}{\kappa},$$

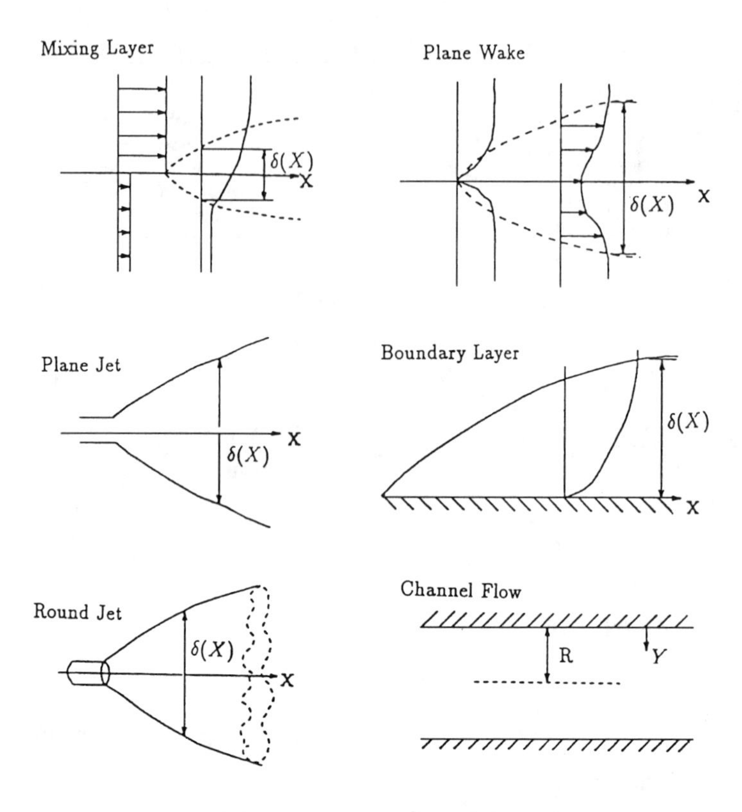

Figure 3.15 Various problems for showing dependence of problem functions.

where $\lambda = 0.09$ and $\kappa = 0.43$.

- for pipe and channel,

$$\frac{l_m}{R} = 0.14 - 0.08g^2 - 0.06g^4,$$

where

$$g = \left(1 - \frac{Y}{R}\right).$$

These problems are shown in Fig. 3.15. It can be easily seen that the l_m function is highly problem dependent.

3.3.3 One-Equation Model (Length-Scale Function)

For the one-equation model, or the k equation model, developed by Rodi (1980)[135], in general, there is still a strong problem-dependent (length-scale) function to be prescribed before a problem is solved. Thus, predictability is reduced.

3.3.4 k-ϵ Model

This model introduces correction functions for free-shear flows. Because both the k-ϵ-A and the k-ϵ-E (see Section 3.2.6) models are shown to be identical in thin-shear flows, let us consider only the k-ϵ-E model. For thin-shear flow, we have

$$-\overline{uv} = \nu_t \frac{\partial U}{\partial Y}, \tag{3.72}$$

where

$$\nu_t = C_\mu \frac{k^2}{\epsilon} \tag{3.73}$$

$$\frac{Dk}{Dt} = C_k \frac{\partial}{\partial Y}\left(\frac{k^2}{\epsilon}\frac{\partial k}{\partial Y}\right) - \overline{uv}\frac{\partial U}{\partial Y} - \epsilon \tag{3.74}$$

$$\frac{D\epsilon}{Dt} = C_\epsilon \frac{\partial}{\partial Y}\left(\frac{k^2}{\epsilon}\frac{\partial \epsilon}{\partial Y}\right) - C_{\epsilon 1}\frac{\epsilon}{k}\overline{uv}\frac{\partial U}{\partial Y} - C_{\epsilon 2}\frac{\epsilon^2}{k} \tag{3.75}$$

$$\frac{DU}{Dt} = -\frac{\partial \overline{uv}}{\partial Y} \tag{3.76}$$

$$\frac{\partial U}{\partial X} + \frac{\partial V}{\partial Y} = 0. \tag{3.77}$$

The model constants are $C_\mu = C_k = 0.09$, $C_{\epsilon 1} = 1.44$, and $C_{\epsilon 2} = 1.92$. For a round jet flow, the prediction for the slope at the half width is $dy_{.5}/dx = 0.12$, whereas the experimental value is $dy_{.5}/dx = 0.086$. Therefore, this model overpredicts by 30 percent for a round jet flow. For a plane jet flow, the prediction is $dy_{.5}/dx = 0.11$, and the experimental value is $dy_{.5}/dx = 0.11$. For a two-dimensional wake, the prediction is $S = 0.067$, whereas the experimental value is $S = 0.098$. Here,

$$S = \frac{U_c}{2W_0}\frac{dY_{.5}}{dX}, \tag{3.78}$$

where U_c is the centerline velocity, and $W_0 = U_\infty - U_c$. Therefore, this model underpredicts for the wake flow. Figure 3.16 depicts the rate of spread for free-shear flows. We find that the model underpredicts the spread of a wake and overpredicts the spread of a round jet. Furthermore, the predictions for a round jet and for a two-dimensional wake are not good, and a problem function has to be introduced to correct the deficiencies. A problem function for a jet (two-dimensional, round) can be introduced (Rodi (1972)[134]) for the k-ϵ-A model as follows:

$$C_\mu = C_k = 0.09 - 0.04f \tag{3.79}$$

$$C_{\epsilon 2} = 1.92 - 0.0667f \tag{3.80}$$

$$f = \frac{\delta}{\Delta U_{\max}}\left(\frac{\partial U_c}{\partial X} - \left|\frac{\partial U_c}{\partial X}\right|\right)^{0.2}, \tag{3.81}$$

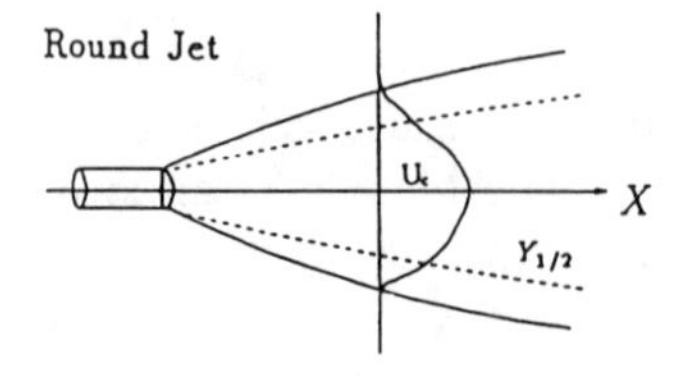

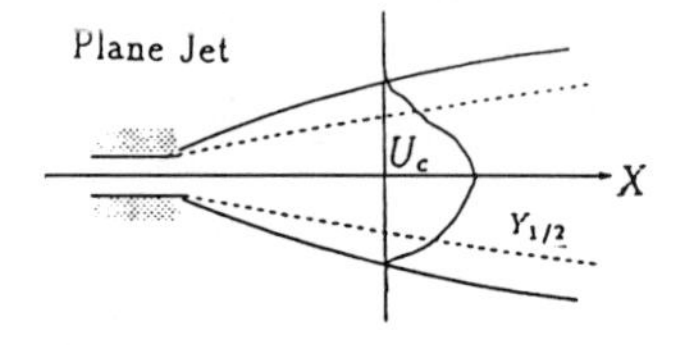

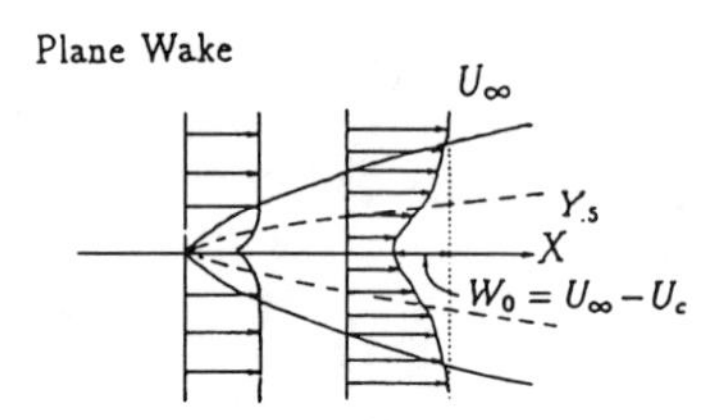

Figure 3.16 Rate of spread of free-shear flows.

where $\Delta U_{\max}$ is the maximum velocity difference, $\delta = y_e - y_c$, and the suffixes e and c mean ambient and center, respectively.

Another problem function for far wakes and weak jets can be introduced (Rodi, 1972 [134]); (far jets or coflowing stream) for the k-ϵ-E model as follows:

$$C_\mu = f\left(\frac{P}{\epsilon}\right)$$

$$C_{\epsilon 1} = 1.44$$

$$C_{\epsilon 2} = 1.92.$$

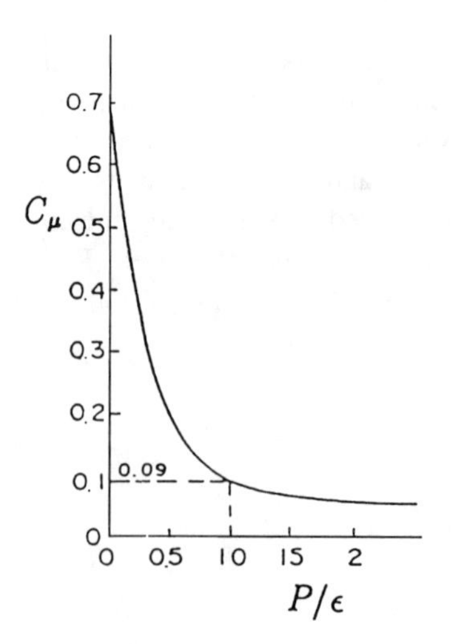

Figure 3.17 Experimental function C_μ as a function of P/ϵ. (Adapted from Rodi (1972).)

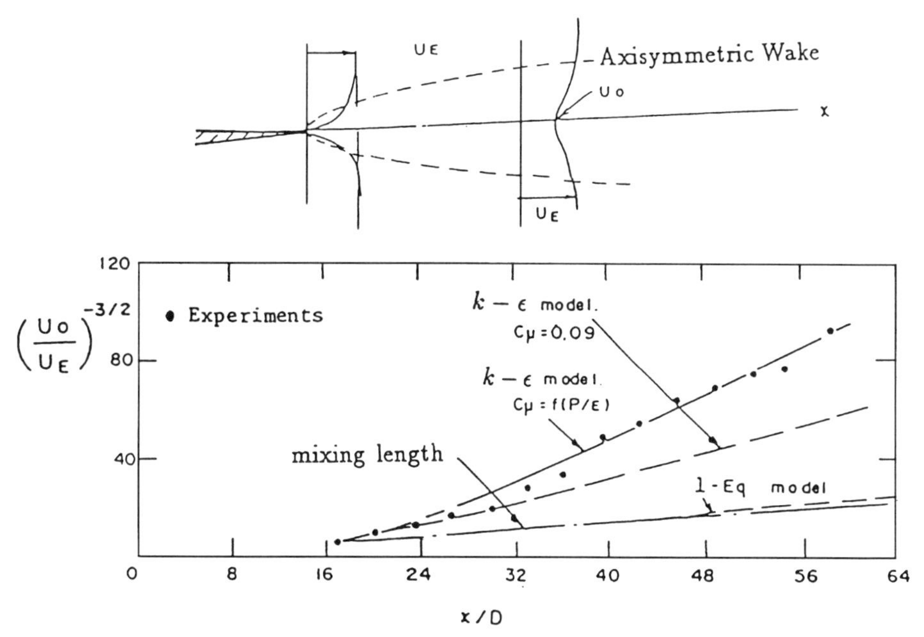

Figure 3.18 Development of maximum deficit velocity, U_0, in an axisymmetric wake. (Adapted from Patel and Scheuerer (1982).)

The C_μ shown in Fig. 3.17 is calibrated with much experimental data and many calculations. It improves prediction of an axisymmetric wake or a round jet. It still underpredicts the asymptotic spreading rate of a plane wake. An approximate derivation of $C_\mu = f(P/\epsilon)$ was given by Rodi (1972)[134]. P is the production term of k equation. The performance of the modified k-ϵ model with $C_\mu = f(P/\epsilon)$ is as shown in Fig. 3.18, whereas Fig. 3.19 (Patel and Scheuerer, 1982 [122]) shows the comparison of predicted and experimental asymmetric wake (k-ϵ) velocity and shear stress profiles, again with $C_\mu = f(P/\epsilon)$.

3.4 TWO-SCALE SECOND-ORDER TURBULENCE MODEL

3.4.1 Rationale

The need to change $C_{\epsilon 1}$, $C_{\epsilon 2}$, and C_μ in predicting simple shear flow is a disappointment after the effort involved in modeling $\overline{u_i u_j}$, k, and ϵ equations. Nevertheless, the second-order turbulence model has greatly improved the accuracy and predictability of the turbulence model. We should consider more fundamental improvements.

The weakest point of the $\overline{u_i u_j}$, k, and ϵ modeling is in the ϵ equation. The modeling of the destruction term in the ϵ equation is questionable. Omission of the production terms in the ϵ equation is also questionable (see Section 2.2.4). The PS term in the $\overline{u_i u_j}$ equation can also be improved (see Section 2.2.2).

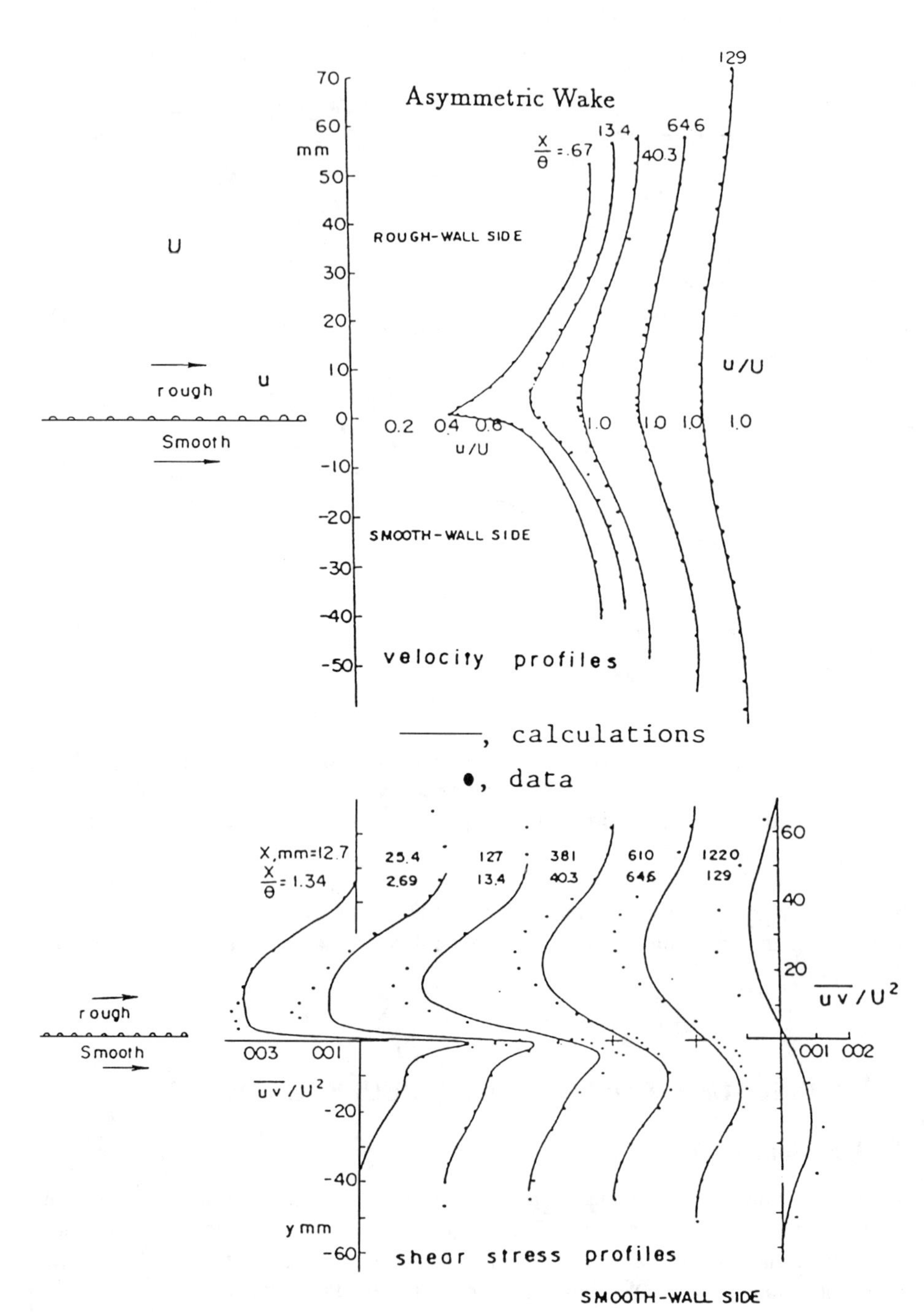

Figure 3.19 Velocity and shear stress profiles for an axisymmetric wake behind plate with one smooth and one rough wall. (Adapted from Patel and Scheuerer (1982).)

Chen and Singh (1986) reasoned that a model for the ϵ equations should include a small-eddy scale (i.e., Kolmogorov scale), and they proposed to modify the seventh postulate of the turbulence closure postulations (see Section 2.1.3) to include two scales, as follows. For the energy-containing large eddies, the (k, ϵ) scale

$$u = \sqrt{k}$$
$$l = \frac{k^{3/2}}{\epsilon}$$
$$t = \frac{k}{\epsilon},$$

and for the energy-dissipating small eddies, the (ν, ϵ) scale

$$u = (\nu\epsilon)^{1/4}$$
$$l = \left(\frac{\nu^3}{\epsilon}\right)^{1/4}$$
$$t = \sqrt{\frac{\nu}{\epsilon}}.$$

3.4.2 Two-Scale Modeling of the ϵ Equation

Consider the exact ϵ equation again,

$$\begin{aligned}\frac{D\epsilon}{Dt} &= \frac{\partial}{\partial X_l}\left(-\overline{\epsilon' u_l} - \frac{2\nu}{\rho}\overline{\frac{\partial u_l}{\partial X_j}\frac{\partial p}{\partial X_j}} + \nu\frac{\partial \epsilon}{\partial X_l}\right)\\ &\quad - 2\nu\frac{\partial U_i}{\partial X_j}\left(\overline{\frac{\partial u_l}{\partial X_i}\frac{\partial u_l}{\partial X_j}} + \overline{\frac{\partial u_i}{\partial X_l}\frac{\partial u_j}{\partial X_l}}\right) - 2\nu\overline{u_l\frac{\partial u_i}{\partial X_j}}\frac{\partial^2 U_i}{\partial X_l \partial X_j}\\ &\quad - 2\nu\overline{\frac{\partial u_i}{\partial X_j}\frac{\partial u_i}{\partial X_l}\frac{\partial u_j}{\partial X_l}} - 2\overline{\left(\nu\frac{\partial^2 u_i}{\partial X_l \partial X_j}\right)^2}.\end{aligned}$$

If we model the ϵ equation as before, we have

$$\frac{D\epsilon}{Dt} = \frac{\partial}{\partial X_l}\left[C_\epsilon\left(\frac{l^2}{t}\right)\frac{\partial \epsilon}{\partial X_l} + \nu\frac{\partial \epsilon}{\partial X_l}\right] - C_{\epsilon 1}\frac{1}{t}\overline{u_i u_l}\frac{\partial U_i}{\partial X_l} - C_{\epsilon 2}\frac{1}{t}\epsilon,$$

where l and t are the characteristic length and time, respectively, needed in the modeling. In the one-scale concept, we reason from the (k, ϵ) scale that

$$t = \frac{k}{\epsilon}$$
$$l = \frac{k^{3/2}}{\epsilon}.$$

Thus, we have

$$\frac{D\epsilon}{Dt} = \frac{\partial}{\partial X_l}\left[C_\epsilon\left(\frac{k^2}{\epsilon} + \nu\right)\frac{\partial \epsilon}{\partial X_l}\right] - C_{\epsilon 1}\frac{\epsilon}{k}\overline{u_i u_l}\frac{\partial U_i}{\partial X_l} - C_{\epsilon 2}\frac{\epsilon^2}{k}.$$

The experiments of Frisch, Sulem, and Nelkim (1978)[48] revealed that large eddies possess most of the turbulent kinetic energy in the flow and do not play any significant role in the dissipation of turbulent kinetic energy. On the other hand, Kolmogorov (1941)[77] found that the characteristics of small eddies are a function of (ν, ϵ) and that small eddies are responsible for dissipation of turbulent kinetic energy. Chen and Singh (1986, 1990)[21, 22] in addition reasoned that because eddies of all sizes exist simultaneously, the eddies with the size between the large and small eddies carry out the transfer of the turbulent kinetic energy possessed by large eddies to the small eddies before it is consumed by viscous dissipation and converted into thermal energy. It is natural to consider both large and small scales together in turbulence modeling. Chen and Singh (1990)[22] suggested that the Kolmogorov scale (ν, ϵ) should be used for the dissipation, or destruction, term of the k and ϵ equations. Because the exact expression of ϵ appears as the dissipation in the k equation, no modeling is required. For the ϵ equation, the turbulent diffusion term was modeled by the (k, ϵ) scale as before, whereas the Kolmogorov time scale, $(t) = (\nu/\epsilon)^{1/2}$, was introduced into the production and destruction terms. On the basis of the two-scale concept, we may reason that the diffusion phenomenon is governed by large-scale motion (k, ϵ), whereas the destruction of ϵ is controlled by small-scale motion (ν, ϵ).

If $(t) = (\nu/\epsilon)^{1/2}$ is used to model the destruction term in the ϵ equation, we have an improved model for the ϵ equation:

$$\frac{D\epsilon}{Dt} = \frac{\partial}{\partial X_l}\left[C'_\epsilon\left(\frac{k^2}{\epsilon} + \nu\right)\frac{\partial \epsilon}{\partial X_l}\right] - C'_{\epsilon 1}\sqrt{\frac{\epsilon}{\nu}}\overline{u_i u_j}\frac{\partial U_i}{\partial X_l} - C'_{\epsilon 2}\sqrt{\frac{\epsilon}{\nu}}\epsilon. \tag{3.82}$$

The constants C'_ϵ, $C'_{\epsilon 1}$, and $C'_{\epsilon 2}$ are determined in the same manner as the constants C_ϵ, $C_{\epsilon 1}$, and $C_{\epsilon 2}$—that is, through experiments (isotropic grid turbulence and homogeneous shear flow)—and are found to be $C'_\epsilon = 2.19$ and $C'_{\epsilon 1} = C'_{\epsilon 2} = 18.70/(Re)^{1/2}$.

3.4.3 Prediction of Two-Scale k-ϵ Model

For free-shear flow (two-scale k-ϵ model), the governing equations are

$$\frac{Dk}{Dt} = C_k\frac{\partial}{\partial Y}\left(\frac{k^2}{\epsilon}\frac{\partial k}{\partial Y}\right) - \overline{uv}\frac{\partial U}{\partial Y} - \epsilon \tag{3.83}$$

$$\frac{D\epsilon}{Dt} = C'_\epsilon\frac{\partial}{\partial Y}\left(\frac{k^2}{\epsilon}\frac{\partial \epsilon}{\partial Y}\right) - C'_{\epsilon 1}\sqrt{\frac{\epsilon}{\nu}}\overline{uv}\frac{\partial U}{\partial y} - C'_{\epsilon 2}\sqrt{\frac{\epsilon}{\nu}}\epsilon \tag{3.84}$$

$$U\frac{\partial U}{\partial X} + V\frac{\partial U}{\partial Y} = -\frac{\partial \overline{uv}}{\partial Y} \tag{3.85}$$

$$\frac{\partial U}{\partial X} + \frac{\partial V}{\partial Y} = 0 \tag{3.86}$$

$$-\overline{uv} = C_\mu\frac{k^2}{\epsilon}\frac{\partial U}{\partial Y}, \tag{3.87}$$

where $C_k = 0.55 \sim 0.90$, $C_\mu = 0.09$, $C'_\epsilon = 2.00$, and $C'_{\epsilon 1} = C'_{\epsilon 2} = 18.70/Re^{1/2}$.

The predictions of the two-scale k-ϵ-E model, when compared with those for the one-scale k-ϵ-E model, yield satisfactory improvement in prediction capability. Now

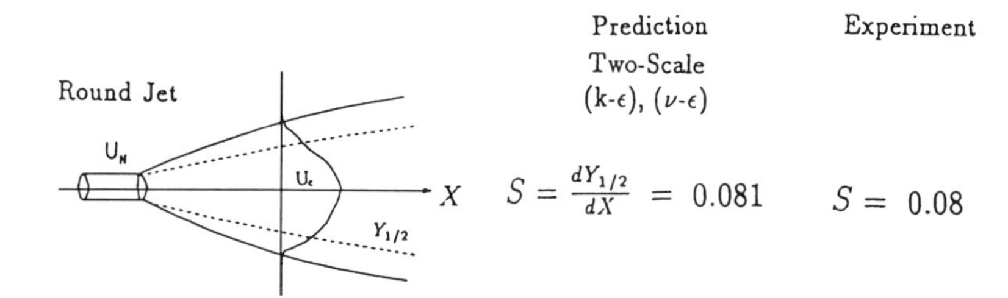

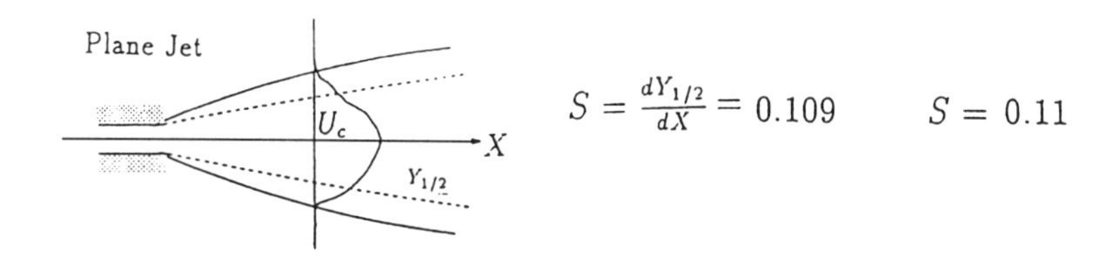

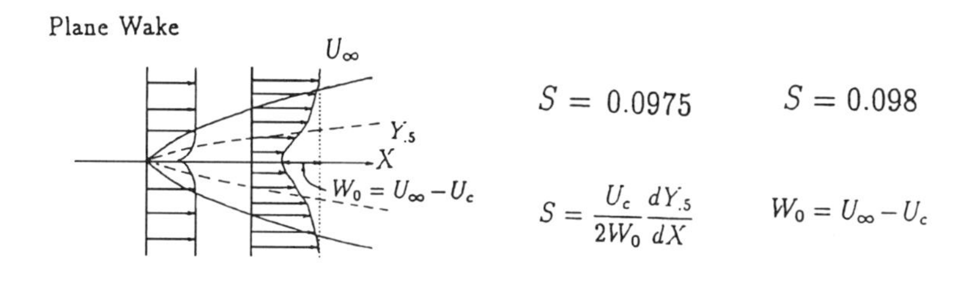

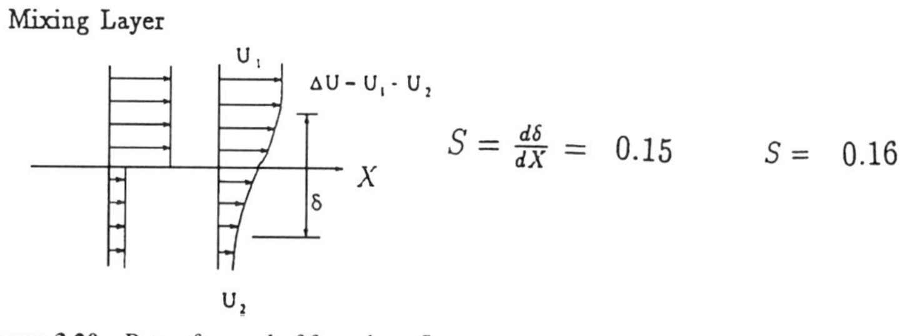

Figure 3.20 Rate of spread of free-shear flows.

the same set of constants can be used to predict two-dimensional and axisymmetric jets, wakes, and mixing-layer flows as shown in Fig. 3.20, Table 3.10, and Fig. 3.21.

3.4.4 Intermittent, Fractal Scale

Jaw and Chen (1990)[68] tried to incorporate fractal dynamics into turbulence modeling by adopting the intermittency model, β model, presented by Frisch et al. (1978)[48]. The main idea of the Frisch et al. model is that the small-scale structures of turbulence become less and less space filling as the scale size decreases.

Table 3.10 Comparison of one-scale and two-scale models with experimental spread rates

Flow	Spread parameter	Experimental data	One scale (k-ϵ)	Two scale (k-ϵ), (ν-ϵ)
Round jet	$S = \frac{dY_{1/2}}{dX}$	0.08	0.1186 (bad)	0.081 (ok)
Plane jet	$S = \frac{dY_{1/2}}{dX}$	0.11	0.1125	0.109
Plane wake	$S = \frac{U_\infty}{2W_o} \frac{dY_{1/2}}{dX}$	0.098	0.067 (bad)	0.0975 (ok)
Mixing layer	$S = \frac{d\delta}{dX}$	0.16	0.159	0.15

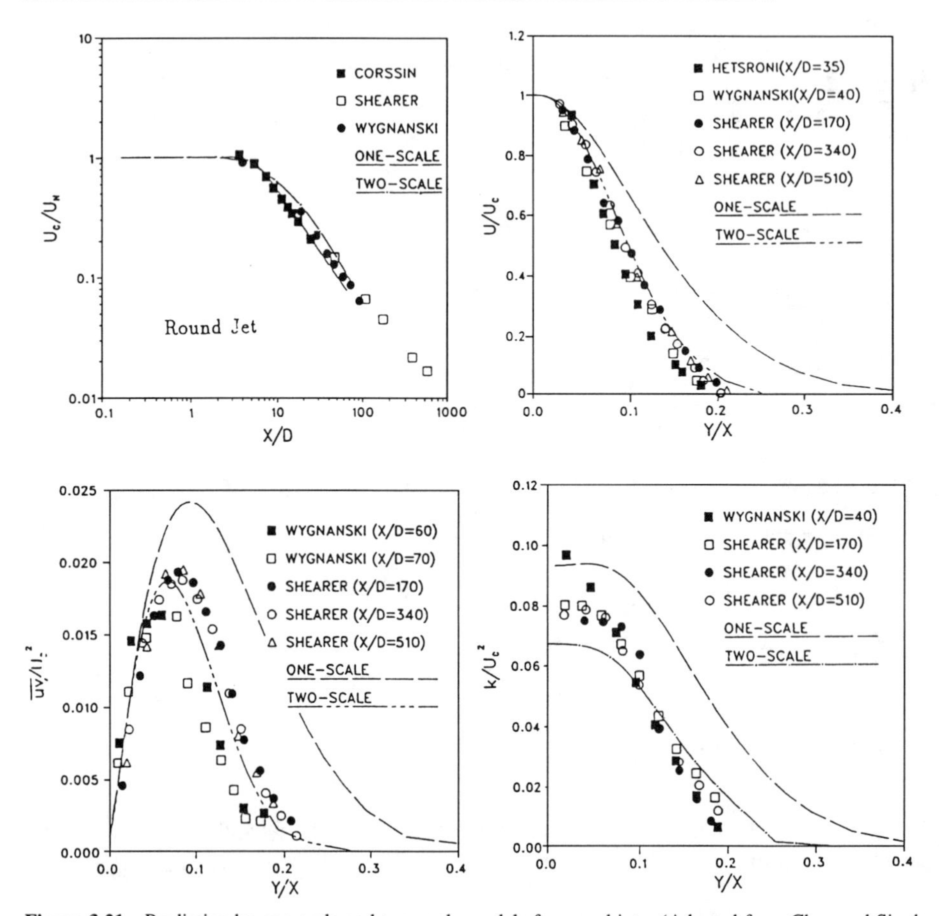

Figure 3.21 Prediction by one-scale and two-scale models for round jets. (Adapted from Chen and Singh (1990).)

Consider a discrete sequence of scales or eddies, $l_n = l_o 2^{-n}$, where $n = 0, 1, 2, \ldots$. If the largest eddies, l_o, are space filling, after n generations only a fraction $\beta_n = (N/2^3)^n$ of the space will be occupied by active fluid. Here N is the average number of eddies formed by each eddy of the preceding generation and is related to the fractal dimension D by $N = 2^D$, where $N \leq 8$. For instance, let v_n denote a typical velocity difference over a distance $\sim l_n$ in an active region. Then the kinetic energy per unit mass on the scale $\sim l_n$ is given by $k_n \sim \beta_n v_n^2$.

By introducing a simple dynamic argument, Frisch et al. (1978)[48] obtained the following relationships:

$$v_n \sim \epsilon^{1/3} l_n^{1/3} \left(\frac{l_n}{l_o} \right)^{-1/3(3-D)} \tag{3.88}$$

$$t_n \sim \epsilon^{-1/3} l_n^{2/3} \left(\frac{l_n}{l_o} \right)^{1/3(3-D)} \tag{3.89}$$

$$k_n \sim \epsilon^{2/3} l_n^{2/3} \left(\frac{l_n}{l_o} \right)^{1/3(3-D)}, \tag{3.90}$$

where t_n is the eddy turnover time, ϵ is the energy dissipation rate, and k_n is the turbulent kinetic energy. From these relationships, new turbulent scales, including the fractal dimension D, can be derived as

$$(l) = \left[\frac{k^{\frac{3(D-3)}{2}}}{\nu^{-3} \epsilon^{D-2}} \right]^{\frac{1}{(1+D)}} \tag{3.91}$$

$$(t) = \left[\frac{k^{3(D-3)}}{\nu^{D-5} \epsilon^{2D-4}} \right]^{\frac{1}{(1+D)}} \tag{3.92}$$

$$(u) = \frac{[l]}{[t]}. \tag{3.93}$$

Note that when D is equal to 3, then $l = (\nu^3/\epsilon)^{1/4}$ and $t = \sqrt{(\nu/\epsilon)}$, which is the classical dissipation scale, the Kolmogorov scale. However, Sreenivasan and Meneveau (1986)[168] showed from experimental observation that the fractal dimension for turbulent dissipation is $D \doteq 2.7$; thus the time scale is $(t) = \nu^{7/11}/(k^{3/11}\epsilon^{4/11})$. The new turbulent scales, with $D \doteq 2.7$, imply that a reasonable second turbulent scale is approximately a Kolmogorov microscale (ν, ϵ) with a fractal dimension of $D = 3$.

The two-scale turbulence model has been applied to predict turbulent free-shear flows (Jaw and Chen, 1990)[68]. Improved results have been obtained, including for the round jet flow, which cannot be predicted satisfactorily by the existing one-scale turbulence models. However, it was found that the predictions were sensitive to the boundary conditions applied. For practical applications, further study is needed.

CHAPTER

FOUR

NEAR-WALL TURBULENCE

4.1 INTRODUCTION

So far we have examined the performance of the second-order turbulence model in free-shear flows. When there is a wall, the turbulence becomes more complex because of the no-slip condition at the wall where the flow is reduced to laminar flow or the molecular viscosity dominates. This presents a problem in turbulence modeling because some postulations become questionable, such as (1) isotropic dissipation, (2) isotropic diffusion, and (3) PS modeling.

Also, in the near-wall zone, the velocity as well as other transport properties vary rapidly a short distance from the wall. Therefore, numerically it requires many grid nodes in this region to resolve properly the variation and the physics.

In the following, we attempt to find some approximate expressions for each variable in the near-wall region in turbulent flows. The derivation is made for the flows without separation, such as shown in Fig. 4.1, and is done on a two-dimensional basis. Flows with separation, such those as shown in Fig. 4.2, require further research. At present, in many practical problems, turbulence modeling for separated flows near the wall still assumes a near-wall function on the basis of a nonseparated near-wall function. In the future, the wall function for separated flows must be studied.

4.2 WALL FUNCTIONS

A wall function is a distribution function that describes the variation of U_i, T, $\overline{u_i u_j}$, k, and ϵ between, as shown in Fig. 4.3, a wall and the turbulent zone near it. It is often used to bypass the necessity of detailed numerical treatment and the uncertain validity of a turbulence model.

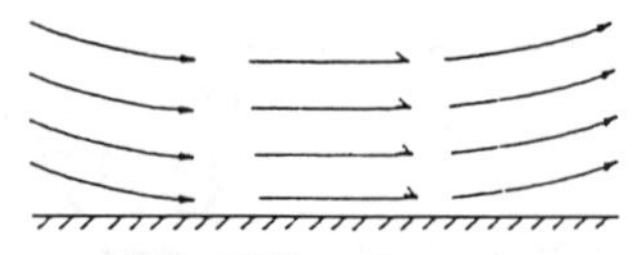

Figure 4.1 Wall shear flow without separation.

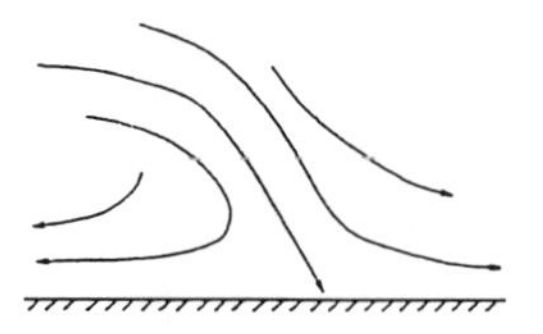

Figure 4.2 Wall shear flow with separation.

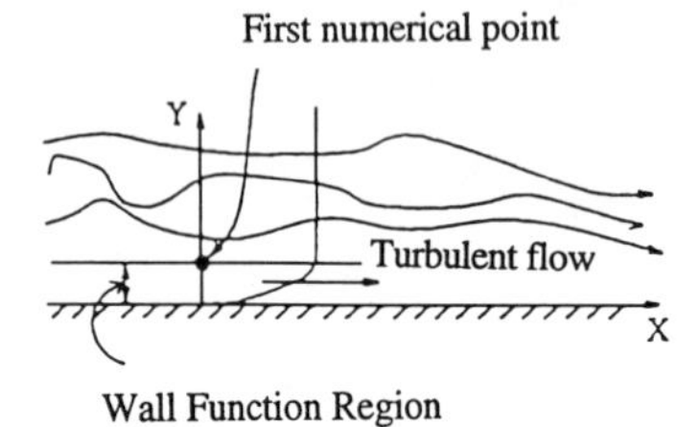

Figure 4.3 Wall function area of application.

4.2.1 Near-Wall Velocity

Consider flows without separation and stagnation. The two-dimensional, turbulent flow near the wall can be approximated by a fully developed or parallel flow assumption (i.e., $\partial/\partial X\,(\text{velocity}) = 0$). We have, in general, for incompressible flow under this assumption,

- the continuity equation yields

$$\frac{\partial V}{\partial Y} = 0, \tag{4.1}$$

- the x-momentum equation yields

$$0 = -\frac{1}{\rho}\frac{\partial P}{\partial X} + \frac{\partial}{\partial Y}\left(\nu\frac{\partial U}{\partial Y} - \overline{uv}\right), \tag{4.2}$$

- the y-momentum equation yields

$$0 = -\frac{1}{\rho}\frac{\partial P}{\partial Y} + \frac{\partial}{\partial Y}(-\overline{v^2}). \tag{4.3}$$

Hence, from Eq. 4.1, $V = f_v(X)$. Applying the boundary condition that V at $Y = 0$ is zero implies that $f_v(X) = 0$. Therefore, $V = 0$. From Eq. 4.3,

$$\frac{P}{\rho} + \overline{v^2} = f_p(X). \tag{4.4}$$

Because $\partial \overline{v^2}/\partial X = 0$, we conclude that the pressure gradient along X near the wall is a function of X only or

$$\frac{\partial}{\partial X}\left(\frac{P}{\rho}\right) = \frac{df_p}{dX}. \tag{4.5}$$

From Eq. 4.2, we have

$$\nu\frac{\partial U}{\partial Y} - \overline{uv} = \frac{d\left(\frac{P}{\rho}\right)}{dX}Y + f_u(X). \tag{4.6}$$

Because at $Y = 0$, $\overline{uv} = 0$, we have

$$\nu\left.\frac{\partial U}{\partial Y}\right|_{Y=0} = \frac{\tau_w}{\rho} = f_u(X). \tag{4.7}$$

Let

$$u^*(\text{friction velocity}) = \sqrt{\frac{\tau_w}{\rho}}$$

$$u^+ = \frac{U}{u^*}$$

$$y^+ = \frac{Yu^*}{\nu}$$

$$p^+ = \frac{P}{\rho u^{*2}}$$

$$x^+ = \frac{Xu^*}{\nu}$$

$$\overline{uv}^+ = \frac{\overline{uv}}{u^{*2}}.$$

Then, Eq. 4.6 becomes

$$\frac{du^+}{dy^+} - \overline{uv}^+ = \frac{dp^+}{dx^+}y^+ + 1. \tag{4.8}$$

The velocity inner law can be stated on the basis of Eq. 4.8 as

$$u^+ = f\left(\overline{uv}^+, \frac{dp^+}{dx^+}, y^+\right),$$

or

$$U = f\left(\overline{uv}, \nu, Y, \frac{\tau_w}{\rho}, \frac{dP}{dX}, \text{boundary conditions}\right).$$

In order to find an approximate velocity profile, we assume

$$-\overline{uv} \sim \frac{dU}{dY},$$

or

$$-\overline{uv}^+ = \kappa y^+\frac{du^+}{dy^+}.$$

The proportionality of κy^+ is assumed to give a reasonable behavior at the boundary because at $y^+ = 0, du^+/dy^+ \neq 0$, but $\overline{uv}^+$ must become zero. With the above-indicated assumption, Eq. 4.8 becomes

$$\frac{du^+}{dy^+} = \left(\frac{dp^+}{dx^+}y^+ + 1\right) \Big/ (1 + \kappa y^+), \tag{4.9}$$

where we see that the 1 in the term $1 + \kappa y^+$ represents the influence of molecular viscosity. Furthermore, for a mild pressure gradient near the wall, we have

$$\frac{dp^+}{dx^+}y^+ < 1.$$

Thus, approximately,

$$\frac{du^+}{dy^+} = \frac{1}{1 + \kappa y^+}. \tag{4.10}$$

Integrating, we have

$$u^+ = \frac{1}{\kappa}\ln(1 + \kappa y^+) + \text{constant}. \tag{4.11}$$

The two constants—a model constant (von Kármán constant, κ) and the integration constant—are determined from experiments as shown in Fig. 4.4.

1. For boundary layer flow over smooth wall, we have
   - for $y^+ > 10$, $u^+ = 2.44 \ln y^+ + 4.9$, where $\kappa = 0.4 \sim 0.43$ is the Kármán constant and
   - for $y^+ < 10$ (i.e., the laminar sublayer),

$$u^+ = y^+.$$

2. For pipe flow with a pressure gradient (i.e., $dP/dx \neq 0$), and $y^+ > 10$, we have

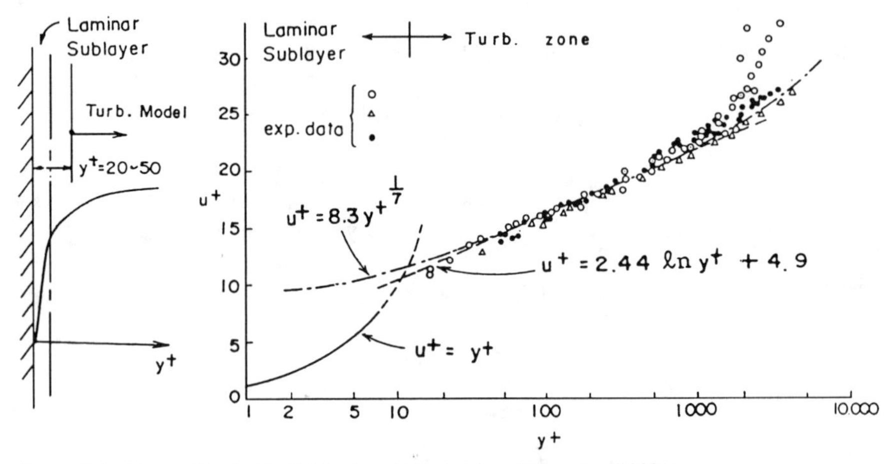

Figure 4.4 Near-wall velocity distribution. (Adapted from Nikuradse (1932))

- for smooth wall,

$$u^+ = 2.5 \ln y^+ + 5.5,$$

and

- for rough wall,

$$u^+ = 2.5 \ln y^+ + 8.5.$$

We see that the Kármán constant appears to be universal near a wall for flow without separation.

In the near-wall region, $0 < y^+ < 300$, u^+ is not sensitive to dp^+/dx^+. However, note that the effect of the pressure gradient is still included in $u^* = (\tau_w/\rho)^{1/2}$ because τ_w is a function of the pressure gradient. Thus, for flow without separation and mild pressure gradients, we may assume that the velocity distribution near the wall is

$$u^+ = \frac{1}{\kappa} \ln E y^+, \tag{4.12}$$

where $\kappa = 0.41$ and $E = 9$. This distribution is used as the wall function for velocity in many turbulent flow computations.

4.2.2 Near-Wall Temperature

Consider a flow that is without separation or stagnation in the near-wall region and is over a wall with a slowly varying temperature along the wall (i.e., $\partial T/\partial X \approx 0$), as shown in Fig. 4.5. In this case, we may assume that the flow is fully developed or

$$\frac{\partial \text{velocity}}{\partial X} = 0$$

and

$$\frac{\partial T}{\partial X} = 0$$

$$\frac{\partial \overline{u\theta}}{\partial X} = 0.$$

From the above-mentioned assumptions, we have the following simplified governing equations:

$$V \equiv 0 \tag{4.13}$$

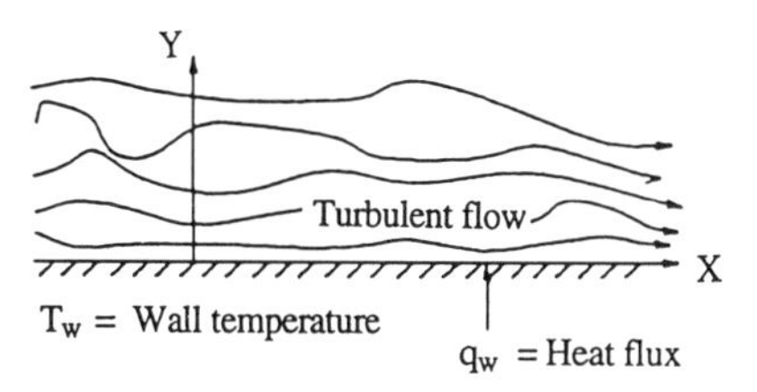

Figure 4.5 Near-wall thermal analysis.

$$0 = -\frac{1}{\rho}\frac{\partial P}{\partial X} + \frac{\partial}{\partial Y}\left(\nu\frac{\partial U}{\partial Y} - \overline{uv}\right) \tag{4.14}$$

$$0 = -\frac{1}{\rho}\frac{\partial P}{\partial Y} + \frac{\partial}{\partial Y}(0 - \overline{v^2}) \tag{4.15}$$

$$0 = \frac{\partial}{\partial Y}\left(\alpha\frac{\partial T}{\partial Y} - \overline{v\theta}\right). \tag{4.16}$$

Integrating Eq. 4.16, we have

$$\alpha\frac{\partial T}{\partial Y} - \overline{v\theta} = f_T(X). \tag{4.17}$$

Because at $Y = 0$,

$$\overline{v\theta} = 0,$$

we have

$$\alpha\left.\frac{\partial T}{\partial Y}\right|_{Y=0} = f_T(X) = -\frac{\alpha}{k}q_w.$$

Here we define

$$f_T = -\frac{\alpha}{k}q_w(X) = -u^*T^*$$

$$T^+ = \frac{T_w - T}{T^*}$$

$$y^+ = \frac{u^*Y}{\nu}$$

$$\overline{v\theta}^+ = \frac{\overline{v\theta}}{u^*T^*}$$

and

$$T^* = \frac{\alpha q_w}{ku^*},$$

where k is the thermal conductivity. Thus, Eq. 4.17 becomes

$$-\frac{1}{Pr}\frac{dT^+}{dy^+} - \overline{v\theta}^+ = -1. \tag{4.18}$$

The temperature inner law, which is based on Eq. 4.18, states that

$$T^+ = f(y^+, u^+, Pr),$$

or

$$T = f(\overline{v\theta}, Y, \tau_w, q_w, U, Pr, \text{boundary condition}).$$

If

$$-\overline{v\theta}^+ \sim -\frac{dT^+}{dy^+}, \tag{4.19}$$

or

$$-\overline{v\theta}^+ = -\frac{\kappa y^+}{Pr_t}\frac{dT^+}{dy^+}, \tag{4.20}$$

then Eq. 4.18 becomes

$$\frac{dT^+}{dy^+} = \left(\frac{1}{\frac{1}{Pr} + \frac{\kappa y^+}{Pr_t}} \right), \tag{4.21}$$

where Pr_t is known as the turbulent Prandtl number, or $Pr_t = \nu_t/\alpha_t$. ν_t and α_t are respectively the eddy viscosity and eddy thermal diffusivity defined in Section 3.1.4 by Eqs. 3.16 and 3.17. Thus,

$$T^+ = \int_0^{y^+} \frac{d\gamma}{\left(\frac{1}{Pr} + \frac{\kappa\gamma}{Pr_t}\right)}, \tag{4.22}$$

where $\kappa = 0.40 \sim 0.43$, $Pr_t = 0.8 \sim 1.3$, and γ is a dummy variable. From Eq. 4.22 and Fig. 4.6, for small Prandtl numbers (i.e., $Pr \ll 1$) and $y^+ < 10$, one may set $T^+ = Pr \cdot y^+$. For large Prandtl numbers (i.e., $Pr \gg 1$) and $y^+ > 10$,

$$T^+ = \frac{Pr_t}{\kappa} \ln E \cdot y^+$$

with $E \equiv 9.0$. From experiments (Kays and Crawford, 1980 [72]), the constants for the near-wall temperature distribution are determined to be

$$T^+ = 2.195 \ln y^+ + 13.2Pr - 5.66 \tag{4.23}$$

for $y^+ > 10$. This temperature distribution may be used as a temperature wall function in turbulent flow computations.

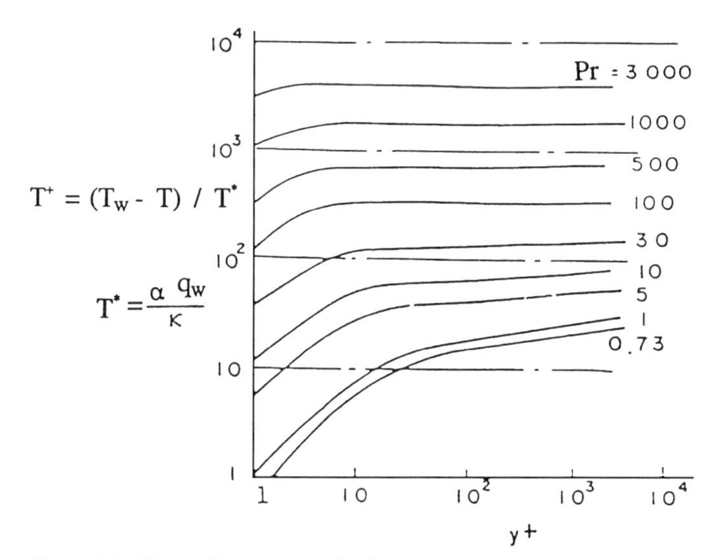

Figure 4.6 Near-wall temperature distribution.

4.2.3 Near-Wall Reynolds-Stress Equation ($\overline{u_iu_j}$)

As discussed in chapter 2, the exact equation for the Reynolds stress term, $\overline{u_iu_j}$, is given by

$$\frac{D\overline{u_iu_j}}{Dt} = \frac{\partial}{\partial X_l}\left(-\overline{u_iu_ju_l} - \overline{\frac{p}{\rho}(\delta_{jl}u_i + \delta_{il}u_j)} + \nu\frac{\partial\overline{u_iu_j}}{\partial X_l}\right)$$
$$-\left(\overline{u_iu_l}\frac{\partial U_j}{\partial X_l} + \overline{u_ju_l}\frac{\partial U_i}{\partial X_l}\right) - 2\nu\overline{\frac{\partial u_i}{\partial X_l}\frac{\partial u_j}{\partial X_l}} + \overline{\frac{p}{\rho}\left(\frac{\partial u_i}{\partial X_j} + \frac{\partial u_j}{\partial X_i}\right)}$$

with the modeled equation being given by Eq. 2.17 as

$$\frac{D\overline{u_iu_j}}{Dt} = \frac{\partial}{\partial X_l}\left(C_k\frac{k^2}{\epsilon}\frac{\partial\overline{u_iu_j}}{\partial X_l} + \nu\frac{\partial\overline{u_iu_j}}{\partial X_l}\right) + P_{ij} - \frac{2}{3}\delta_{ij}\epsilon$$
$$-C_1\frac{\epsilon}{k}\left(\overline{u_iu_j} - \frac{2}{3}\delta_{ij}k\right) - C_2\left(P_{ij} - \frac{2}{3}\delta_{ij}P_k\right).$$

In the near-wall region, we expect that the turbulence even on a small scale is highly anisotropic. Several modifications of the $\overline{u_iu_j}$ equation can be considered. The main modifications considered are in the pressure-strain (PS) term and in the diffusion term. To introduce anisotropic effect in the diffusion term, instead of using

$$\frac{\partial}{\partial X_l}\left(C_k\frac{k^2}{\epsilon}\frac{\partial\overline{u_iu_j}}{\partial X_l}\right),$$

we have near the wall, the term with an isotropic diffusion,

$$\frac{\partial}{\partial X_l}\left(C_k'\frac{k}{\epsilon}\overline{u_lu_k}\frac{\partial\overline{u_iu_j}}{\partial X_k}\right),$$

where $C_k' = 0.235$. For the dissipation term, instead of using the idea of isotropic dissipation (Postulation 3 in Section 2.1.3)

$$\frac{2}{3}\delta_{ij}\epsilon,$$

we may have near the wall the term

$$\frac{\overline{u_iu_j}}{k}\epsilon.$$

For the PS term, from Eq. 1.14, we have

$$\frac{Du_i}{Dt} = -\frac{1}{\rho}\frac{\partial p}{\partial X_l} + \nu\frac{\partial^2 u_i}{\partial X_l\partial X_l} - u_l\frac{\partial U_i}{\partial X_l} - u_l\frac{\partial u_i}{\partial X_l} + \frac{\partial\overline{u_iu_l}}{\partial X_l}.$$

Taking the divergence of above equation, we obtain,

$$\nabla^2\frac{p}{\rho} = -\left(\frac{\partial^2(u_lu_m - \overline{u_lu_m})}{\partial X_l\partial X_m} + 2\frac{\partial U_l}{\partial X_m}\frac{\partial u_m}{\partial X_l}\right)$$
$$= -F.$$

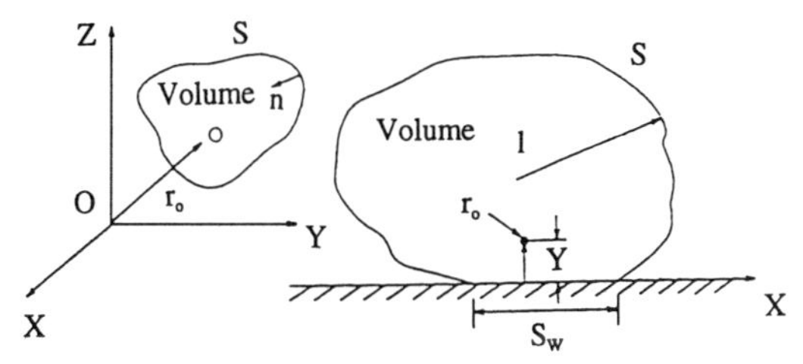

Figure 4.7 Near-wall integration.

Consider Fig. 4.7. From Green's theorem, we have

$$\int_{\text{vol}} \frac{1}{r} \nabla^2 \left(\frac{p}{\rho}\right) \text{dvol} = -\int_s \left[\frac{p}{\rho}\frac{\partial}{\partial n}\left(\frac{1}{r}\right) - \frac{1}{r}\frac{\partial}{\partial n}\left(\frac{p}{\rho}\right)\right] ds - 4\pi \left.\frac{p}{\rho}\right|_o .$$

Hence,

$$\frac{p}{\rho} = \frac{1}{4\pi}\int_{\text{vol}} F\frac{\text{dvol}}{r} - \frac{1}{4\pi}\int_s \frac{p}{\rho}\frac{\partial}{\partial n}\left(\frac{1}{r}\right) ds + \frac{1}{4\pi}\int_s \frac{1}{r}\frac{\partial}{\partial n}\left(\frac{p}{\rho}\right) ds.$$

When the wall (or the surface of the integral S) is away from r_o, then

$$\frac{p}{\rho} = \frac{1}{4\pi}\int_{\text{vol}} F\frac{\text{dvol}}{r}$$

because the integration over the surface S that is far from r_o is equal to zero. However, when the wall (S_w) is close to r_o, then the integration over the surface S is zero everywhere except over S_w. In this case,

$$\frac{p}{\rho} \equiv \frac{1}{4\pi}\int_{\text{vol}} F\frac{\text{dvol}}{r} - \frac{1}{4\pi}\int_{S_w} \frac{p}{\rho}\frac{\partial}{\partial n}\left(\frac{1}{r}\right) ds - 0.$$

Here it is assumed that the variation of the pressure in the normal direction is negligible, that is,

$$\begin{aligned}\frac{\partial}{\partial n}\left(\frac{p}{\rho}\right) &= \frac{\partial}{\partial Y}\left(\frac{p}{\rho}\right) \\ &= 0.\end{aligned}$$

Therefore,

$$\begin{aligned}\overline{\frac{p}{\rho}\left(\frac{\partial u_i}{\partial X_j} + \frac{\partial u_j}{\partial X_i}\right)} &\doteq \frac{1}{4\pi}\int_{\text{vol}} \overline{F^*\left(\frac{\partial u_i}{\partial X_j} + \frac{\partial u_j}{\partial X_i}\right)}\frac{\text{dvol}}{r} \\ &\quad - \frac{1}{4\pi}\int_{S_w} \overline{\frac{p^*}{\rho}\left(\frac{\partial u_i}{\partial X_j} + \frac{\partial u_j}{\partial X_i}\right)}\left(\frac{-1}{r^2}\right) ds \\ &\doteq \frac{1}{4\pi}\overline{F\left(\frac{\partial u_i}{\partial X_j} + \frac{\partial u_j}{\partial X_i}\right)}l^2 + \frac{1}{4\pi}\overline{\frac{p}{\rho}\left(\frac{\partial u_i}{\partial X_j} + \frac{\partial u_j}{\partial X_i}\right)}\frac{l^2}{Y^2},\end{aligned}$$

where the second term on the right represents the new model near the wall, whereas the first term on the right is identical to that discussed in chapter 2 as $\Phi_{ij,1}$. Chen (1983) modeled $l^2 = k^3/\epsilon^2$ on the basis of dimensional reasoning about k and ϵ; hence, we have

$$\overline{\frac{p}{\rho}\left(\frac{\partial u_i}{\partial X_j}+\frac{\partial u_j}{\partial X_i}\right)} = -C_1\frac{\epsilon}{k}\left(\overline{u_i u_j}-\frac{2}{3}\delta_{ij}k\right) - C_2\left(P_{ij}-\frac{2}{3}\delta_{ij}P_k\right) + C_w\overline{\frac{p}{\rho}\left(\frac{\partial u_i}{\partial X_j}+\frac{\partial u_j}{\partial X_i}\right)}\frac{k^3}{\epsilon^2 Y^2}.$$

Thus,

$$\overline{\frac{p}{\rho}\left(\frac{\partial u_i}{\partial X_j}+\frac{\partial u_j}{\partial X_i}\right)} = -\tilde{C}_1\frac{\epsilon}{k}\left(\overline{u_i u_j}-\frac{2}{3}\delta_{ij}k\right) - \tilde{C}_2\left(P_{ij}-\frac{2}{3}\delta_{ij}P_k\right), \tag{4.24}$$

where

$$\tilde{C}_1 = C_1\Big/\left(1 - C_w\frac{k^3}{\epsilon^2 Y^2}\right) \doteq C_1\left(1 + C_w\frac{k^3}{\epsilon^2 Y^2}\right)$$

and

$$\tilde{C}_2 = C_2\Big/\left(1 - C_w\frac{k^3}{\epsilon^2 Y^2}\right) \doteq C_2\left(1 + C_w\frac{k^3}{\epsilon^2 Y^2}\right).$$

Simpler coefficients were proposed by Launder et al. (1975)[88] for Eq. 4.25 to set

$$\tilde{C}_1 = C_1 + 0.125\frac{k^{3/2}}{\epsilon Y}$$

and

$$\tilde{C}_2 = C_2 + 0.015\frac{k^{3/2}}{\epsilon Y}.$$

Here C_1 and C_2 are the same constants that were described in the modeling of the PS term in chapter 2 (i.e., $C_1 = 1.5$ and $C_2 = 0.4 \sim 0.6$). The difference between Chen's (1983) derivation and that of Launder et al.'s is that the latter assume a linear relation of l/Y instead of Chen's $(l/Y)^2$ relation for the surface integral.

Note that as Y becomes large, the model is reduced to the original model.

4.2.4 Near-Wall Turbulence Kinetic Energy (k) and Stress ($\overline{uv}$)

Because PS $= 0$ in the k equation, there is no particular difficulty with the k equation. Near the wall, turbulent fluctuation is small, whereas production and dissipation are large. Therefore, the k equation is given by

$$P_k - \epsilon = 0,$$

where

$$P_k = -\overline{u_n u_m}\frac{\partial U_n}{\partial X_m}.$$

Then,

$$-\overline{uv}\frac{\partial U}{\partial Y} = \epsilon. \tag{4.25}$$

To derive an approximate Reynolds stress $-\overline{uv}$ near the wall, we see that the momentum equation in the X direction near the wall is given by

$$0 = -\frac{1}{\rho}\frac{dP}{dX} + \frac{\partial}{\partial Y}\left(\nu\frac{\partial U}{\partial Y} - \overline{uv}\right). \tag{4.26}$$

Then

$$\begin{aligned} \nu\frac{\partial U}{\partial Y} - \overline{uv} &= f_u(X) \\ &= \frac{\tau_w(X)}{\rho}. \end{aligned} \tag{4.27}$$

In the turbulent zone,

$$\nu\frac{\partial U}{\partial Y} \ll -\overline{uv}.$$

Hence, we may approximate Eq. 4.27 as

$$\begin{aligned} -\overline{uv} &= \frac{\tau_w(X)}{\rho} \\ &= u^{*2}, \end{aligned} \tag{4.28}$$

where u^* is the friction velocity. The analysis of the momentum equation in the Y direction Eq. 4.3 shows that near the wall, $-\overline{uv}$ is not a function of Y. From the $\overline{uv}$ equation analysis, we have from the eddy viscosity model, Eq. 3.16

$$-\overline{uv} = C_\mu\frac{k^2}{\epsilon}\frac{\partial U}{\partial Y}. \tag{4.29}$$

Hence, we may combine Eqs. 4.25, 4.28, and 4.29 to obtain the following equation:

$$-\overline{uv} = C_\mu\left(k^2\frac{\partial U}{\partial Y}\right)\bigg/\left(-\overline{uv}\frac{\partial U}{\partial Y}\right).$$

Hence,

$$(-\overline{uv})^2 = C_\mu k^2.$$

Therefore, near the wall with $-\overline{uv} = u^{*2}$,

$$\begin{aligned} k &= \frac{\tau_w}{\rho}\frac{1}{\sqrt{C_\mu}} \\ &= \frac{u^{*2}}{\sqrt{C_\mu}}. \end{aligned} \tag{4.30}$$

In summary, for $y^+ > 10$, near the wall, the wall function for k and $\overline{uv}$ is

$$k = \frac{u^{*2}}{\sqrt{C_\mu}}$$

$$-\overline{uv}_{\text{wall}} = u^{*2} = \frac{\tau_w}{\rho}.$$

These two equations may be used as wall functions or computational boundary conditions for k and $\overline{uv}$.

4.2.5 Near-Wall Dissipation Function (ϵ)

Using the same reasoning as in the previous section, we have from the kinetic energy equation

$$\epsilon = -\overline{uv}\frac{\partial U}{\partial Y}. \tag{4.31}$$

The near-wall velocity profile can be approximated, from the momentum equation in the X direction, by

$$u^+ = \frac{1}{\kappa} \ln y^+ + \text{constant}.$$

Differentiating the above equation, we have

$$\frac{\partial U}{\partial Y} = \frac{u^*}{\kappa Y}, \tag{4.32}$$

noting that $u^+ = U/u^*$. Furthermore, from the $\overline{uv}$ equation near the wall, we have

$$-\overline{uv} = \frac{\tau_w}{\rho} = u^{*2}.$$

Combining Eqs. 4.31 and 4.32, we have for $y^+ > 10$,

$$\epsilon = \frac{u^{*3}}{\kappa Y}.$$

This may be used as a wall dissipation function.

4.2.6 Summary

To avoid sharp gradients in which many grid points in computation are needed, and because the turbulence model is invalid too close to the wall, the computation may be started in the turbulent zone. Between the turbulent zone and the wall, a wall function is used, as shown in Fig. 4.8. The wall functions for each variable are summarized in

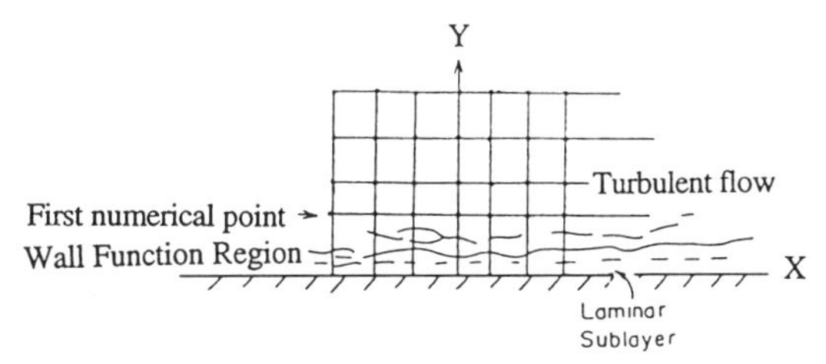

Figure 4.8 Computational region of wall functions.

the following way:

- From the momentum equations, respectively,

$$u^+ = y^+, \quad y^+ < 10$$
$$u^+ = \frac{1}{\kappa} \ln E y^+, \quad y^+ > 10,$$

  where $\kappa = 0.41$ and $E = 9.00$.
- From the energy equation, respectively,

$$T^+ = Pr \cdot y^+, \quad y^+ < 10, \ Pr < 10,$$
$$T^+ = \frac{Pr_t}{\kappa} \ln E y^+, \quad y^+ > 10.$$

  where $Pr_t \doteq 0.9$ is the turbulent Prandtl number.
- From the $\overline{uv}$ equation, we have for $y^+ > 10$,

$$-\overline{uv} = u^{*2} = \frac{\tau_w}{\rho}.$$

- From the kinetic energy, k, equation, we have for $y^+ > 10$,

$$k = \frac{u^{*2}}{\sqrt{C_\mu}},$$

  where $C_\mu = 0.09$.
- From the energy dissipation, ϵ, equation, we have for $y^+ > 10$,

$$\epsilon = \frac{u^{*3}}{\kappa Y}.$$

4.3 LOW-REYNOLDS-NUMBER TURBULENCE MODELS

There are a number of instances in which the modeled turbulence equations and the wall function approach are not adequate, and, hence, one has to modify the wall function approach in predicting, for example, turbulent boundary layers with low and transitional Reynolds numbers, unsteady and separated flows, or flows over spinning surfaces, and so forth (Patel, Rodi, and Schenerer, 1984 [121]). Also, wall functions do not allow prediction of turbulent wall-shear flows directly from the wall. From the practical stand point, if the momentum and the continuity equations are solved from the wall, such a prediction capability would provide complex turbulent flows near the wall without invoking wall functions.

Over the years, many substitutions for the wall function approach have been made. Some researchers tried to develop a low-Reynolds-number model by incorporating either a wall damping effect or a direct effect of molecular viscosity, or both, on the empirical constants and functions in the turbulence transport equations devised originally for high-Reynolds-number, fully turbulent flows remote from the wall. Others tried the idea of a two-layer model by resolving the viscosity-affected near-wall layer with a simpler and numerically more stable algebraic model. The two-layer model often involves a length-scale model replacing the inconsistent and improperly behaved ϵ equation in the near-wall layer and uses the k-ϵ model in the region away from the wall.

A variety of low-Reynolds-number k-ϵ models involving different functions and additional terms have been proposed, which in general can be written in the form

$$-\overline{u_i u_j} = \nu_t \left(\frac{\partial U_i}{\partial X_j} + \frac{\partial U_i}{\partial X_j} \right) - \frac{2}{3}\delta_{ij} k \tag{4.33}$$

$$\nu_t = C_\mu f_\mu \frac{k^2}{\bar{\epsilon}} \tag{4.34}$$

$$\epsilon = \bar{\epsilon} + D \tag{4.35}$$

$$\frac{Dk}{Dt} = \frac{\partial}{\partial X_l}\left(C_k f_\mu \frac{k^2}{\bar{\epsilon}} \frac{\partial k}{\partial X_l} + \nu \frac{\partial k}{\partial X_l} \right) - \overline{u_i u_l}\frac{\partial U_i}{\partial X_l} - \epsilon \tag{4.36}$$

$$\frac{D\bar{\epsilon}}{Dt} = \frac{\partial}{\partial X_l}\left(C_\epsilon f_\mu \frac{k^2}{\bar{\epsilon}} \frac{\partial \bar{\epsilon}}{\partial X_l} + \nu \frac{\partial \bar{\epsilon}}{\partial X_l} \right) - C_{\epsilon 1} f_1 \frac{\bar{\epsilon}}{k}\overline{u_i u_l}\frac{\partial U_i}{\partial X_l} - C_{\epsilon 2} f_2 \frac{\overline{\epsilon^2}}{k} + E \tag{4.37}$$

$$R_T = \frac{k^2}{\nu \bar{\epsilon}} \tag{4.38}$$

$$R_y = \frac{\sqrt{k} y}{\nu} \tag{4.39}$$

$$U^* = \sqrt{\frac{\tau_w}{\rho}} \tag{4.40}$$

$$y^+ = \frac{YU^*}{\nu}. \tag{4.41}$$

Several low-Reynolds-number models have been reviewed by Patel et al. (1984)[121], Nagano and Hishida (1987)[112], Shih and Mansour (1990)[153], and Michelassi and Shih (1991)[107]. Ten of those models, namely, those of Chien (1982)[30], Dutoya and Michard (1981)[45], Hassid and Porch (1975)[59], Hoffman (1975)[61], Lam and Bremhorst (1981)[81], Launder and Sharma (1974)[90], Reynolds (1970)[129], Nagano and Hishida (1987)[112], Michelassi and Shih (1991)[107], and Wilcox and Rubesin (1980)[186], have been considered here. The first nine models are variants of the k-ϵ model. The tenth is a k ω model. Wilcox and Rubesin employ an equation for the turbulent kinetic energy together with a transport equation for a pseudovorticity, ω.

Table 4.1 Constants for the k-ϵ group of models

Model	Code	D	$\overline{\epsilon_\omega}$– B.C.	c_μ	$c_{\epsilon 1}$	$c_{\epsilon 2}$	σ_k	σ_k
Standard	HR	0	Wall functions	0.09	1.44	1.92	1.00	1.30
Launder–Sharma	LS	$2\nu\left(\frac{\partial\sqrt{k}}{\partial Y}\right)^2$	0	0.09	1.44	1.92	1.00	1.30
Hassid–Poreh	HP	$2\nu\frac{k}{Y^2}$	0	0.09	1.45	2.00	1.00	1.30
Hoffman	HO	$\frac{\nu}{Y}\frac{\partial k}{\partial Y}$	0	0.09	1.81	2.00	2.00	3.00
Dutoya–Michard	DM	$2\nu\left(\frac{\partial\sqrt{k}}{\partial Y}\right)^2$	0	0.09	1.35	2.00	0.90	0.95
Chien	CH	$2\nu\frac{k}{Y^2}$	0	0.09	1.35	1.80	1.00	1.30
Reynolds	RE	0	$\nu\frac{\partial^2 k}{\partial Y^2}$	0.084	1.00	1.83	1.69	1.30
Lam–Bremhorst	LB	0	$\nu\frac{\partial^2 k}{\partial Y^2}$	0.09	1.44	1.92	1.00	1.30
Lam–Bremhorst	LB1	0	$\frac{\partial\epsilon}{\partial Y}=0$	0.09	1.44	1.92	1.00	1.30
Nagano–Hishida	NH	$2\nu\left(\frac{\partial\sqrt{k}}{\partial Y}\right)^2$	0	0.09	1.45	1.90	1.00	1.30
Michelassi–Shih	MS	$\frac{\partial}{\partial Y}\left(\frac{0.004\mu_t}{1.3f_\mu^2}\frac{\partial k}{\partial Y}\right)$	$\frac{\nu}{2k}\left(\frac{\partial k}{\partial Y}\right)^2$	0.09	1.45	2.00	1.00	1.30

The k-ω model is

$$U\frac{\partial k}{\partial X}+V\frac{\partial k}{\partial Y}=\frac{\partial}{\partial Y}\left[\left(\nu+\frac{\nu_t}{\sigma_k}\right)\frac{\partial k}{\partial Y}\right]+\nu_t\left(\frac{\partial U}{\partial Y}\right)^2-C_\mu k\omega \tag{4.42}$$

$$U\frac{\partial\omega^2}{\partial X}+V\frac{\partial\omega^2}{\partial Y}=\frac{\partial}{\partial Y}\left[\left(\nu+\frac{\nu_t}{\sigma_\omega}\right)\frac{\partial\omega^2}{\partial Y}\right]+c_{\omega1}f_1\omega\left(\frac{\partial U}{\partial Y}\right)^2-c_{\omega2}\omega^3+E, \tag{4.43}$$

where $\nu_t=f_\mu(k/\omega)$, $R_T=\sqrt{k}l/\nu$, and $l=\sqrt{k}/\omega$. Tables 4.1, 4.2, and 4.3 summarize the low-Reynolds-number constants and functions for the k-ϵ group of models. The first line in the tables is the parent high-Reynolds-number model (HR). The first line in Table 4.1 also contains the five basic constants. Table 4.3 summarizes the model of Wilcox and Rubesin (WR). In these equations, x is in the streamwise direction, y is normal to the wall, and U and V are the corresponding mean velocity components. In the interest of brevity, the various models will henceforth be referred to by the letter codes indicated in Tables 4.1, 4.2, and 4.3 (e.g., Chien is CH, and so forth). The models in the k-ϵ group differ from their basic version by the inclusion of the viscous diffusion terms and of functions, f, to modify the constants, c. Also, extra terms, denoted by D and E, are added in some cases to better represent the near-wall behavior.

Table 4.2 Functions for the k-ϵ group of models

Code	f_μ	f_1	f_2	E
HR	1.0	1.0	1.0	0.0
LS	$\exp\left[\frac{-3.4}{\left(1+\frac{R_T}{50}\right)^2}\right]$	1.0	$1-0.3\exp\left(-R_T^2\right)$	$2\nu\nu_t\left(\frac{\partial^2 U}{\partial Y^2}\right)^2$
HP	$1-\exp(-0.0015R_T)$	1.0	$1-0.3\exp\left(-R_T^2\right)$	$-2\nu\left(\frac{\partial\sqrt{\tilde{\epsilon}}}{\partial Y}\right)^2$
HO	$\exp\left(\frac{-1.75}{1+R_T/50}\right)$	1.0	$1-0.3\exp\left(-R_T^2\right)$	0
DM	$1-0.86\exp\left[-\left(\frac{R_T}{600}\right)^2\right]$	$1-0.04\exp\left[-\left(\frac{R_T}{50}\right)^2\right]$ $+0.25\left(\frac{\lambda}{Y}\right)^2$	$1-0.3\exp\left[-\left(\frac{R_T}{50}\right)^2\right]$ $-0.08\left(\frac{\lambda}{Y}\right)^2$	$-c_{\epsilon 2}f_2(\tilde{\epsilon}D/k)$
CH	$1-\exp(-0.0115y^+)$	1.0	$1-0.22\exp[-(R_T/6)^2]$	$-2\nu(\tilde{\epsilon}/Y^2)$ $\times\exp(-0.5Y)$
RE	$1-\exp(-0.0198R_y)$	1.0	$[1-0.3\exp[-(R_T/3)^2]$ $\times h(R_y)$	0
LB	$[1-\exp(-0.0165R_y)]^2$ $\times\left(1+\frac{20.5}{R_T}\right)$	$1+(0.05/f_\mu)^3$	$1-\exp\left(-R_T^2\right)$	0
LB1	$[1-\exp(-0.0165R_y)]^2$ $\times\left(1+\frac{20.5}{R_T}\right)$	$1+(0.05/f_\mu)^3$	$1-\exp\left(-R_T^2\right)$	0
NH	$[1-\exp(-R_T/26.5)]^2$	1.0	$1-0.3\exp\left(-R_T^2\right)$	$\nu\nu_t(1-f_\mu)$ $\times\left(\frac{\partial^2 U}{\partial Y^2}\right)^2$
MS	$1-\frac{\exp\left[-0.0004\exp\left(1.2R_L^{1/4}\right)\right]}{\exp[-0.0004]}$	1.0	$1-0.22\exp[-(R_T/6)^2]$	$-\nu\nu_t\left(\frac{\partial^2 U}{\partial Y^2}\right)^2$

The new nondimensional parameter R_L, which appeared in MS's model, is defined as $R_L = L/L_\nu$, where $L = k^{3/2}/\epsilon$ and $L_\nu = \nu/|U|$. $|U|$ is the amplitude of the relative mean velocity in a frame fixed to the solid boundary. R_L represents the ratio of the turbulent length-scale L to the viscous length scale L_ν. This ratio approaches zero at the wall and asymptotically reaches a maximum in the log-law region. With the introduction of the R_L parameter in the damping function f_μ, MS's model is independent of the wall distance, y or y^+, and thus does not give unphysical results in the case of separation or reattachment points.

Table 4.3 Constants and functions for the WR model

Constant	Value	Function	Value
c_μ	0.09	σ_ω	2.0
$c_{\omega 1}$	1.11	f_μ	$1-0.992\exp(-R_T)$
$c_{\omega 2}$	0.15	f_l	$1-0.992\exp(-R_T/2)$
σ_k	2.00	E	$-\frac{2}{\sigma_\omega}\left(\frac{\partial l}{\partial Y}\right)^2\omega^3$

All versions have been tested mainly in boundary layer, pipe, and channel flows, except the model of NH (1987)[112], which has also been tested by a diffuser flow. Depending somewhat on the version considered, low-Reynolds-number k-ϵ models were found to require rather high numerical resolution near the wall, mainly, because of the steep gradient of the dissipation rate, ϵ, in which the distribution is determined by solving a transport equation for this quantity. Typically, 60 to 100 grid points across boundary layers are required for proper numerical resolution. Further, k-ϵ models have been found to perform rather poorly in boundary layers with adverse pressure gradients (Rodi and Scheuerer, 1986 [139]). In separated flows, the low-Reynolds-number k-ϵ models are little tested so far. Initial calculations indicate that the damping functions developed for attached boundary layers are not always well behaved in these flows.

4.4 TWO-LAYER MODEL

In order to save grid points and hence computer storage and time, to increase the robustness of the method, and also to introduce the fairly well established length-scale distribution very near walls into the model, a recent trend has been to use the two-layer model. That is, use the k-ϵ model only away from the wall and resolve the near-wall region with a simpler model involving a length-scale prescription to replace the ϵ equation, which is known as the most troublesome and the weakest modeled equation of the turbulence model.

In the two-layer model, the near-wall, viscous-affected regions are resolved with a one-equation turbulence model; that is, turbulent kinetic energy k is determined from the model while its dissipation rate ϵ is determined from a prescribed length-scale distribution, l. Specifically, the eddy viscosity relation is rewritten as

$$\nu_t = C_\mu \sqrt{k} l_\mu,$$

and ϵ is determined from

$$\epsilon = \frac{k^{3/2}}{l_\epsilon}.$$

The length-scale l_μ, adopted by Chen and Patel (1988)[28] and Iacovides and Launder (1990)[65], is from the model by Wolfshtein (1969)[187], whereas Rodi and his co-workers (1988)[137] adopted the length-scale model of Norris and Reynolds (1975)[119] because the Norris–Reynolds model was found to perform well in boundary layers with an adverse pressure gradient and transpiration. Both the Wolfshtein and the Norris–Reynolds model employ the following relation for l_μ

$$l_\mu = C_1 Y \left[1 - \exp\left(-\frac{R_y}{A_\mu}\right)\right],$$

which involves the argument R_y defined as before, $R_y = \sqrt{k}y/\nu$. R_y varies relatively slowly along lines parallel to the wall, does not vanish at separation, and remains well defined in regions of flow reversal. For the length-scale l_ϵ, Wolfshtein proposed an

Table 4.4 Coefficients for the two-layer model

Authors	A_μ	A_ϵ
Chen and Patel (1988)	70.0	5.08
Iacovides and Launder (1990)	62.5	3.80
Rodi et al. (1988)	50.5	

exponential damping function similar to the one used in l_μ;

$$l_\epsilon = C_1 Y \left[1 - \exp\left(-\frac{R_y}{A_\epsilon}\right)\right].$$

Norris and Reynolds proposed a different damping function, and their l_ϵ prescription reads

$$l_\epsilon = \frac{C_1 Y}{\left(1 + \frac{5.3}{R_y}\right)}.$$

The constant C_1 is chosen as $C_1 = \kappa C_\mu^{-3/4}$, where κ is the von Kármán constant while A_μ and A_ϵ are model coefficients. Different values for these coefficients have been proposed by various modelers, and these are listed in Table 4.4 for comparison. Chen and Patel (1988)[28] determined the parameter A_μ from numerical tests to recover the additive constant $B = 5.45$ in the logarithmic law in the case of a flat-plate boundary layer. Hence, their constants are somewhat different from those reported in Wolfshtein (1969)[187] and Iacovides and Launder (1990)[65].

In the two-layer modeling, the two models have to be matched at some location, and this should be placed near the edge of the viscous sublayer; that is, in a region where viscous effects have become negligible. Chen and Patel (1988)[28] matched along a grid line where the minimum of R_y is in the order of 250. They found that the results were not sensitive to the matching criterion so long as the minimum value of R_y was greater than 200. It should be noted that in normal boundary layer flow, $R_y = 250$ corresponds roughly to $y^+ = 135$. Iacovides and Launder (1990)[65] used ten nodes in this region and reported that the matching took place in a y^+ region of 80 to 120. Instead of matching the two models at preselected grid lines, Rodi and his co-workers (1988)[137] tested two different matching criteria. The first one was to match the models at a location where the ratio of eddy viscosity to molecular viscosity has a certain value. ν_t/ν ratios in the range of twelve to forty-eight have been tested (Cordes, 1991)[33], and the calculation results were found independent of the exact matching criterion when the ratio was greater than thirty. The second criterion was to match the models in which the damping function in the length-scale relation had a value close to unity so that the viscous effects were small. In their application, Rodi and his co-workers chose a value of 0.95 for the damping function. In boundary layers, these matching criteria effectively led to a switching between the models at $y^+ = 80 \sim 90$.

A variety of two-dimensional boundary layer and separated flows have been tested with these two-layer models; for example, adverse pressure gradient boundary layer flows or steady separated flows (Chen and Patel, 1988 [28]; Choi and Chen, 1988 [31];

Rodi, 1991 [138]). In general, two-layer models have been reported to predict more promising results than low-Reynolds-number models or the wall function approach. Jaw (1991)[67] noticed from prediction of different backward facing step flows that two-layer models seem to be sensitive to adverse pressure gradients, whereas predictions with low-Reynolds-number models are quite consistent.

4.5 DIRECT NUMERICAL SIMULATION (DNS)

Direct Numerical Simulations (DNS; Kim et al., 1987 [74]) of turbulent flows provide a complete database to develop and test turbulence models. In order to investigate the performance of each modeling term of the k and ϵ transport equations, we compute the low-Reynolds-number channel flow simulated by DNS. The simulated flow fields are a channel flow at Reynolds number $Re_\tau = u_\tau \delta / \nu = 180$, which is based on the kinematic viscosity ν, wall shear velocity u_τ, and channel half-width δ. Computations are also performed by using the low Reynolds number model of Lam and Bremhorst (1981)[81] and the two-layer model of Chen and Patel (1988)[28]. But for the two-layer model, only budgets of the k equation can be presented. Because the whole computational domain is below $R_y = 250$ and is dominated by the algebraic ϵ equation, no differential ϵ equation is adopted and hence no individual distribution of the modeled ϵ equation is available.

The following figures compare the DNS budget of exact turbulence kinetic energy with modeled k budgets. Figure 4.9(a) and (b) show that for viscous diffusion,

$$\frac{\partial}{\partial X_l}\left(\nu \frac{\partial k}{\partial X_l}\right),$$

and for turbulent diffusion,

$$\frac{\partial}{\partial X_l}\left(-\overline{u_l k} - \frac{\overline{p u_l}}{\rho}\right) = \frac{\partial}{\partial X_l}\left(C_k f_\mu \frac{k^2}{\epsilon}\frac{\partial k}{\partial X_l}\right).$$

Both models yield similar trends as the DNS data, but the low-Reynolds-number model predicts larger diffusion effects and is closer to the DNS data. Figure 4.9(c) shows the dissipation budget of k. It is found that the two-layer model overpredicts its peak value, $\epsilon = -0.195$, whereas the low Reynolds number model predicts $\epsilon(\text{peak}) = -0.16$, closer to the DNS data prediction of $\epsilon = -0.162$. Both profiles shift their peak position from the wall to about $Y^+ = 16$, the location of the second peak of the DNS data. It is interesting to point out that the dissipation budget deduced from experimental data for a flat-plate boundary layer (Patel et al., 1984 [121]) also presents such a shift, but with a peak of 0.2, which is much closer to the two-layer profile. This is anticipated because the model coefficients of the two-layer model were adjusted to fit the experimental data of a flat-plate boundary layer. Figure 4.9(d) shows that for the production term $-\overline{u_i u_l}\partial U_i/\partial X_l$, both the low-Reynolds-number model and the two-layer model overpredict their peak values, with $P(\text{peak}) = 0.24$ for the low Reynolds number model, and $P(\text{peak}) = 0.25$ for the two-layer model. The DNS peak value is about $P(\text{peak}) = 0.20$.

It should be pointed out that in the k equation, only the turbulent diffusion term requires a model. From Fig. 4.9(b), it is known that the modeled diffusion term fits pretty well with DNS data. Hence the k equation is considered to be relatively accurate so that the prediction of k directly from the wall can be achieved with only slight modification.

The following figures compare the budgets of the modeled ϵ equation with DNS data. Figure 4.10(a) and (b) present the distribution of the viscous diffusion,

$$\frac{\partial}{\partial X_l}\left(\nu \frac{\partial \epsilon}{\partial X_l}\right),$$

and for turbulent diffusion,

$$\frac{\partial}{\partial X_l}\left(C_\epsilon f_\mu \frac{k^2}{\epsilon} \frac{\partial \epsilon}{\partial X_l}\right)$$

of ϵ versus DNS data. It is clear that for $Y^+ > 18$, the modeled diffusion terms match

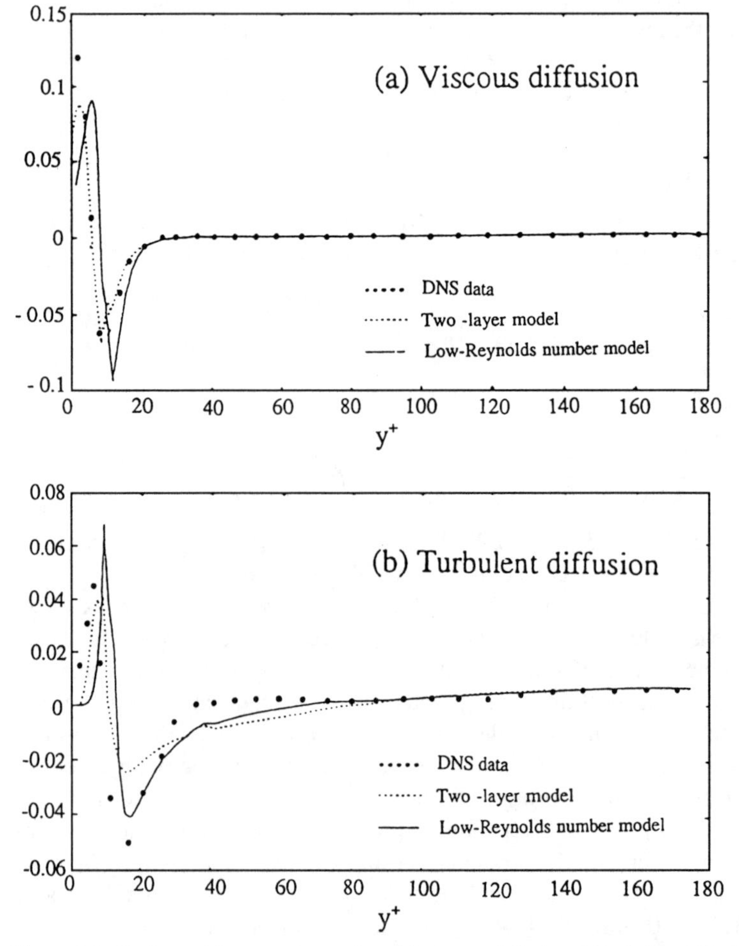

Figure 4.9 Budgets of k-transport equation: (a) viscous diffusion, (b) turbulent diffusion.

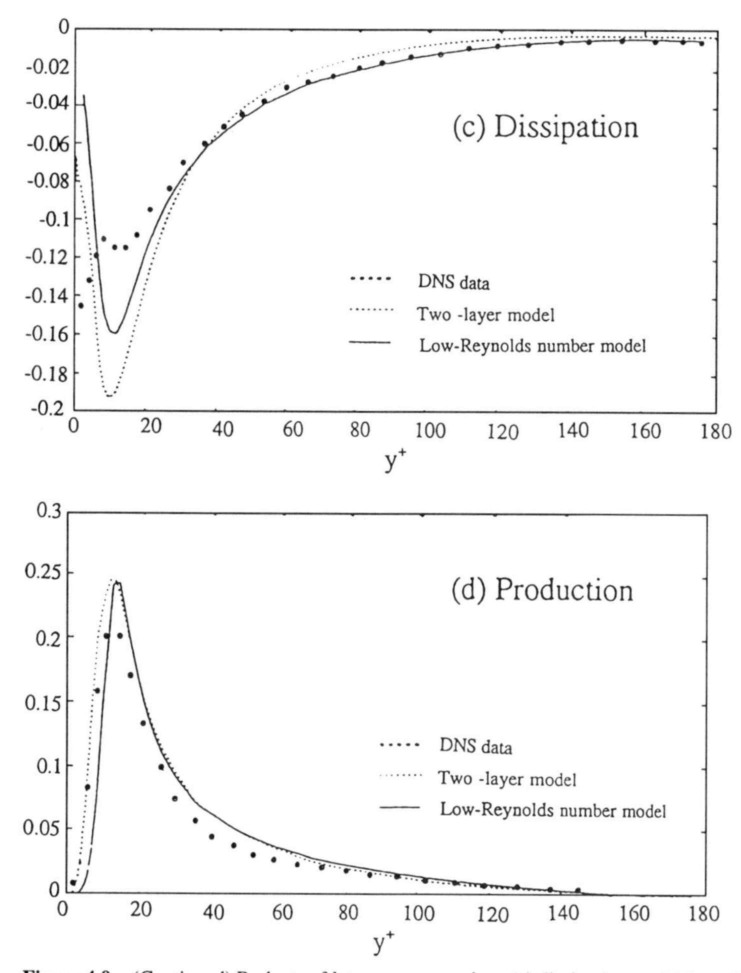

Figure 4.9 (Continued) Budgets of k-transport equation: (c) dissipation, and (d) production. (Adapted from Kim et al. (1987).)

correctly with DNS data. For $Y^+ < 18$, obvious disagreement exists in the turbulent diffusion term, especially in the region close to the wall, $Y^+ < 10$. However, the disagreement is small compared with the errors of the other terms because the magnitude of the diffusion is about an order smaller than other modeling terms, as shown in the figures. Figure 4.10(c) shows that the modeled destruction of ϵ,

$$-2\left(\nu\overline{\frac{\partial^2 u_i}{\partial X_j \partial X_l}}\right)^2 = -C_{\epsilon 2} f_2 \frac{\epsilon^2}{k}$$

is overpredicted. It should be remarked that the trend of the profile is only similar in the outer region and is substantially different in the inner region, $Y^+ < 10$. This is an indication that the destruction term in the ϵ equation as modeled is not proper because it gives an inconsistent trend.

Figure 4.11(a) and (b) plot the modeled production term

$$-C_{\epsilon 1} f_1 \frac{\epsilon}{k} \overline{u_i u_l} \frac{\partial U_i}{\partial X_l}$$

with respect to the triple correlation,

$$-2\nu \overline{\frac{\partial u_i}{\partial X_j} \frac{\partial u_i}{\partial X_l} \frac{\partial u_j}{\partial X_l}}$$

and the exact production term,

$$-2\nu \frac{\partial U_i}{\partial X_j} \left(\overline{\frac{\partial u_i}{\partial X_l} \frac{\partial u_j}{\partial X_l}} + \overline{\frac{\partial u_l}{\partial X_i} \frac{\partial u_l}{\partial X_j}} \right),$$

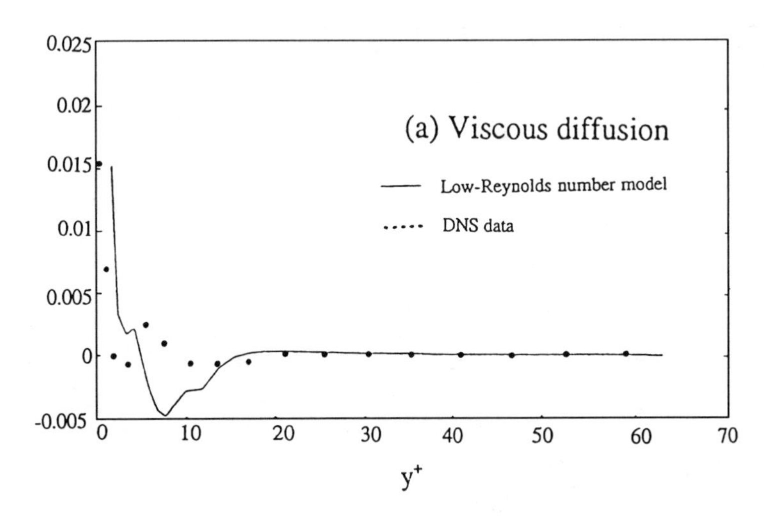

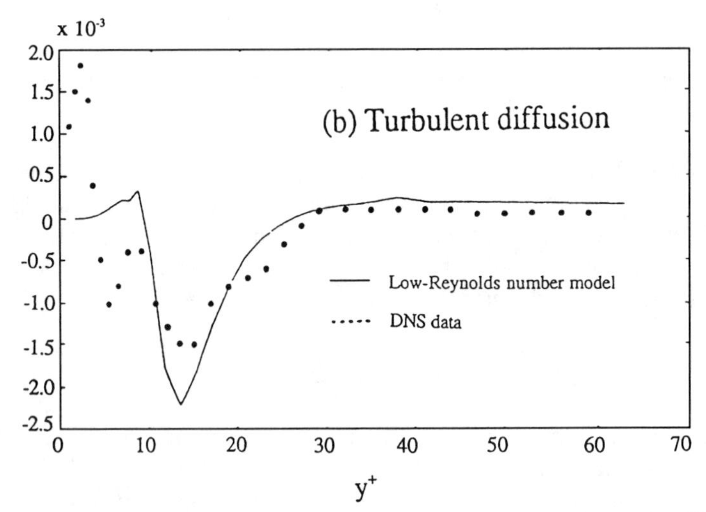

Figure 4.10 Budgets of ϵ-transport equation: (a) viscous diffusion, (b) turbulent diffusion.

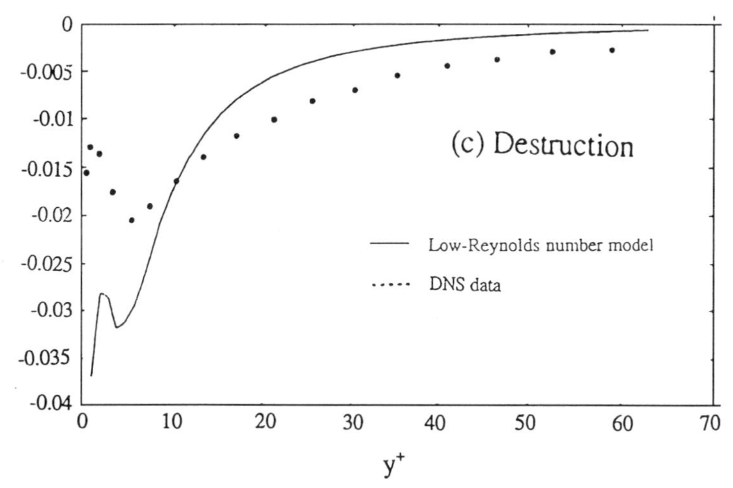

Figure 4.10 (Continued) Budgets of ϵ-transport equation: (c) destruction. (Adapted from Kim et al. (1987).)

respectively, to examine the relation of the modeled production with these two terms. From these two figures, it is known that the modeled production fits neither of these two terms. However, if we plot the summation of modeled production and destruction with respect to the summation of exact production, triple correlation, and destruction terms, the curve fits much better with DNS data, as shown in Fig. 4.11(c). The summation of errors is an order of magnitude smaller than the error in each individual term. This may explain why an inaccurately modeled ϵ equation may still predict fairly well some simple turbulent flows. Because the modeled production and destruction terms yield errors with similar magnitude but opposite in sign, their errors are canceled by each other. This may not be true, however, for more complex flows. In that case, the k-ϵ model may produce greater error in predicting turbulent flows. From the above-indicated examination, it is

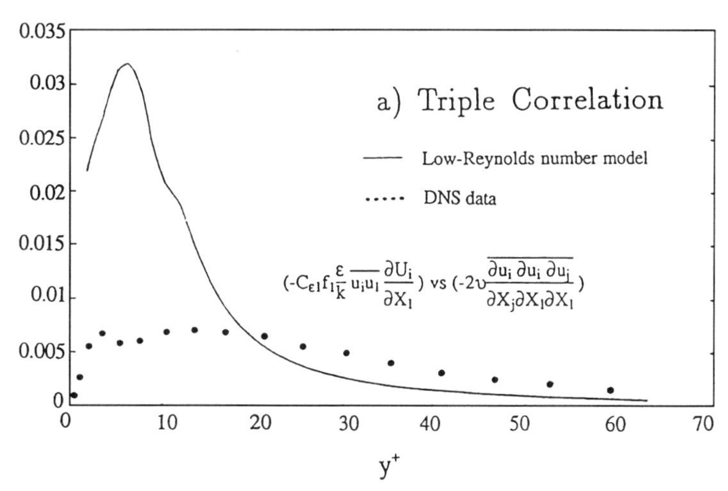

Figure 4.11 Budgets of ϵ-transport equation: (a) triple correlation.

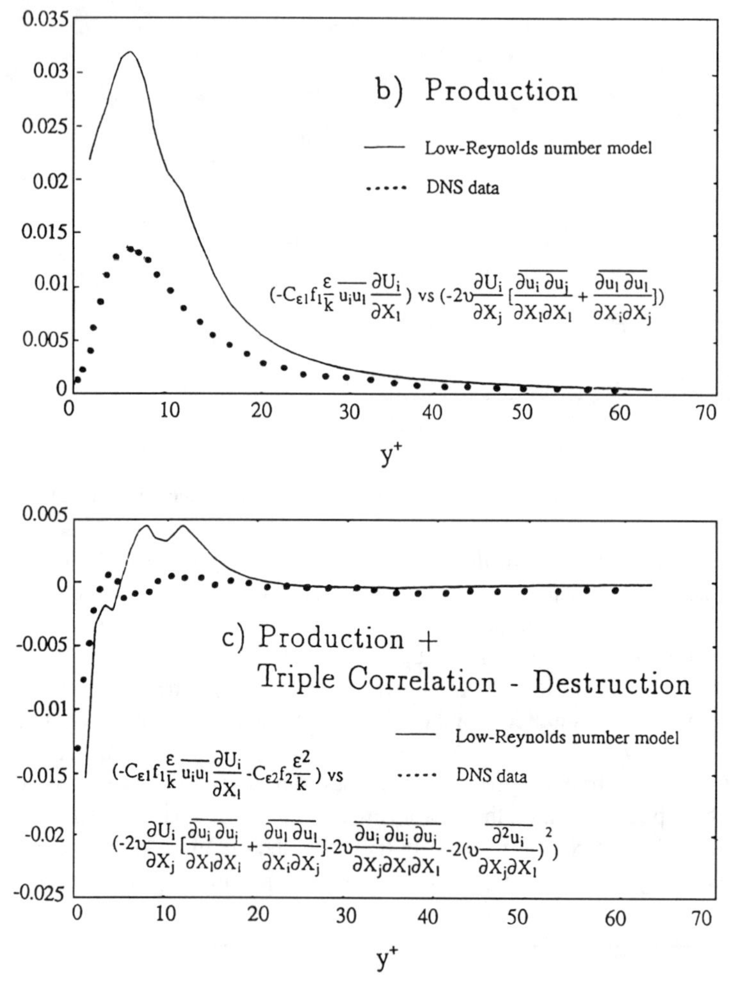

Figure 4.11 (Continued) Budgets of ϵ-transport equation: (b) production, (c) production + triple correlation–destruction. (Adapted from Kim et al. (1987).)

clear that modeling of the ϵ equation requires a more fundamental revision than the use of a damping function or an algebraic length equation for ϵ in the inner wall region.

4.6 TURBULENT FLOW PREDICTIONS: TWO (WALL-SHEAR FLOWS)

4.6.1 Examples of Wall-Shear Flows

Wall shear flows are flows in which shear is created near the wall. This includes boundary layer flow, channel flow, and wall jets. We shall examine the performance of the second-order turbulence model in predicting these flows, in which in two-dimensional flows,

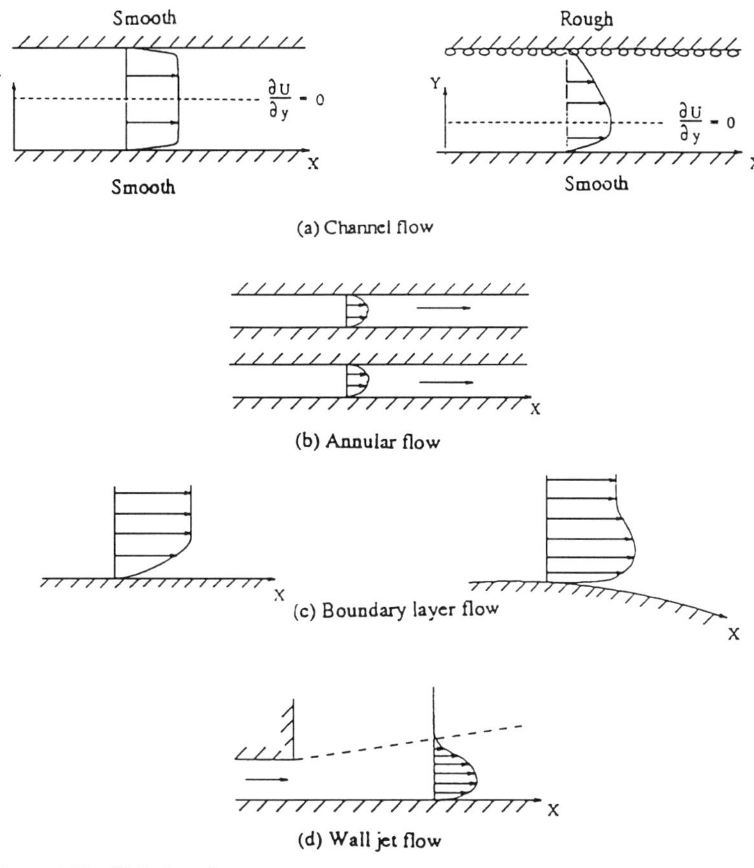

Figure 4.12 Wall-shear flows.

$U \gg V$ and $\partial/\partial Y \gg \partial/\partial X$. Examples are given in Fig. 4.12 for the following:

- fully developed channel flow, that is, $\partial \text{velocity}/\partial X = 0$;
- fully developed annular flow, that is, $\partial \text{velocity}/\partial X = 0$;
- boundary layer flow, that is, $\partial/\partial Y \gg \partial/\partial X$; and
- wall jet, that is, $\partial/\partial Y \gg \partial/\partial X$.

4.6.2 Three Models (RSM, *k*-ϵ-*A/E*, and Wall Function)

For shear flows such as boundary layer or jet flows, the continuity equation is given by

$$\frac{\partial U}{\partial X} + \frac{\partial V}{\partial Y} = 0, \tag{4.44}$$

and the momentum equation in the X direction is given by

$$U\frac{\partial U}{\partial X} + V\frac{\partial U}{\partial Y} = -\frac{1}{\rho}\frac{dP}{dX} + \frac{\partial}{\partial Y}\left(\nu\frac{\partial U}{\partial Y} - \overline{uv}\right). \tag{4.45}$$

For fully developed flows, however, the continuity equation is given by

$$\frac{\partial \text{velocity}}{\partial X} = 0, \tag{4.46}$$

whereas the momentum equation in the X direction is given by

$$0 = -\frac{1}{\rho}\frac{dP}{dX} + \frac{\partial}{\partial Y}\left(\nu \frac{\partial U}{\partial Y} - \overline{uv}\right). \tag{4.47}$$

The RSM is given by

- the $\overline{uv}$ equation,

$$\frac{D\overline{uv}}{Dt} = \frac{\partial}{\partial Y}\left[\left(C_k \frac{k^2}{\epsilon} + \nu\right)\frac{\partial \overline{uv}}{\partial Y}\right] - \overline{v^2}\frac{\partial U}{\partial Y} - \tilde{C}_1 \frac{\epsilon}{k}\overline{uv} + \tilde{C}_2 \overline{v^2}\frac{\partial U}{\partial Y},$$

- the $\overline{v^2}$ equation is given by

$$\frac{D\overline{v^2}}{Dt} = \frac{\partial}{\partial Y}\left[\left(C_k \frac{k^2}{\epsilon} + \nu\right)\frac{\partial \overline{v^2}}{\partial Y}\right] - \frac{2}{3}\epsilon - \tilde{C}_1 \frac{\epsilon}{k}\left(\overline{v^2} - \frac{2}{3}k\right) - \tilde{C}_2 \frac{2}{3}\overline{uv}\frac{\partial U}{\partial Y},$$

or

$$\overline{v^2} = 0.5k.$$

- The kinetic energy, k, equation is given by

$$\frac{Dk}{Dt} = \frac{\partial}{\partial Y}\left[\left(C_k \frac{k^2}{\epsilon} + \nu\right)\frac{\partial k}{\partial Y}\right] - \overline{uv}\frac{\partial U}{\partial Y} - \epsilon.$$

- The rate of dissipation, ϵ, equation is given by

$$\frac{D\epsilon}{Dt} = \frac{\partial}{\partial Y}\left[\left(C_\epsilon \frac{k^2}{\epsilon} + \nu\right)\frac{\partial \epsilon}{\partial Y}\right] - C_{\epsilon 1}\frac{\epsilon}{k}\overline{uv}\frac{\partial U}{\partial Y} - C_{\epsilon 2}\frac{\epsilon^2}{k},$$

where $C_k = 0.064$, $\tilde{C}_1 = C_1 + 0.125\frac{k^{3/2}}{\epsilon Y}$, $\tilde{C}_2 = C_2 + 0.05\frac{k^{3/2}}{\epsilon Y}$, $C_1 = 1.5$, $C_2 = 0.4 \sim 0.6$, $C_\epsilon = 0.065$, $C_{\epsilon 1} = 1.45$, and $C_{\epsilon 2} = 1.90 \sim 2.0$ from Hanjalic and Launder (1972)[52].

The k-ϵ-A and k-ϵ-E model are given by the following:

- The kinetic energy, k, equation is given by

$$\frac{Dk}{Dt} = \frac{\partial}{\partial Y}\left[\left(C_k \frac{k^2}{\epsilon} + \nu\right)\frac{\partial k}{\partial Y}\right] - C_\mu \frac{k^2}{\epsilon}\left(\frac{\partial U}{\partial Y}\right)^2 - \epsilon.$$

- The rate of dissipation, ϵ, equation is given by

$$\frac{D\epsilon}{Dt} = \frac{\partial}{\partial Y}\left[\left(C_\epsilon \frac{k^2}{\epsilon} + \nu\right)\frac{\partial \epsilon}{\partial Y}\right] - C_{\epsilon 1}C_\mu \frac{k^2}{\epsilon}\left(\frac{\partial U}{\partial Y}\right)^2 - C_{\epsilon 2}\frac{\epsilon^2}{k},$$

- The $\overline{uv}$ equation is given by

$$-\overline{uv} = C_\mu \frac{k^2}{\epsilon}\frac{\partial U}{\partial Y}. \tag{4.29}$$

where $C_k = 0.09$, $C_\mu = 0.09 \sim 0.11$, $C_\epsilon = 0.07$.

The wall function model is given by

- the X-momentum equation

$$u^+ = \frac{1}{\kappa} \ln Ey^+$$

- the $\overline{uv}$ equation

$$-\overline{uv} = u^{*2}$$

- the k equation

$$k = \frac{u^{*2}}{\sqrt{C_\mu}}$$

- the ϵ equation

$$\epsilon = \frac{u^{*3}}{\kappa Y},$$

where $\kappa = 0.41 \sim 0.43$, $E = 9.00$, $u^* = (\tau_w/\rho)^{1/2}$, and $y^+ > 10$.

4.6.3 Asymmetric Channel Flows (RSM)

Let us examine the flow in an asymmetric channel predicted by full-stress turbulence model (RSM; Hanjalic and Launder, 1972 [55]). The U, $\overline{uv}$, and k predictions are plotted in Fig. 4.13. With the full differential model (RSM), we find that the locations in which $-\overline{uv} = 0$ and $\partial U/\partial Y = 0$ do not coincide. It should be emphasized here that this cannot be predicted with the k-ϵ model. Therefore, the RSM is in general better than the k-ϵ model.

The rough wall function used is

$$\frac{U}{u^*} = \frac{1}{0.42} \ln \frac{Yu^*}{\nu} + 3.5,$$

whereas the smooth wall function used is

$$\frac{U}{u^*} = \frac{1}{0.42} \ln \frac{Yu^*}{\nu} + 5.45.$$

The wall conditions used are

$$\overline{u^2} = 5.1u^{*2}$$

$$\overline{v^2} = u^{*2}$$

$$\overline{w^2} = 2.3u^{*2}$$

$$\overline{uv} = -u^{*2} + y\frac{\partial P}{\partial X},$$

and

$$\epsilon = -\overline{uv}\frac{dU}{dY},$$

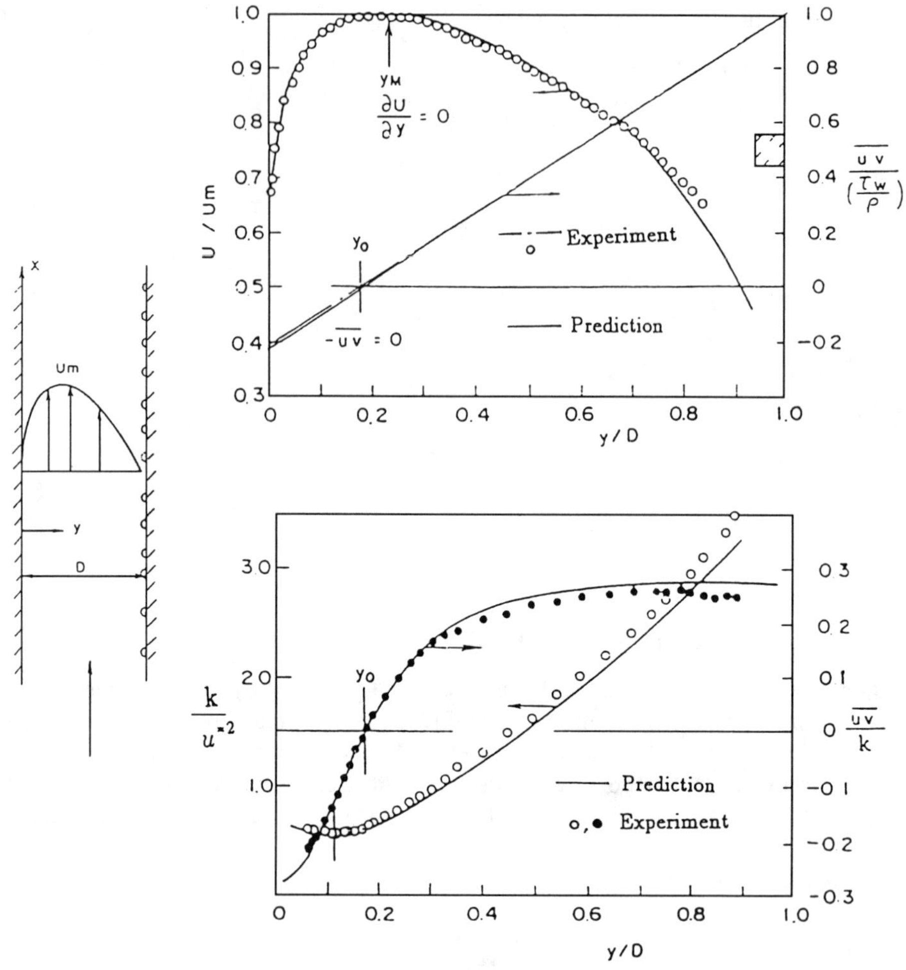

Figure 4.13 U, $\overline{uv}$, and k predictions for asymmetric channel flow. (Adapted from Hanjalic and Launder (1972).)

where

$$u^* = \sqrt{\frac{\tau_w}{\rho}},$$

and

$$\tau_w = \mu \left. \frac{\partial U}{\partial Y} \right|_{y=0}.$$

The turbulent energy equation balance can be also checked. We denote the k equation as

$$\frac{Dk}{Dt} = D_k + P_k - \epsilon.$$

We note from Figs. 4.14 and 4.15 that near the wall the balance of the k equation shows near-turbulent equilibrium, or $P_k \cong \epsilon$.

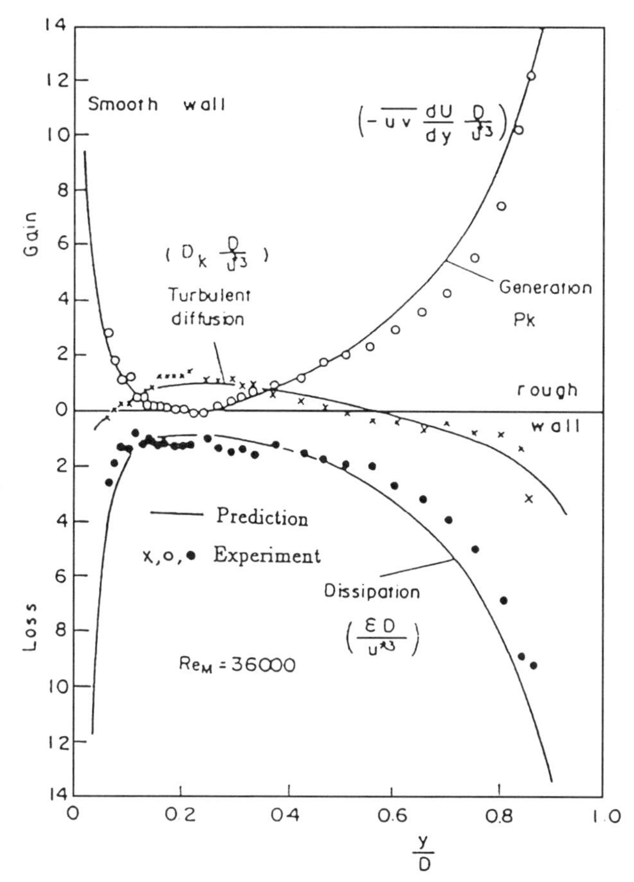

Figure 4.14 Variation of turbulent diffusion, generation, and dissipation for smooth- and rough-walled asymmetric channel flow. (Adapted from Hanjalic and Launder (1972).)

4.6.4 Boundary Layer Flows (RSM)

Let us examine the influence of different constants on the prediction of boundary layer flows. Figure 4.16 compares the predictions that are based on two different models. Model 1 uses $C_1 = 1.5$ and $C_2 = 0.4$, whereas Model 2 uses $C_1 = 1.5$ and $C_2 = 0.6$. The comparison shows that Model 1 predicts boundary layer velocity profiles better than Model 2.

4.6.5 Wall Jet Flows (RSM)

Let us examine the prediction of wall jet flow by the RSM (Hanjalic and Launder, 1972 [55]). Figure 4.17 presents the velocity and shear stress profiles in plane wall jets as predicted by Hanjalic and Launder, as compared with the experimental data from Tailland and Mathieu (1967)[171]. Hanjalic and Launder made the following

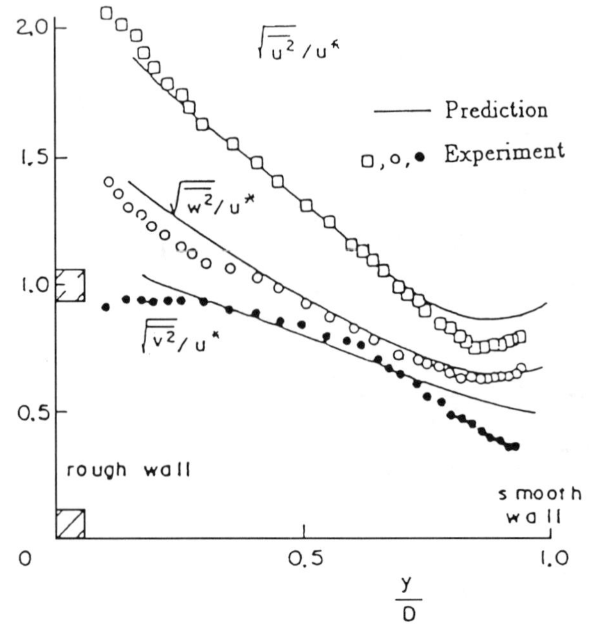

Figure 4.15 Kinetic energy balance near the wall for asymmetric channel flow. (Adapted from Hanjalic and Launder (1972).)

observation: "We found the spread rate of the wall jet to be about 20% greater than what the measurements suggest. At present, we are uncertain whether these differences represent essential deficiencies of the present model or merely indicate that the boundary conditions used are not quite appropriate to the flow in question."

The discrepancy may be due to improper modeling of the ϵ equation. Figure 4.18 presents the predictions for the rate of spread of wall jet flow. Chen and Singh (1990)[22] showed that the two-scale k-ϵ model can perform well in free-shear flows. The two-scale k-ϵ model may be used to improve the prediction of wall jet flow.

4.6.6 Boundary Layer Flow (k-ϵ)

Let us examine the performance of the k-ϵ model in predicting boundary layer flow (Rodi, 1972)[134], as shown in Fig. 4.19. Figure 4.20 presents the predictions for the friction coefficient, C_f, and for H, where $H = \delta^*/\theta$; for zero pressure gradient. δ^* and θ are defined as

$$\delta^* = \int_0^\delta \left(1 - \frac{U}{U_E}\right) \cdot dY$$

and

$$\theta = \int_0^\delta \left(1 - \frac{U}{U_E}\right) \cdot \frac{U}{U_E} \cdot dY,$$

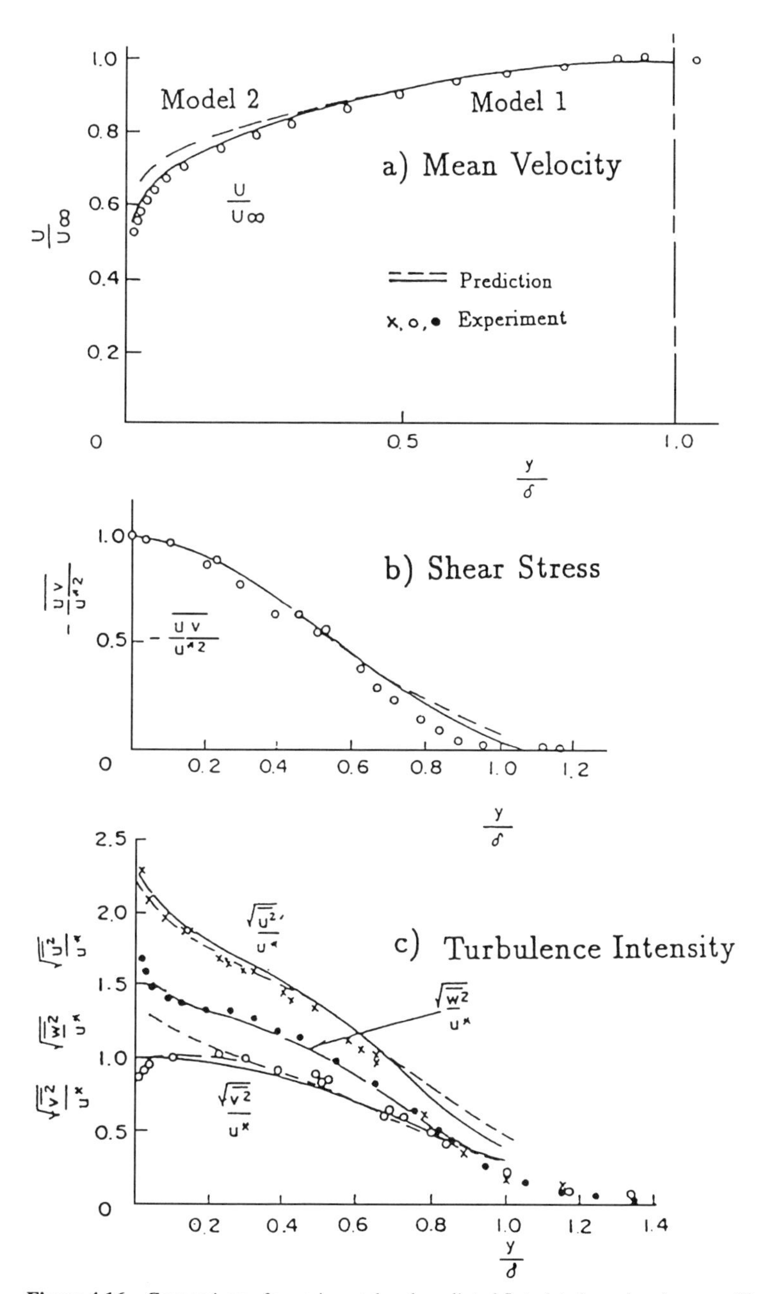

Figure 4.16 Comparison of experimental and predicted flat-plate boundary layer profiles: (a) mean velocity, (b) shear stress, and (c) turbulence intensity. (Adapted from Hanjalic and Launder (1972).)

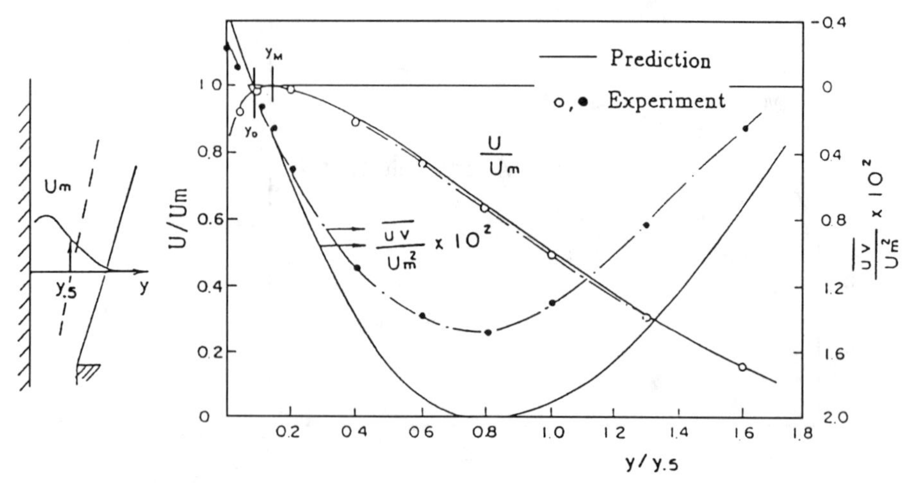

Figure 4.17 Comparison of the experimental and predicted velocity and shear stress profiles in plane wall jet. (Adapted from Hanjalic and Launder (1972).)

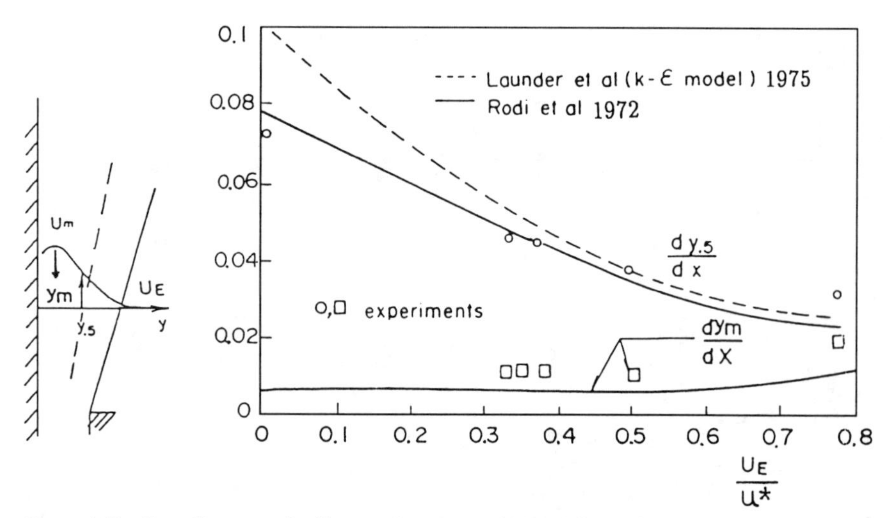

Figure 4.18 Spreading rates of self-preserving plane wall jet. (Adapted from Rodi (1972).)

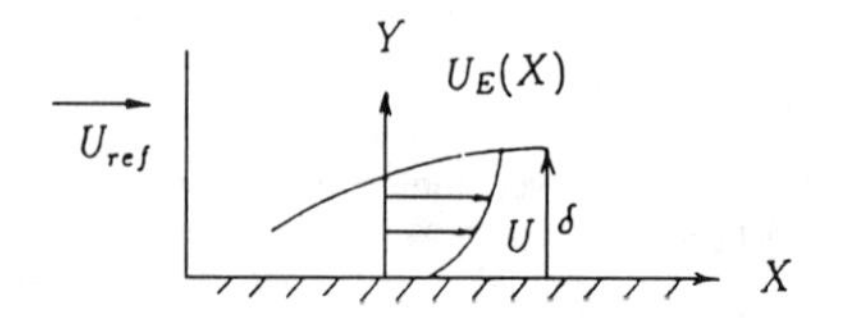

Figure 4.19 Boundary layer flow parameters.

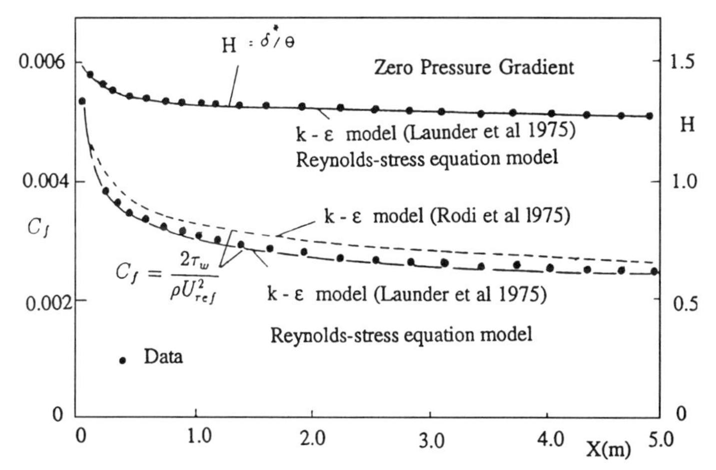

Figure 4.20 Predictions for boundary layer flow with zero pressure gradient. (Adapted from Rodi (1972).)

where δ and U and U_E are defined as shown in Fig. 4.19. Figure 4.21 shows the predictions for C_f and U_E/U_{ref} for adverse pressure gradient, where

$$C_f = \frac{2\tau_w}{\rho U^2_{\text{ref}}},$$

whereas Fig. 4.22 shows the velocity profile $U/U_E(X)$ at $X = 3.4$ m.

These predictions showed that within moderate pressure gradients, the friction coefficient and velocity profiles are predicted within fifteen.

4.6.7 Low-Reynolds-Number Flow

When the flow is turbulent but at relatively low Reynolds number, the turbulence model performs less satisfactorily. Jones and Launder (1972)[70] proposed a modification of

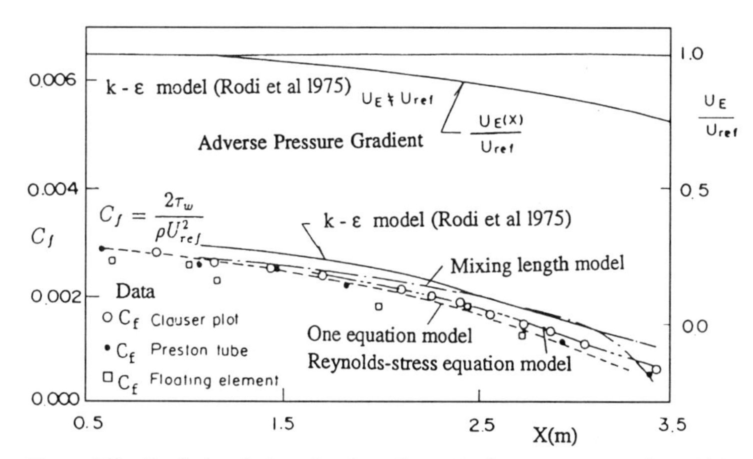

Figure 4.21 Predictions for boundary layer flow with adverse pressure gradient. (Adapted from Rodi (1972).)

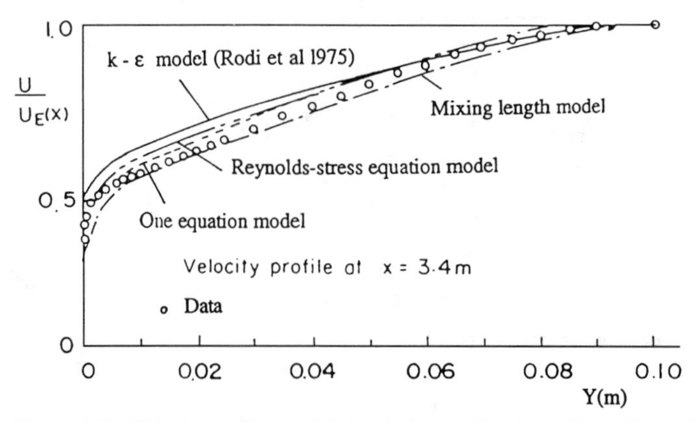

Figure 4.22 Velocity profile at $x = 3.4$ m in the boundary layer flow. (Adapted from Rodi (1972).)

the k-ϵ-E model for low Reynolds numbers. It is summarized as follows:

- Their mean equations are
  the continuity equation,

$$\frac{\partial U}{\partial X} + \frac{\partial V}{\partial Y} = 0, \tag{4.48}$$

and the momentum equation in the X-direction,

$$U\frac{\partial U}{\partial X} + V\frac{\partial U}{\partial Y} = -\frac{1}{\rho}\frac{\partial P}{\partial X} + \frac{\partial}{\partial Y}\left[(\nu + \nu_t)\frac{\partial U}{\partial Y}\right]. \tag{4.49}$$

- Their turbulent transport equations are
  the k equation,

$$\frac{Dk}{Dt} = \frac{\partial}{\partial Y}\left[(\nu + \nu_t)\frac{\partial k}{\partial Y}\right] + \nu_t\left(\frac{\partial U}{\partial Y}\right)^2 - \epsilon - 2\nu\left(\frac{\partial\sqrt{k}}{\partial Y}\right)^2, \tag{4.50}$$

the ϵ equation,

$$\frac{D\epsilon}{Dt} = \frac{\partial}{\partial Y}\left[\left(\nu + \frac{\nu_t}{\sigma\epsilon}\right)\frac{\partial \epsilon}{\partial Y}\right] + C_{\epsilon 1}\frac{\epsilon}{k}\nu_t\left(\frac{\partial U}{\partial Y}\right)^2 - C_{\epsilon 2} f_\epsilon \frac{\epsilon^2}{k} + 2\nu\nu_t\left(\frac{\partial^2 U}{\partial Y^2}\right), \tag{4.51}$$

with

$$\nu_t = C_\mu \frac{k^2}{\epsilon} f_u,$$

where the problem functions f_u and f_ϵ are defined as

$$f_u = \exp\left[-2.5/\left(1 + R_t/50\right)\right]$$

and

$$f_\epsilon = 1 - 3.0\exp\left(-R_t^2\right).$$

R_t is given by

$$R_t = \frac{k^2}{\epsilon \nu}.$$

Note that the last term in the k equation has been added so that ϵ may be set equal to zero at the wall. Launder and Jones have also added the last term in the ϵ equation but are unable to provide any physical argument for its adoption. Its inclusion simply appears to be to reduce the peak of the turbulent kinetic energy close to the wall.

4.6.8 Turbulent Flow Past Axisymmetric Bodies

Here we consider the work of Choi and Chen (1988, 1991)[31, 32]. The fully elliptic numerical solutions of the turbulent flow past axisymmetric bodies are obtained by using the k-ϵ turbulence model and the wall function method. Complex physical domains are resolved by the use of numerically generated, body-fitted coordinates. The finite analytic (FA) method proposed by Chen (1986)[13] is adopted for the numerical discretization scheme. Calculations are performed for the two axisymmetric bodies that have different afterbodies and are considered to be the benchmark geometries in ship design.

The dimensionless continuity equation is given by

$$\frac{\partial u}{\partial x} + \frac{1}{r}\frac{\partial (rv)}{\partial r} = 0. \tag{4.52}$$

The generic dimensionless transport equation for the FA method is

$$\frac{\partial \phi}{\partial t} + u\frac{\partial \phi}{\partial x} + v\frac{\partial \phi}{\partial r} = \frac{1}{R_\phi}\left(\frac{\partial^2 \phi}{\partial x^2} + \frac{1}{r}\frac{\partial \phi}{\partial r} + \frac{\partial^2 \phi}{\partial r^2}\right) + S_\phi, \tag{4.53}$$

where ϕ represents any dynamic variable, such as u, v, and so forth. The variables in these equations are made dimensionless by their respective reference value of free-stream velocity, U_∞, and the body length, L. The values of R_ϕ and S_ϕ for different representations of ϕ are as indicated in Table 4.5. Figure 4.23 indicates the geometry of various afterbodies and their computational domains. In the study of Choi and Chen (1988)[31], the domain of investigation consists of a near-wall region and an outer region

Table 4.5 Generic variables in FA method

ϕ	R_ϕ	S_ϕ
u	$R_u = \frac{1}{(1/Re+\nu_t)}$	$-\frac{\partial p}{\partial x} + \frac{\partial \nu_t}{\partial y}\left(\frac{\partial u}{\partial y} + \frac{\partial v}{\partial x}\right) + 2\frac{\partial \nu_t}{\partial x}\frac{\partial u}{\partial x} - \frac{2}{3}\frac{\partial k}{\partial x}$
v	$R_v = \frac{1}{(1/Re+\nu_t)}$	$-\frac{\partial p}{\partial y} + \frac{\partial \nu_t}{\partial x}\left(\frac{\partial u}{\partial y} + \frac{\partial v}{\partial x}\right) + 2\frac{\partial \nu_t}{\partial y}\frac{\partial u}{\partial y} - \frac{2}{3}\frac{\partial k}{\partial y}$
k	$R_k = \frac{\sigma_k}{(\sigma_k/Re+\nu_t)}$	$\frac{1}{\sigma_k}\left(\frac{\partial \nu_t}{\partial x}\frac{\partial k}{\partial x} + \frac{\partial \nu_t}{\partial y}\frac{\partial k}{\partial y}\right) + G - \epsilon$
ϵ	$R_\epsilon = \frac{\sigma_k}{(\sigma_k/Re+\nu_t)}$	$\frac{1}{\sigma_\epsilon}\left(\frac{\partial \nu_t}{\partial x}\frac{\partial \epsilon}{\partial x} + \frac{\partial \nu_t}{\partial y}\frac{\partial \epsilon}{\partial y}\right) + C_{\epsilon 1} G\frac{\epsilon}{k} - C_{\epsilon 2}\frac{\epsilon^2}{k}$

where $G = \nu_t\left[2\left(\frac{\partial u}{\partial x}\right)^2 + 2\left(\frac{\partial v}{\partial y}\right)^2 + \left(\frac{\partial u}{\partial y} + \frac{\partial v}{\partial x}\right)^2\right]$; $\nu_t = c_\mu \frac{k^2}{\epsilon}$

and $c_\mu = 0.09$, $\sigma_k = 1.00$, $\sigma_\epsilon = 1.30$, $C_{\epsilon 1} = 1.44$, $C_{\epsilon 2} = 1.92$.

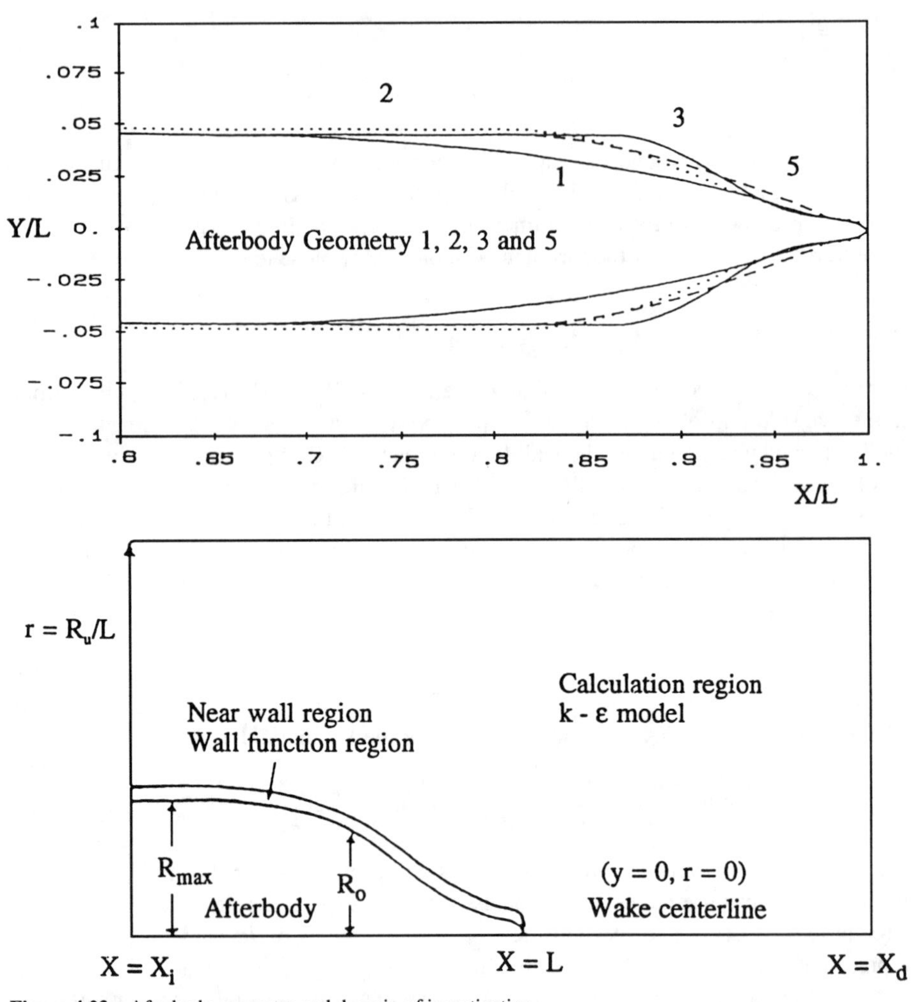

Figure 4.23 Afterbody geometry and domain of investigation.

as shown in Fig. 4.23. The near-wall region is a computational control volume on the body normally extending from $y^+ = 0$ to $y^+ = 150$. The boundary conditions for the calculation region are as follows:

- At the upstream inlet ($x = x_i$), U, k, and ϵ are specified and $V_x = 0$.
- At the downstream outlet ($x = x_d$), (the derivative is denoted by the subscript)

$$U_x = V_x = k_x = \epsilon_x = P_x = 0.$$

- Along the wake centerline ($y = 0$ or $r = 0$),

$$U_y = k_y = \epsilon_y = 0; V = 0.$$

- Along the upper boundary ($r = R_u/L$), $U = 1$, $P = 0$, and $k_y = \epsilon_y = 0$.
- In the wall function region ($x_i < x < 1; 0 < y^+ < 150$), U, V, k, and ϵ, are specified.

Because the upstream inlet conditions for U, k, and ϵ are not readily available, they are approximately specified by the flat-plate correlations with boundary layer thickness δ and friction velocity U_τ given by experimental data as

$$U = \left(\frac{y}{\delta}\right)^{1/7},$$

$$k = c_\mu^{-1/2} U_\tau^2 \left(1 - \frac{y}{\delta}\right),$$

and

$$\epsilon = c_\mu^{3/4} \frac{k^{3/2}}{\kappa y},$$

where κ is the von Kármán constant, $\kappa = 0.418$, and $c_\mu = 0.090$.

On the wall $R = R_0$, the wall function is imposed. In the wall function region, U, V, k, and ϵ are calculated by the following wall functions. The pressure gradient effects on the flow in the wall region are taken into consideration by the use of the generalized form of the law of the wall,

$$\frac{q}{U_\tau} = \frac{1}{\kappa}\left[\ln\left(\frac{4}{\Delta\tau}\frac{\sqrt{1+\Delta\tau y^+} - 1}{\sqrt{1+\Delta\tau y^+} + 1}\right) + 2\left(\sqrt{1+\Delta\tau y^+} - 1\right)\right] + 5.45 + 3.7\Delta p,$$

where $y^+ = ReU_\tau y$ is the dimensionless normal distance from the surface, $\Delta p = \nabla p / ReU_\tau^3$ is the dimensionless pressure gradient on the body surface, $\Delta\tau$ is the dimensionless shear stress gradient and is assumed to be $0.5\Delta p$, and q is the magnitude of the velocity, or $q = (U^2 + V^2)^{1/2}$.

Choi and Chen performed the calculations for two axisymmetric bodies: Afterbody 1 and Afterbody 5 of the David Taylor Naval Ship Research and Development Center where extensive experimental data are available. These bodies have the same forebodies and middlebodies but different afterbodies, as shown in Fig. 4.23. The detailed geometric parameters of these bodies can be found in reference by Huang et al. (1979, 1980)[64, 63]. The Reynolds numbers for these two bodies, which are based on the free-stream velocity and body length are, respectively, $Re = 6.6 \cdot 10^6$ for Afterbody 1 and $Re = 9.3 \cdot 10^6$ for Afterbody 5. The change in the surface curvature in the stern and propeller hub region induces strongly favorable and adverse pressure gradients but no flow separation was measured or predicted in the model. Figures 4.24 and 4.25 show the computed and experimentally obtained pressure distribution and friction velocity profiles for Afterbody 1 and Afterbody 5, respectively, whereas Fig. 4.26 displays the turbulent kinetic energy profile for Afterbody 1 and Afterbody 5 as predicted by the code versus the experimentally obtained values. Figure 4.27 shows the velocity distributions for Afterbody 1 and Afterbody 5 at different x/L locations as predicted by the code versus the experimentally obtained data.

Chen and Patel (1988)[28] investigated uniform flow past the Axisymmetric Afterbody 3 of the David Taylor Naval Research and Development Center. Experimental data are available from Huang et al. (1980) reported that a small separation bubble exists near the inflection point of Afterbody 3, although experimental data cannot reveal such detailed information. Figures 4.28(a)–4.29(b) present the results predicted by the two-layer model of Chen and Patel [28]. Figure 4.28(a) is the computation grid generated by

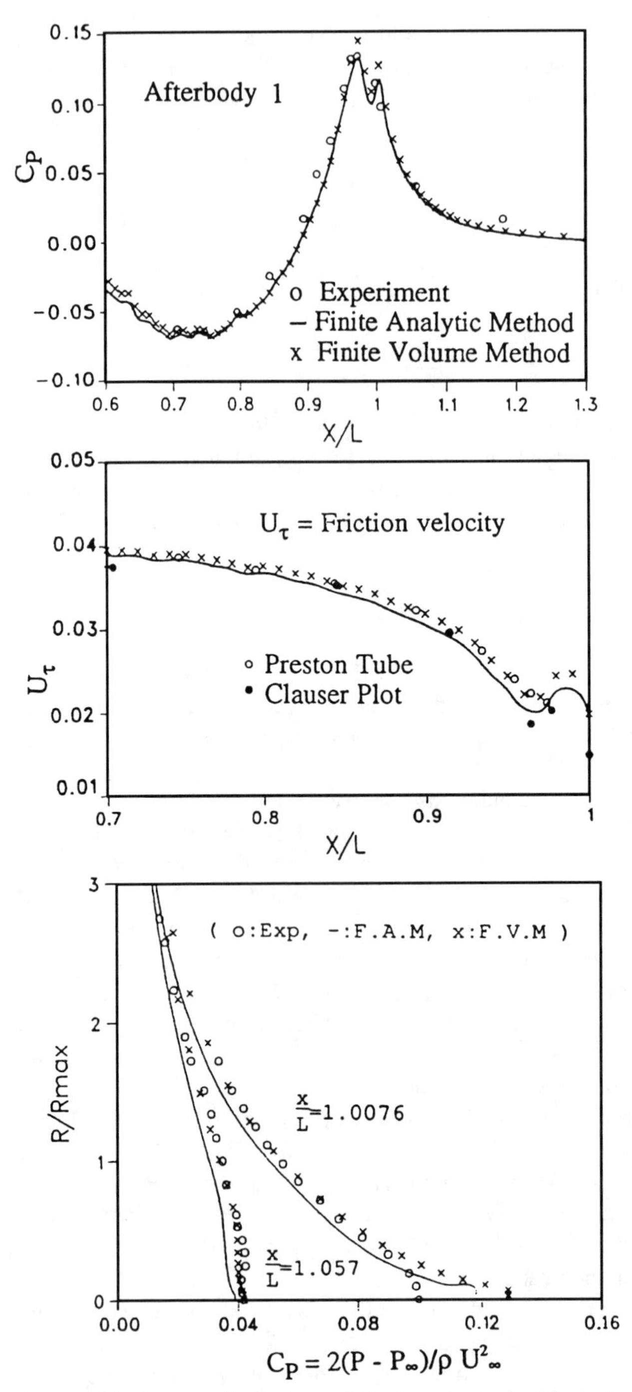

Figure 4.24 Pressure distribution and friction velocity for Afterbody 1. (Adapted from Choi and Chen (1972).)

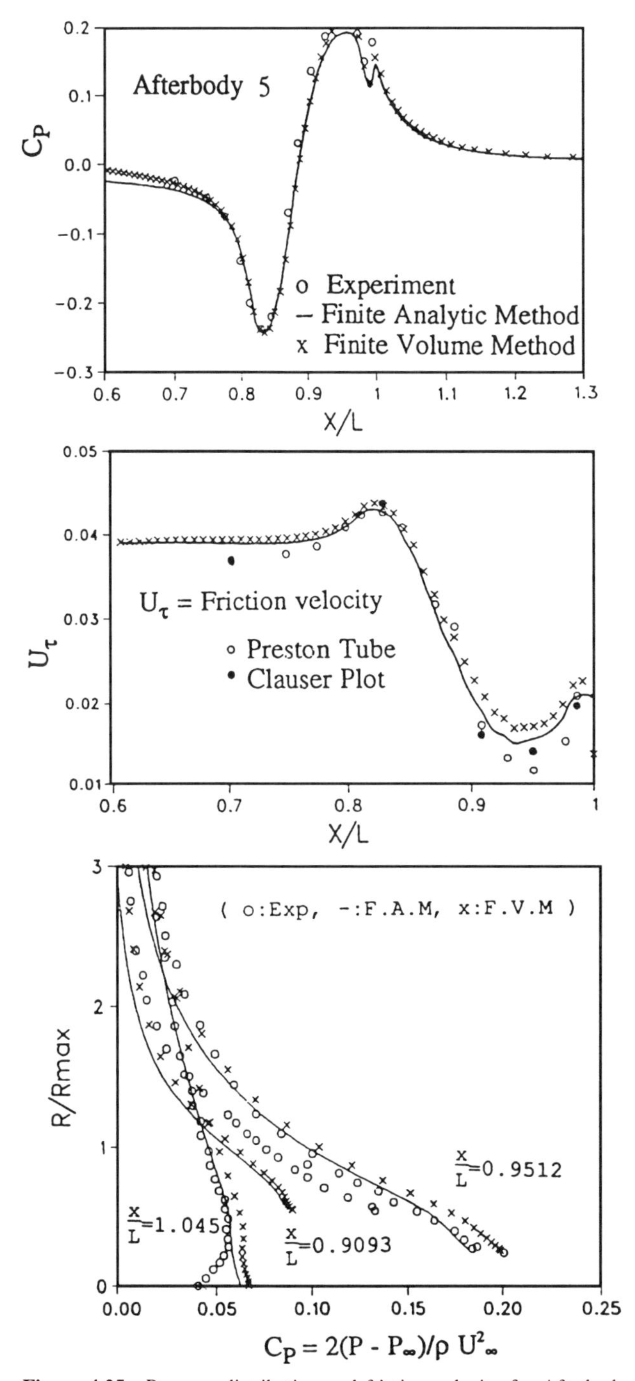

Figure 4.25 Pressure distribution and friction velocity for Afterbody 5. (Adapted from Choi and Chen (1988).)

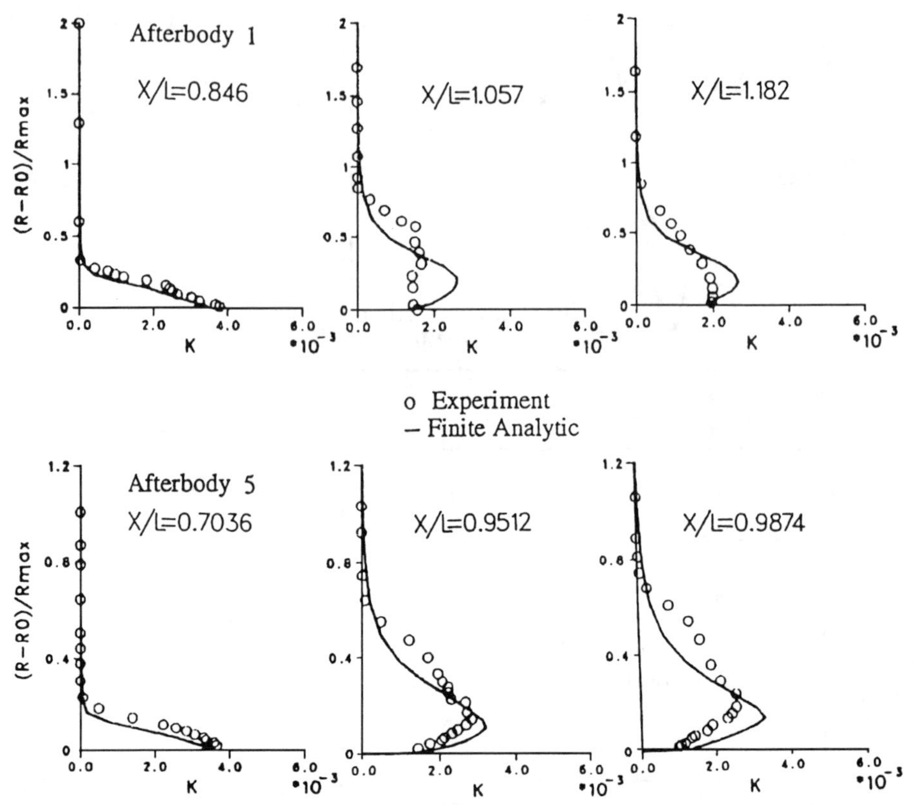

Figure 4.26 Turbulent kinetic energy distribution of Afterbody 1 and Afterbody 5. (Adapted from Choi and Chen (1988).)

body-fitted coordinates. Figures 4.28(b) and 4.29(a) are the surface pressure and shear velocity distributions along the axial direction. Figure 4.29(b) is the velocity distribution at five different locations on the afterbody. All of the predicted results fit pretty well to the experimental data; the separation bubble near the inflection point can also be predicted by using the two-layer model. In Fig. 4.29(a), the surface shear velocities are less than zero within $X/L = 0.91 \sim 0.93$, which implies that separation flow occurs in this range. The two-layer model satisfactorily predicts the flow past an axisymmetric body.

4.7 OTHER NEAR-WALL TURBULENCE MODELS

4.7.1 General Remarks

Most of the models presented so far are applicable only to flows or flow regions with high turbulent Reynolds numbers, ν_t/ν, and are not accurate in predicting turbulence near the wall region, where viscous effects become dominant. Although the so-called *wall functions* (Chen, 1983)[12] that related surface boundary conditions to points in fluid away from the boundaries and, thereby, avoid the problem of modeling the direct

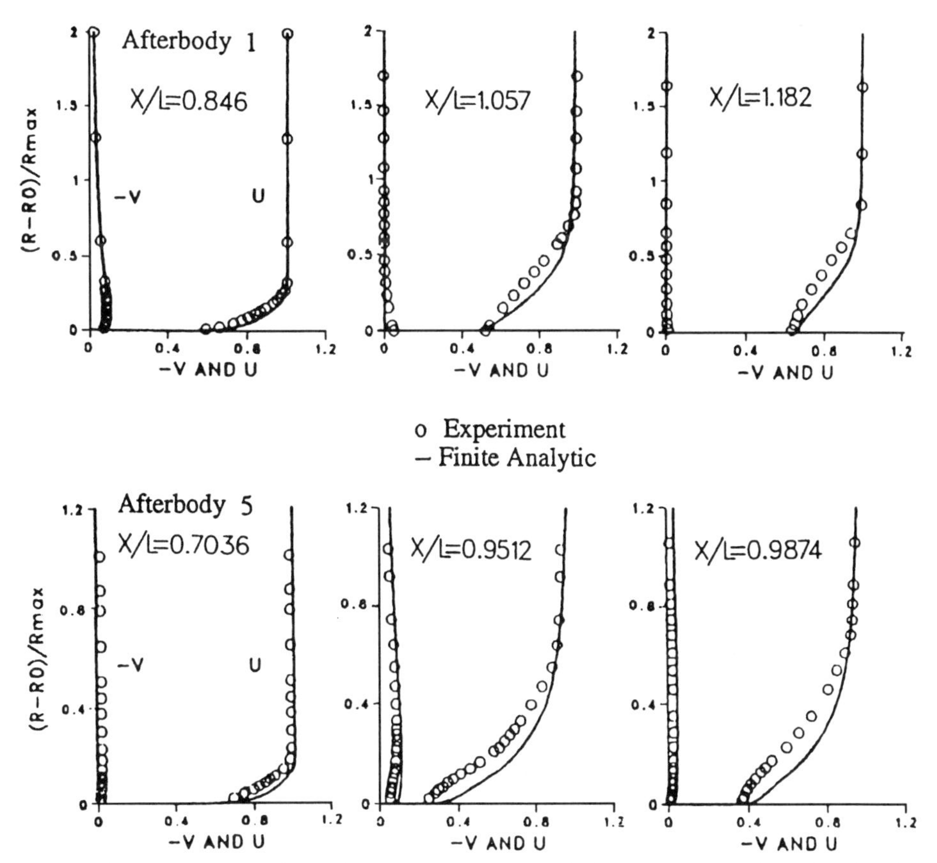

Figure 4.27 Velocity distribution of Afterbody 1 and Afterbody 5. (Adapted from Choi and Chen (1988).)

influence of viscosity, are often used in order to bridge the viscous sublayer, there are a number of instances in which the wall function approach is not adequate, and, hence, one has to abandon the wall function approach in predicting, for example, turbulent boundary layers with low and transitional Reynolds number, unsteady and separated flows, or flow over spinning surfaces, and so forth (Patel et al., 1984 [121]).

A rigid boundary exerts many different effects on turbulence (Launder, 1989 [85]), the most important of which are (1) it reduces the length scales of the fluctuation raising the dissipation rate; (2) it reflects the pressure fluctuations, thereby inhibiting the transfer of turbulence energy into fluctuations normal to the wall; and (3) it enforces a no-slip condition, thus ensuring that within a wall-adjacent sublayer, turbulent stresses are negligible and viscous effects on transport processes become of vital importance.

Accordingly, the extensions of high Reynolds number closures to the near-wall region all involve modifying the viscous diffusion, dissipation, and pressure redistribution. In a recent study, So, Lai, and Zhang (1991)[161] evaluated several of the available second-order near-wall closure models. Among them, four more popularly adopted

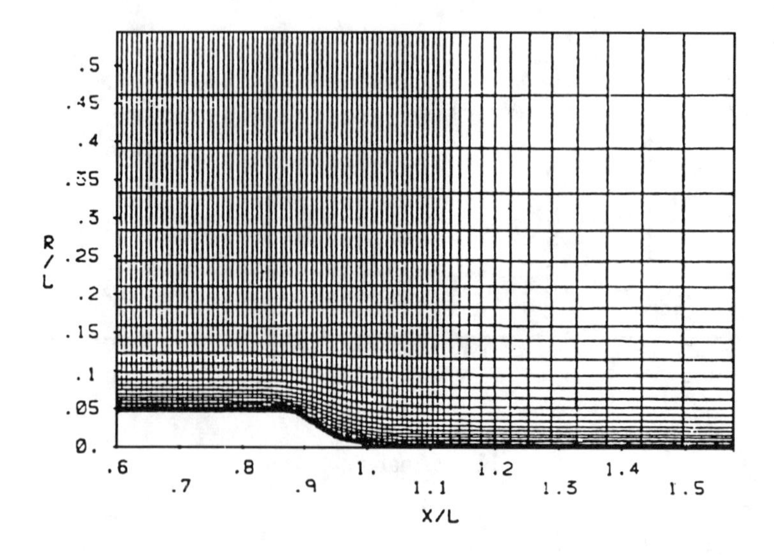

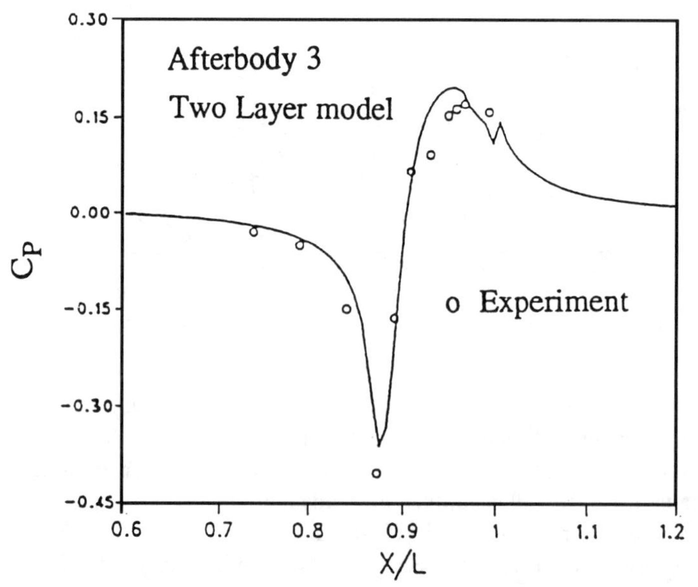

Figure 4.28 Uniform flow past axisymmetric Afterbody 3: (a) computational grid, and (b) axial pressure distribution. (Adapted from Chen and Patel (1988).)

models (viz., that of Hanjalic and Launder (1976)[56], Shima (1988)[154], Launder and Shima (1989)[91], and Lai and So (1990)[80] are introduced. It should be remarked first that the high Reynolds number versions of these near-wall closures are very similar. They all use the isotropic dissipation model for ϵ_{ij}, Rotta's model for $\Phi_{ij,1}$, Launder et al.'s model for $\Phi_{ij,2}$, except that Launder and Shima use Daly and Harlow's model for D^t_{ij}, whereas the other three use the model of Hanjalic and Launder. However, as already pointed out, the difference in the modeling of turbulent diffusion is slight and is

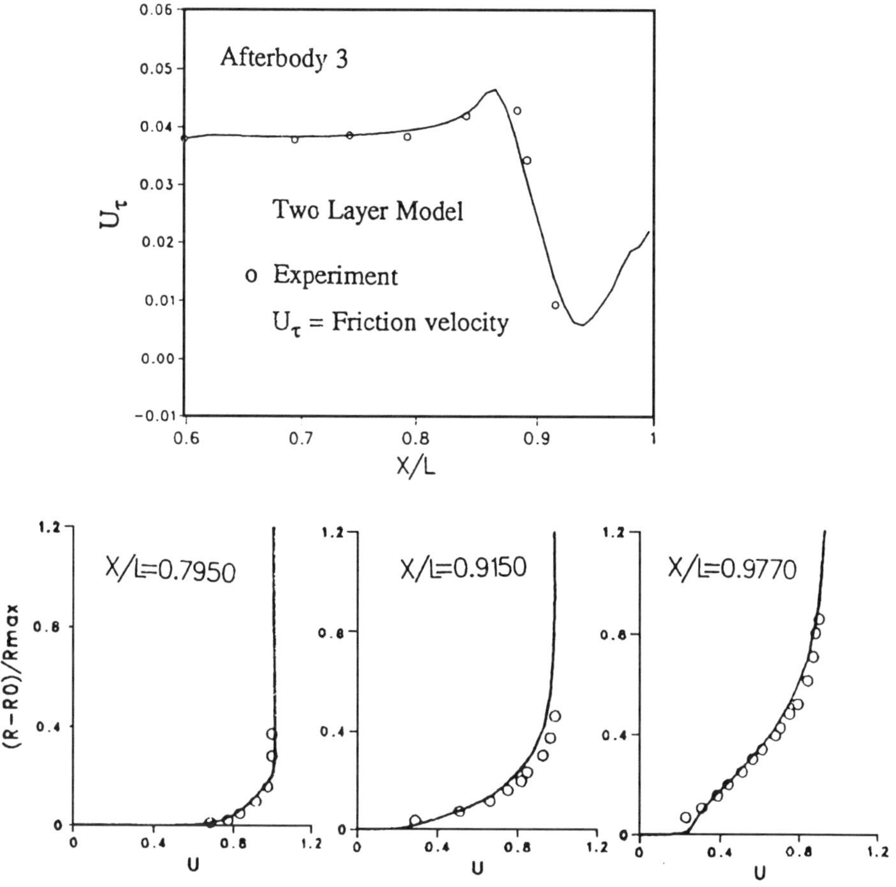

Figure 4.29 Uniform flow past axisymmetric Afterbody 3: (a) axial shear velocity distribution, and (b) velocity profiles. (Adapted from Chen and Patel (1988).)

insignificant. As a result, all of these closures can be expected to perform similarly far away from a wall.

4.7.2 Near-Wall Turbulence Models for ϵ and $\overline{u_i u_j}$

The near-wall ϵ-transport equation can be written in the form of

$$\frac{D\epsilon}{Dt} = \frac{\partial}{\partial X_l}\left(C_\epsilon \frac{k}{\epsilon}\overline{u_l u_m}\frac{\partial \epsilon}{\partial X_m} + \nu\frac{\partial \epsilon}{\partial X_l}\right) + C_{\epsilon 1}\frac{\epsilon}{k}P + \Psi - C_{\epsilon 2} f_\epsilon \frac{\epsilon}{k}\tilde{\epsilon} + \zeta. \qquad (4.54)$$

Here Ψ denotes the additional production term, ζ is the near-wall modification term, f_ϵ is the near-wall damping function, and $\tilde{\epsilon}$ is defined as

$$\tilde{\epsilon} = \epsilon - 2\nu\left(\frac{\partial\sqrt{k}}{\partial Y}\right)^2.$$

For the Reynolds-stress transport equations, a near-wall modification of Φ_{ij} is added to

Table 4.6 Additional production term and near-wall modification term

Authors	Ψ	ζ
Hanjalic and Launder (1976)[56]	$2\nu\frac{k}{\epsilon}\overline{u_j u_k}\frac{\partial^2 U_i}{\partial X_j \partial X_l}\frac{\partial^2 U_i}{\partial X_k \partial X_l}$	0
Shima (1988)[154]	$C_{\epsilon 1} f_{w,1}\frac{\epsilon}{k}P$	$[(-2+7/9C_{\epsilon 2})\epsilon\tilde{\epsilon}/k - 0.5\tilde{\epsilon}^2/k]f_{w,1}$ $\tilde{\epsilon} = \epsilon - \nu(\partial^2 k/\partial y^2)$
Launder and Shima (1989)[91]	$\frac{\epsilon}{k}(\Psi_1 + \Psi_2)P$ $\Psi_1 = 2.5A(P/\epsilon - 1)$ $\Psi_2 = 0.3(1 - 0.3a_{ik}a_{ki})f_{w,1}$	0
Lai and So (1990)	$C_{\epsilon 1}\sigma\frac{\epsilon}{k}f_{w,1}P$ $\sigma = 1 - 0.6\exp[-10^{-4}Re]$	$[(-2+7/9C_{\epsilon 2})\epsilon\tilde{\epsilon}/k - 0.5\epsilon^{*2}/k]f_{w,1}$ $\epsilon^* = \epsilon - 2\nu k/y^2$

Eq. (3.1) and reads as

$$\frac{D\overline{u_i u_j}}{Dt} = \left(D_{ij}^t + \Phi_{ij,p} + D_{ij}^\nu\right) + P_{ij} - \epsilon_{ij} + \Phi_{ij} + \Phi_{ij,w}$$
$$= \left(D_{ij}^t + \Phi_{ij,p} + D_{ij}^\nu\right) + P_{ij} - \epsilon_{ij} + \left(\Phi_{ij,1} + \Phi_{ij,2} + \Phi_{ij,w}\right). \quad (4.55)$$

Here $\Phi_{ij,w}$ represents the PS wall-echo effects. Hanjalic and Launder (1976)[56] argued that ν does not appear in Φ_{ij}. Therefore, the model for Φ_{ij} need not be modified for near-wall viscous effects. They also argued that for $y^+ < 15$, the effects of diffusive transport on the stress budgets are insignificant, thus the diffusion model can be used without modification. Near-wall modifications to the ϵ-transport equation, Eq. 3.3, anisotropic dissipation ϵ_{ij}, and models of PS wall-echo effects $\Phi_{ij,w}$ are listed in Tables 4.6–4.8.

Hanjalic and Launder (1972)[55] argued that as the Reynolds number approaches zero, the energy containing and dissipating range of motions overlap and ϵ_{ij} could be approximated by $\epsilon_{ij} = (\epsilon/k)\overline{u_i u_j}$. Consequently, they proposed to modify ϵ_{ij} by

$$\epsilon_{ij} = \frac{2}{3}\epsilon\left[(1 - f_s)\delta_{ij} + f_s\frac{\overline{u_i u_j}}{2k/3}\right]. \quad (4.56)$$

This ϵ_{ij} contracts to 2ϵ everywhere in the flow. They further modified their ϵ equation by defining the destruction term as $C_{\epsilon 2} f_\epsilon \tilde{\epsilon}\epsilon/k$. By introducing $\tilde{\epsilon}$ to replace one of the ϵs in

Table 4.7 Damping functions and model coefficients

Authors	f_ϵ	$f_{w,1}$	C_ϵ	$C_{\epsilon 1}$	$C_{\epsilon 2}$
Hanjalic and Launder (1976)[56]	$1 - \frac{2}{9}\exp\left[-\left(\frac{R_T}{6}\right)^2\right]$		0.15	1.275	1.80
Shima (1988)[154]	$1 - \frac{2}{9}\exp\left[-\left(\frac{R_T}{6}^2\right)\right]$	$\exp\left[-\left(\frac{0.015\sqrt{k}y}{\nu}\right)^4\right]$	0.15	1.35	1.80
Launder and Shima (1989)[91]	1	$\exp[-(0.002R_T)^2]$	0.18	1.45	1.90
Lai and So (1990)[80]	$1 - \frac{2}{9}\exp\left[-\left(\frac{R_T}{6}^2\right)\right]$	$\exp\left[-\left(\frac{R_T}{64}^2\right)\right]$	0.15	1.35	1.80

Table 4.8 Near-wall modifications of the Reynolds-stress-transport equations

Equations
Hanjalic and Launder (1972)[55]
$\epsilon_{ij} = \frac{2}{3}\epsilon\left\{(1-f_s)\delta_{ij} + f_s\frac{\overline{u_iu_j}}{2k/3}\right\},\ f_s = 1/(1+0.1R_T)$
$\Phi_{ij,w} = \left\{0.125\frac{\epsilon}{k}\left(\overline{u_iu_j} - \frac{2}{3}\delta_{ij}k\right) + 0.015(P_{ij} - D_{ij})\right\}(k^{3/2}/y\epsilon)$
y = distance from the wall
Shima (1988)[154]
$\epsilon_{ij} = \frac{2}{3}\delta_{ij}\epsilon,\ f_{w,1} = \exp\left[-\left(\frac{0.015\sqrt{k}y}{\nu}\right)^4\right],\ S_{ij} = \left(\frac{\partial U_i}{\partial X_j} + \frac{\partial U_j}{\partial X_i}\right)$
$\Phi_{ij,w} = \left[0.45\left(P_{ij} - \frac{2}{3}\delta_{ij}P\right) - 0.03\left(D_{ij} - \frac{2}{3}\delta_{ij}P\right) + 0.08kS_{ij}\right] f_{w,1}$
$\Phi_{ij,1} = -C_1^*\frac{\epsilon}{k}\left(\overline{u_iu_j} - \frac{2}{3}\delta_{ij}k\right),\ C_1^* = C_1\left[1 - \left(1 - \frac{1}{C_1}\right)\right] f_{w,1}$
Launder and Shima (1989)[91]
$\epsilon_{ij} = \frac{2}{3}\delta_{ij}\epsilon,\ \Phi_{ij,2} = -\xi\left(P_{ij} - \frac{2}{3}\delta_{ij}P\right),\ \xi = 0.75\sqrt{A}$
$\Phi_{ij,w} = C_1^*\frac{\epsilon}{k}\left[\overline{u_ku_m}n_kn_m\delta_{ij} - \left(\frac{3}{2}\right)\overline{u_ku_i}n_kn_j - \left(\frac{3}{2}\right)\overline{u_ku_j}n_kn_i\right]\left(\frac{0.4k^{3/2}}{y\epsilon}\right)$
$+ C_2^*\left[\Phi_{km,2}n_kn_m\delta_{ij} - \left(\frac{3}{2}\right)\Phi_{ik,2}n_kn_j - \left(\frac{3}{2}\right)\Phi_{jk,2}n_kn_i\right]\left(\frac{0.4k^{3/2}}{y\epsilon}\right)$
$C_1^* = -\frac{2}{3}C_1 + 1.67,\ C_2^* = \max\left[\left(\frac{2}{3}C_2 - \frac{1}{6}\right)/C_2, 0\right]$
Lai and So (1991)
$\epsilon_{ij} = \frac{2}{3}\epsilon(1-f_w)\delta_{ij} + f_w\frac{\epsilon}{k}[\overline{u_iu_j} + \overline{u_ku_i}n_kn_j + \overline{u_ku_j}n_kn_i$
$+ n_in_j\overline{u_ku_l}n_kn_l]/(1 + 3\overline{u_ku_l}n_kn_l/2k)$
$\Phi_{ij,w} = f_w\left[C_1\frac{\epsilon}{k}\left(\overline{u_iu_j} - \frac{2}{3}\delta_{ij}k\right) - \frac{\epsilon}{k}(\overline{u_iu_k}n_kn_j + \overline{u_ju_k}n_kn_i) + 0.45\left(P_{ij} - \frac{2}{3}\delta_{ij}P\right)\right]$
$f_w = \exp[-(R_T/150)^2]$

the destruction term, the term approaches zero at the wall. In addition, they introduced an extra production term in the ϵ equation to account for mean-strain generation of ϵ, or

$$\Psi = 2\nu\frac{k}{\epsilon}\overline{u_ku_j}\frac{\partial^2 U_i}{\partial X_j\partial X_l}\frac{\partial^2 U_i}{\partial X_k\partial X_l}. \tag{4.57}$$

Shima (1988)[154] pointed out that the redistribution model of Hanjalic and Launder is not consistent with the term in the exact equation. For the exact equation, the redistribution term vanishes in the $\overline{u_1^2}$, $\overline{u_2^2}$, and $\overline{u_3^2}$ equations and vanishes in the $\overline{u_iu_j}$ equation. To remove the defect, Shima adopted Lumley's (1980)[99] arrangement of the dissipation and redistribution terms. Instead of modeling these two terms separately, the treatment rewrites them as

$$-\epsilon_{ij} + \phi_{ij} = -\frac{2}{3}\delta_{ij}\epsilon + \left[\phi_{ij,1} - \epsilon_{ij} + \frac{2}{3}\delta_{ij}\epsilon\right] + \phi_{ij,2}. \tag{4.58}$$

This way, Shima does not have to model ϵ_{ij}. He proposes to model the square bracketed term by the Rotta's (1951)[143] model except that C_1 is replaced by $C_1^* = C_1[1-(1-1/C_1)]f_{w.1}$. In addition, Shima proposes a near-wall correction $\phi_{ij,w}$ given by

$$\phi_{ij,w} = \left[0.45\left(P_{ij} - \frac{2}{3}\delta_{ij}P\right) - 0.03\left(D_{ij} - \frac{2}{3}\delta_{ij}P\right) + 0.08kS_{ij}\right] f_{w,1}. \tag{4.59}$$

This term vanishes at the wall, as well as in high-Reynolds-number regions. Because the function $f_{w.1}$ depends explicitly on viscosity, this model does not include the nonviscous proximity effect.

Shima (1988)[154] also analyzed the near-wall behavior of the ϵ-transport equation proposed by Hanjalic and Launder (1972)[55]. He found that an additional term

$$\zeta = \left[\left(-2 + \frac{7}{9}C_{\epsilon 2}\right)\frac{\epsilon}{k}\tilde{\epsilon} - 0.5\frac{\tilde{\epsilon}^2}{k}\right] f_{w,1} \tag{4.60}$$

has to be introduced into the ϵ equation to fulfill the coincidence of $\partial\epsilon/\partial t$ and $\partial(\nu\partial^2 k/\partial x_j \partial x_j)$ at the wall. Note that $\epsilon = \nu\partial^2 k/\partial x_j \partial x_j$ is the wall condition derived from the k-transport equation. Shima also argued that because Ψ essentially enhances ϵ generation in the immediate vicinity of a wall, its complex form can be replaced by

$$\Psi = C_{\epsilon 1} f_{w.1}\frac{\epsilon}{k}P.$$

Launder and Shima (1989)[91] adopted an approach similar to that of Shima (1988)[154] and proposed only to modify Φ_{ij}. A practice suggested by Lumley (1978)[98] is to retain isotropic dissipation for ϵ_{ij} and to regard Φ_{ij} as including not only the true PS contribution but the departure of ϵ_{ij} from isotropic dissipation. This consideration means that, as the wall is approached, C_1 should approach unity. In this way, the limiting value of $(\epsilon_{ij} + \Phi_{ij,1})$ is $-(\epsilon/k)\overline{u_k u_j}$, which is consistent with the requirement that $\overline{(p/\rho)[(\partial u_i/\partial X_j) + (\partial u_j/\partial X_i)]}$, and ϵ_{22} should vanish at the wall. In modeling these processes, the echo effects that arise from the reflection of pressure fluctuations from the rigid wall are

$$\begin{aligned}\Phi_{ij,w} = {} & C_1^*\left(\frac{\epsilon}{k}\right)\left[\overline{u_k u_m} n_k n_m \delta_{ij} - \left(\frac{3}{2}\right)\overline{u_k u_i} n_k n_j - \left(\frac{3}{2}\right)\overline{u_k u_j} n_k n_i\right]\left(\frac{0.4k^{3/2}}{y\epsilon}\right) \\ & + C_2^*\left[\Phi_{km,2} n_k n_m \delta_{ij} - \left(\frac{3}{2}\right)\Phi_{ik,2} n_k n_j - \left(\frac{3}{2}\right)\Phi_{jk,2} n_k n_i\right]\left(\frac{0.4k^{3/2}}{y\epsilon}\right).\end{aligned} \tag{4.61}$$

After systematic tuning, the recommended model coefficients are

$$\begin{aligned}C_1 &= 1 + 2.58A(a_{ik}a_{ki})^{1/4}\{1 - \exp[-(0.0067R_T)^2]\} \\ C_2 &= 0.75\sqrt{A} \\ C_1^* &= -\frac{2}{3}C_1 + 1.67 \\ C_2^* &= \max\left[\left(\frac{2}{3}C_2 - \frac{1}{6}\right)\Big/C_2, 0\right].\end{aligned}$$

As for the ϵ equation, ζ is assumed to be zero, and Ψ is taken to be given by

$$\frac{\epsilon}{k}(\Psi_1 + \Psi_2)P,$$

where $\Psi_1 = 2.5A(P/\epsilon - 1)$, and $\Psi_2 = 0.3(1 - 0.3a_{ik}a_{ki})f_{w.1}$.

Lai and So (1990)[80] pointed out that convection, turbulent diffusion, and production go to zero very rapidly as the wall is approached. On the other hand, viscous diffusion, D^{ν}_{ij}, and ϵ_{ij} are dominant near a wall and their difference has to be compensated for by the velocity-pressure-gradient correlation, $\Phi^*_{ij} = \Phi_{ij.p} + \Phi_{ij}$. If a model for Φ^*_{ij} is chosen such that Φ^*_{ij} is exactly zero at the wall but fails to balance the difference $(\epsilon_{ij} - D^{\nu}_{ij})$, then the model cannot be expected to mimic the anisotropic behavior of near-wall turbulence very well. In general, the pressure diffusion $\phi_{ij.p}$ is small compared with D^t_{ij} for high-Reynolds-number flow and is normally neglected in Reynolds-stress closure. As a result, Φ^*_{ij} with a nonzero trace is now replaced by Φ_{ij} with a zero trace. In order that the model proposed for Φ_{ij} be tensorially correct, it should also have a zero trace. Therefore, the Φ_{ij} model cannot correctly replicate the behavior of Φ^*_{ij} near a wall. To remove the defect, Lai and So proposed to model Φ^*_{ij} as $\Phi^*_{ij} = \Phi_{ij} + \Phi_{ij.w}$ with

$$\Phi_{ij.w} = f_w\left[C_1\frac{\epsilon}{k}\left(\overline{u_iu_j} - \frac{2}{3}\delta_{ij}k\right) - \frac{\epsilon}{k}(\overline{u_iu_k}n_kn_j + \overline{u_ju_k}n_kn_i) + 0.45\left(P_{ij} - \frac{2}{3}\delta_{ij}P\right)\right]. \tag{4.62}$$

Note that f_w is introduced to guarantee the disappearance of $\Phi_{ij.w}$ far away from the wall, and $\Phi_{ij.w}$ is the model for the pressure diffusion term plus the adjustment for the incorrect modeling of Φ_{ij} near a wall.

For ϵ_{ij}, a model proposed by Launder and Reynolds (1983)[89] along the suggestions of Hanjalic and Launder (1976)[56] reads Ok to

$$\epsilon_{ij} = \frac{2}{3}\epsilon(1 - f_w)\delta_{ij} + \frac{f_w\frac{\epsilon}{k}(\overline{u_iu_j} + \overline{u_ku_i}n_kn_j + \overline{u_ku_j}n_kn_i + n_in_j\overline{u_ku_l}n_kn_l)}{1 + 3\overline{u_ku_l}n_kn_l/2k}, \tag{4.63}$$

which guarantees that ϵ_{ij} will asymptote to Kolmogorov's (1942)[78] model far away from the wall and satisfies the near-wall kinematic conditions (Launder and Reynolds, 1983)[89]

$$\frac{\epsilon_{11}}{\overline{u^2}} = \frac{\epsilon_{33}}{\overline{w^2}} = \frac{\epsilon_{13}}{\overline{uw}} = \frac{1}{2}\frac{\epsilon_{23}}{\overline{vw}} = \frac{1}{2}\frac{\epsilon_{12}}{\overline{uv}} = \frac{1}{4}\frac{\epsilon_{22}}{\overline{v^2}} = \frac{\epsilon}{k}.$$

The ϵ equation is modified similarly to that of Shima (1988)[154] so that $\partial\epsilon/\partial t$ is required to have the proper behavior at a wall. Lai and So proposed to model Ψ by $C_{\epsilon 1}\sigma\frac{\epsilon}{k}f_{w.1}P$, where σ depends on the Reynolds number as a result of the direct simulation study of Mansour et al. (1988)[102]. Furthermore, the term $\tilde{\epsilon} = \epsilon - \nu(\partial^2 k/\partial y^2)$ proposed by Shima in the near-wall modification model ζ is replaced by a more convenient and numerically stable form; namely, $\epsilon^* = \epsilon - 2\nu k/y^2$ and $\zeta = [(-2 + 7/9C_{\epsilon 2})\epsilon\tilde{\epsilon}/k - 0.5\epsilon^{*2}/k]f_{w.1}$. The near-wall modification term is later further modified to $\zeta = (-2\epsilon\tilde{\epsilon}/k + 1.5\epsilon^{*2}/k)f_{w.1}$ (So, Zhang, and Speziale, (1991)[159]) with model coefficient $C_{\epsilon 1}$ changed to 1.5 and σ redefined as $\sigma = 1 - 1.5\exp(-10^{-4}Re)$. Thus modified, the calculated properties of a flat-plate boundary layer are in better agreement with measurements and direct simulation results.

Applications of these models are reported for the flat-plate boundary layer, fully developed plane channel flow, and pipe flow. Predictions in complex, nonequilibrium shear flows, such as flow through a curved pipe bend (So, Lai and Hwang, 1991 [160]), flow through a strongly curved, 180° square duct (Sotiropoulos and Patel, 1993 [162]), flow through a circular-to-rectangular transition duct (Sotiropoulos and Patel, 1993 [162]), flow past *Hamburgische Schiffbau-Versuchsanstalt* (Hamburg Ship Model Basin) tanker (Sotiropoulos and Patel, 1994 [163]; Chen, Lin and Weems, 1994), submarine flows (Chen, 1992 [27]), and so forth, have also been reported. It is found that, in general, agreement with experiment is better than when the eddy viscosity model is adopted.

Note that with the appearance of a wall, near-wall modifications and damping functions are required for all Φ_{ij} models if they are applied directly from the wall. Because the basic Φ_{ij} models proposed are based on quasihomogeneous assumptions, they are incorrect in the strongly inhomogeneous near-wall region. Near-wall modifications and damping functions are then introduced to correct the erroneous behavior of the basic model. These modifications adjust the model solution to fit a particular data set but remove the flexibility that the differential equations were meant to provide. Although it may not be obvious how the aspects of inhomogeneity and anisotropy of the near-wall turbulence can be represented by model differential equations, attempts in that direction are certainly warranted. Durbin (1993)[43] suggested that elliptic effects within the flow, which are caused by the proximity of a boundary, might be included by formulating an elliptic relaxation model for Φ_{ij}. The elliptic effects are blocking the normal velocity and pressure reflection from the surface. An elliptic model provides a natural way to let the wall effects appear; they enter through the boundary conditions, and the model relaxes to quasihomogeneous behavior in the interior, far away from the boundary. Wall effects then enter via solution at the governing equations rather than via prescribed damping function profiles.

4.7.3 Near-Wall Turbulence Model with Kolmogorov Scale

Recently, a near-wall second-order closure model that also includes Kolmogorov scale has been proposed by Durbin (1993)[43]. The length scale l and time scale t are formulated as the maximum of $C_L(k^{3/2}/\epsilon, C_\eta(\nu^3/\epsilon)^{1/4})$ and $(k/\epsilon, C_t\nu/\epsilon^{1/2})$, respectively. The primary purpose for using the Kolmogorov scale as a lower bound is to avoid the singularity at $y = 0$ in the first term of these scale representations. For the turbulent diffusion model,

$$D_{ij}^t = \frac{\partial}{\partial X_l}\left(C_k t\overline{u_l u_m}\frac{\partial \overline{u_i u_j}}{\partial X_m}\right) \tag{4.64}$$

is used. The ϵ-transport equation adopted is

$$\frac{D\epsilon}{Dt} = \frac{\partial}{\partial X_l}\left[(C_\epsilon t\overline{u_l u_m} + \nu\delta_{lm})\frac{\partial \epsilon}{\partial X_m}\right] + \frac{C_{\epsilon 1}^* P - C_{\epsilon 2}\epsilon}{t} \tag{4.65}$$

Here $C_{\epsilon 1}^* = C_{\epsilon 1}(1 + 0.1P/\epsilon)$ was made a function of P/ϵ to account for the anisotropic production terms in the near-wall region.

The modeled Reynolds stress equation can be written as

$$\frac{D\overline{u_i u_j}}{Dt} = \frac{\partial}{\partial X_l}\left[(C_k t\overline{u_l u_m} + \nu\delta_{lm})\frac{\partial \overline{u_i u_j}}{\partial X_m}\right] + P_{ij} - \frac{2}{3}\delta_{ij}\epsilon + \Phi_{ij}. \qquad (4.66)$$

As already mentioned in the previous section, Durbin (1991)[42] proposed an elliptic relaxation model for the PS term,

$$\Phi_{ij} = kf_{ij}$$

$$l^2\nabla^2 f_{ij} - f_{ij} = -\frac{1}{k}(\Phi_{ij,1} + \Phi_{ij,2})$$

In the quasihomogeneous limit, the above two equations reduce to

$$\Phi_{ij} = (\Phi_{ij,1} + \Phi_{ij,2}).$$

For $(\Phi_{ij,1} + \Phi_{ij,2})$, any quasihomogeneous model can be used. Durbin adopts the simple model recommended by Launder (1989)[85]

$$\Phi_{ij} = -\frac{C_1}{t}\left(\overline{u_i u_j} - \frac{2}{3}\delta_{ij}k\right) - \xi\left(P_{ij} - \frac{2}{3}\delta_{ij}P\right). \qquad (4.67)$$

The model coefficients adopted are $C_{\epsilon 1} = 1.44$, $C_{\epsilon 2} = 1.9$, $C_k = 0.2$, $C_\epsilon = 0.14$, $C_L = 0.2$, $C_\eta = 80$, $C_t = 6$, $C_1 = 1.22$, and $\xi = 0.6$. This model has been applied to solve channel flow and boundary layers with zero and adverse pressure gradients. Good predictions of Reynolds-stress components, mean flow, skin friction, and displacement thickness are obtained in various comparisons to experimental and direct numerical simulation data.

4.8 SUMMARY

In this chapter, we considered three models; namely,

1. RSM (Reynolds stress model or differential model)
2. k-ϵ-A (kinetic energy-dissipation-algebraic stress model)
3. k-ϵ-E (kinetic energy-dissipation-eddy viscosity model).

With the exception of round jet and far wake flows, all three models perform satisfactorily for nonseparated flows such as free-shear and wall-shear flows. A two-scale k-ϵ model may be introduced to correct the deficiency of the one-scale k-ϵ turbulence model.

For complex flows with no separation and curved geometry, the RSM, in general, will perform better than the other models. The k-ϵ-A model, in turn, will perform better than the k-ϵ-E model. Although problem functions are required in some problems, the second-order turbulence model performs much better than the first-order model in which the Reynolds stresses are assumed on an ad hoc basis.

The weakest modeling is still in the ϵ equation. The PS modeling can be improved. Similarly, the isotropic dissipation and eddy viscosity modeling can be revised.

CHAPTER

FIVE

APPLICATIONS OF TURBULENCE MODELS

5.1 INTRODUCTION

In the Turbulent Flow Predictions 1 (chapter 3, Section 3.2) and 2 (chapter 4, Section 4.6), we examined the validity of the second-order turbulence model in predicting free- and wall-shear flows. In general, the second-order turbulence model performs satisfactorily in providing mean-flow quantities and gives fair results for turbulence transport properties. There are, however, some disappointments as the model fails in some cases to predict even some gross characteristics of the mean flow. In these cases, problem functions or modifying functions are introduced to alter some turbulent constants.

In this chapter we consider further the applications of turbulence models to more complex flows.

5.1.1 Two-Dimensional Separated Flows

In many engineering and practical problems, turbulent flows may separate. The ability to predict separated flows is then of great importance to engineers. In analyzing separated flows, more complex governing equations must be solved compared with those for nonseparated flows. Although it is more desirable to employ the full Reynolds stress model (RSM) for the separated flows, most recent predictions are done with the k-ϵ model for reasons of computational economy. Some two-dimensional separated flows are given in Fig. 5.1. These separate flows present different degrees of difficulty in the prediction. The less difficult cases are those separations that do not involve shedding or unsteadyness. When the flow separates and oscillates, the turbulence model near the wall becomes very complicated. In some instances the turbulence vortex shedding behind a cylinder remains one of most challenging prediction today.

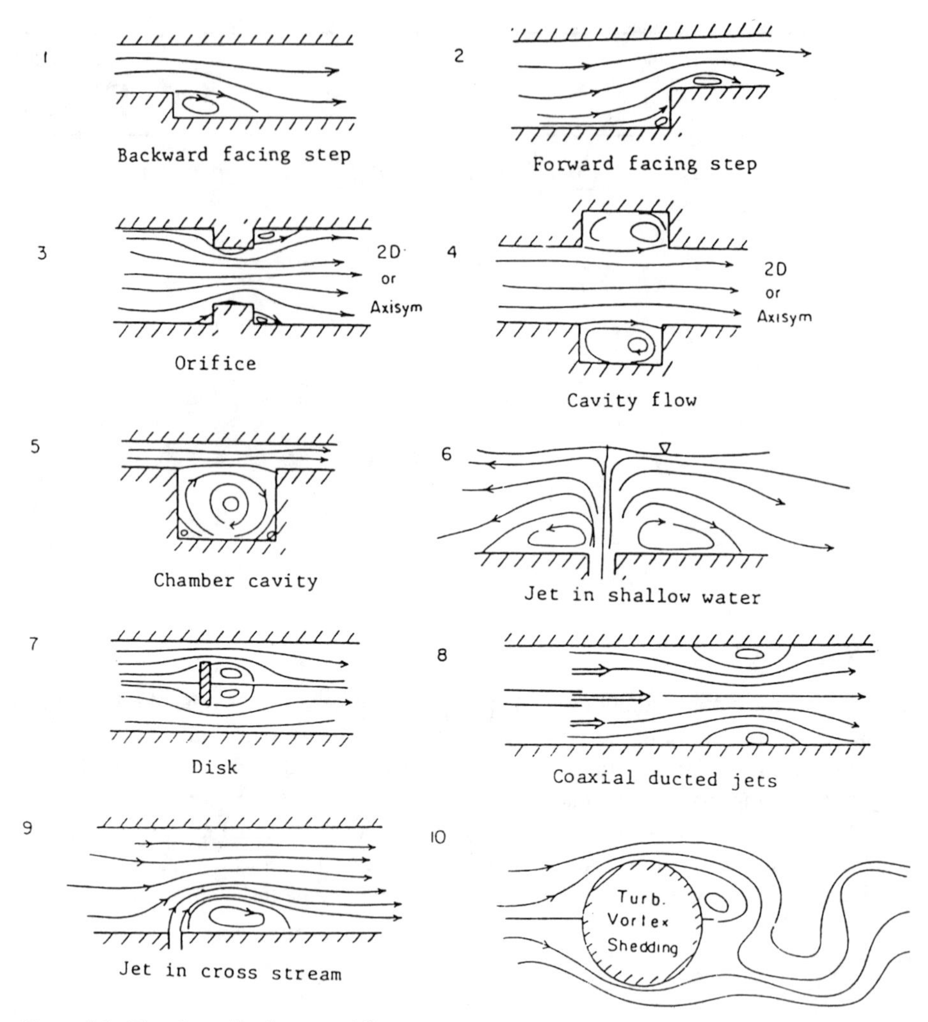

Figure 5.1 Two-dimensional separated flows.

5.1.2 Mean Equations

For two-dimensional unsteady incompressible flow involving flow separation, we have the following governing equations:

- the continuity equation,

$$\frac{\partial U}{\partial X} + \frac{\partial V}{\partial Y} = 0, \tag{5.1}$$

- the X-momentum equation,

$$\frac{\partial U}{\partial t} + U\frac{\partial U}{\partial X} + V\frac{\partial U}{\partial Y} = -\frac{1}{\rho}\frac{\partial P}{\partial X} + \frac{\partial}{\partial Y}\left(\nu\frac{\partial U}{\partial Y} - \overline{uv}\right) + \frac{\partial}{\partial X}\left(\nu\frac{\partial U}{\partial X} - \overline{u^2}\right), \tag{5.2}$$

- the Y-momentum equation,

$$\frac{\partial V}{\partial t}+U\frac{\partial V}{\partial X}+V\frac{\partial V}{\partial Y}=-\frac{1}{\rho}\frac{\partial P}{\partial Y}+\frac{\partial}{\partial Y}\left(\nu\frac{\partial V}{\partial Y}-\overline{v^2}\right)+\frac{\partial}{\partial X}\left(\nu\frac{\partial V}{\partial X}-\overline{uv}\right), \quad (5.3)$$

and

- the energy equation,

$$\frac{\partial T}{\partial t}+U\frac{\partial T}{\partial X}+V\frac{\partial T}{\partial Y}=\frac{\partial}{\partial y}\left(\alpha\frac{\partial T}{\partial X}-\overline{v\theta}\right)+\frac{\partial}{\partial X}\left(\alpha\frac{\partial T}{\partial X}-\overline{u\theta}\right). \quad (5.4)$$

To close the above indicated equations, a turbulence model is needed for the turbulent quantities. In the following two sections, the k-ϵ-A and the k-ϵ-E will be derived for the two-dimensional steady incompressible flow. It should be remarked that the set of equations indicated above are elliptic partial differential equations in space (X, Y). The numerical solution of the elliptic equations is in general more involved than that of a parabolic equation in the boundary layer and free-shear flows.

5.1.3 k-ϵ-A Model

The k-ϵ-A model consists of two differential equations for k and ϵ and algebraic equations for the Reynolds stress. The modeled equations for the k, ϵ, and the algebraic stress model are

- the k equation,

$$\frac{Dk}{Dt}=\frac{\partial}{\partial Y}\left[\left(C_k\frac{k^2}{\epsilon}+\nu\right)\frac{\partial k}{\partial Y}\right]+\frac{\partial}{\partial X}\left[\left(C_k\frac{k^2}{\epsilon}+\nu\right)\frac{\partial k}{\partial X}\right]-\left[\overline{u^2}\frac{\partial U}{\partial X}+\overline{uv}\frac{\partial U}{\partial Y}\right]-\epsilon, \quad (5.5)$$

- the ϵ equation,

$$\frac{D\epsilon}{Dt}=\frac{\partial}{\partial Y}\left[\left(C_\epsilon\frac{k^2}{\epsilon}+\nu\right)\frac{\partial \epsilon}{\partial Y}\right]+\frac{\partial}{\partial X}\left[\left(C_\epsilon\frac{k}{\epsilon}+\nu\right)\frac{\partial \epsilon}{\partial X}\right]$$
$$-C_{\epsilon 1}\frac{\epsilon}{k}\left(\overline{u^2}\frac{\partial U}{\partial X}+\overline{uv}\frac{\partial U}{\partial Y}\right)-C_{\epsilon 2}\frac{\epsilon^2}{k}, \quad (5.6)$$

and

- the algebraic stress equations for $\overline{u_iu_j}$,

$$0=P_{ij}-\frac{2}{3}\delta_{ij}\epsilon-C_1\frac{\epsilon}{k}\left(\overline{u_iu_j}-\frac{2}{3}\delta_{ij}k\right)-C_2\left(P_{ij}-\frac{2}{3}\delta_{ij}P_k\right). \quad (5.7)$$

For $i=j=1$, Eq. 5.7 becomes

$$-\overline{u^2}=\frac{k}{C_1\epsilon}\left[(C_2-1)P_{11}-\frac{2}{3}C_2P_k\right]-\frac{2}{3}\frac{C_1-1}{C_1}k. \quad (5.8)$$

Substituting for P_{11} and P_k, we get

$$-\overline{u^2} = \frac{k}{C_1\epsilon}\left[\left(2-\frac{4}{3}C_2\right)\left(\overline{u^2}\frac{\partial U}{\partial X} + \overline{uv}\frac{\partial U}{\partial Y}\right) + \frac{2}{3}C_2\left(\overline{uv}\frac{\partial V}{\partial X} + \overline{v^2}\frac{\partial V}{\partial Y}\right)\right] - \frac{2\,(C_1-1)}{3C_1}k. \tag{5.9}$$

For $i = j = 2$, Eq. 5.7 becomes

$$-\overline{v^2} = \frac{k}{C_1\epsilon}\left[(C_2-1)\,P_{22} - \frac{2}{3}C_2P_k\right] - \frac{2}{3}\frac{C_1-1}{C_1}k. \tag{5.10}$$

Substituting for P_{22} and P_k, we get

$$-\overline{v^2} = \frac{k}{C_1\epsilon}\left[\left(2-\frac{4}{3}C_2\right)\left(\overline{uv}\frac{\partial V}{\partial X} + \overline{v^2}\frac{\partial V}{\partial Y}\right) + \frac{2}{3}C_2\left(\overline{u^2}\frac{\partial U}{\partial X} + \overline{uv}\frac{\partial U}{\partial Y}\right)\right] - \frac{2\,(C_1-1)}{3C_1}k. \tag{5.11}$$

For $i = 1$ and $j = 2$, Eq. 5.7 yields

$$(1-C_2)\,P_{12} - C_1\frac{\epsilon}{k}\overline{uv} = 0. \tag{5.12}$$

Substituting for P_{12} and P_k, we get

$$-\overline{uv} = \frac{k}{\epsilon}\frac{(1-C_2)}{C_1}\left(\overline{u^2}\frac{\partial V}{\partial X} + \overline{v^2}\frac{\partial U}{\partial Y}\right), \tag{5.13}$$

where $C_1 = 2.3$ and $C_2 = 0.4$.

The algebraic equation for $\overline{u_i\theta}$ is

$$-\overline{u_i\theta} = \frac{k}{C_{T1}\epsilon}\overline{u_iu_l}\frac{\partial T}{\partial X_l} - \frac{(C_{T2}-1)\,k}{C_{T1}\epsilon}\overline{u_l\theta}\frac{\partial U_i}{\partial X_l}. \tag{5.14}$$

The experimental values for the turbulence coefficients are

$C_k = 0.09$, $C_\epsilon = 0.07$, $C_{\epsilon 1} = 1.44$, $C_{\epsilon 2} = 1.92$,
$C_1 = 2.30$, $C_2 = 0.40$, $C_{T1} = 3.20$, and $C_{T2} = 0.50$.

5.1.4 k-ϵ-E Model

In this model, the k and ϵ equations (5.5) and (5.6) remain the same. The Reynolds stress and the Reynolds heat flux are modeled on the basis of the eddy viscosity and eddy diffusivity models, respectively. In this model, the Reynolds stress and the Reynolds heat flux can be written as

- the Reynolds stress equation,

$$-\overline{u_iu_j} = C_\mu\frac{k^2}{\epsilon}\left(\frac{\partial U_i}{\partial X_j} + \frac{\partial U_j}{\partial X_i}\right) - \frac{2}{3}\delta_{ij}k, \tag{5.15}$$

and

- the Reynolds heat flux equation,

$$-\overline{u_i\theta} = \frac{C_\mu k^2}{Pr_t \epsilon}\frac{\partial T}{\partial X_i}, \tag{5.16}$$

where $C_\mu = 0.09$, $Pr_t (\approx 0.8 \sim 1.3)$ is the turbulence Prandtl number.

5.1.5 Comparison of k-ϵ-A and k-ϵ-E Models

Let us examine the $\overline{uv}$ equation. The k-ϵ-A model will give

$$-\overline{uv} = \frac{k\overline{u^2}}{\epsilon}\left(\frac{1-C_2}{C_1}\right)\frac{\partial V}{\partial X} + \frac{k\overline{v^2}}{\epsilon}\left(\frac{1-C_2}{C_1}\right)\frac{\partial U}{\partial y}, \tag{5.17}$$

whereas the k-ϵ-E model will give

$$-\overline{uv} = C_\mu\frac{k^2}{\epsilon}\frac{\partial V}{\partial X} + C_\mu\frac{k^2}{\epsilon}\frac{\partial U}{\partial y}. \tag{5.18}$$

If we set $\overline{v^2} = \overline{u^2} = 0.5k$ in Eq. 5.17 and substitute for the coefficients $C_1 = 2.3$ and $C_2 = 0.4$, we have

$$\frac{k\overline{u^2}}{\epsilon}\left(\frac{1-C_2}{C_1}\right) = \frac{k\overline{v^2}}{\epsilon}\left(\frac{1-C_2}{C_1}\right) = 0.13\frac{k^2}{\epsilon}. \tag{5.19}$$

The k-ϵ-A model gives $C_\mu = 0.13$, whereas the k-ϵ-E model has $C_\mu = 0.09$. Thus the k-ϵ-A model gives a $\overline{uv}$ equation similar to that of the k-ϵ-E model but with a higher diffusion coefficient if $\overline{u^2} = \overline{v^2} = 0.5k$.

Let us consider the normal stress $\overline{u^2}$. The (k-ϵ-A) model gives

$$-\overline{u^2} = \left(\frac{6-4C_2}{3C_1}\right)\frac{k\overline{u^2}}{\epsilon}\frac{\partial U}{\partial X} - \frac{2}{3}\left(\frac{C_1-1}{C_1}\right)k + \left(\frac{6-4C_2}{3C_1}\right)\frac{k\overline{uv}}{\epsilon}\frac{\partial U}{\partial Y}$$
$$+ \frac{2C_2}{3C_1}\frac{k}{\epsilon}\left(\overline{uv}\frac{\partial V}{\partial X} + \overline{v^2}\frac{\partial V}{\partial Y}\right), \tag{5.20}$$

and the (k-ϵ-E) model gives

$$-\overline{u^2} = 2C_\mu\frac{k^2}{\epsilon}\frac{\partial U}{\partial x} - \frac{2}{3}k. \tag{5.21}$$

With $\overline{u^2} = 0.5k$, $C_1 = 2.3$, and $C_2 = 0.4$, the coefficients in Eq. 5.20 become

$$\frac{6-4C_2}{3C_1}\frac{k\overline{u^2}}{\epsilon} = 2(0.16)\frac{k^2}{\epsilon} \tag{5.22}$$

$$\frac{2}{3}(C_1 - 1) = \frac{2}{3}(1.3). \tag{5.23}$$

Thus the k-ϵ-A model gives $C_\mu = 0.16$, which is larger than $C_\mu = 0.09$ from the k-ϵ-E model. It also gives 1.3 times larger magnitude of normal stress in proportion to the kinetic energy. In addition, there are other contributions to $\overline{u^2}$ in the k-ϵ-A model. The normal stress $\overline{v^2}$ in the k-ϵ-A model, likewise, differs from $\overline{v^2}$ in the k-ϵ-E model. The comparison of the Reynolds heat flux also shows a difference between the two models such that in the k-ϵ-A model

$$-\overline{u_i\theta} = \frac{\overline{u_i u_l}}{C_{T1}} \frac{k}{\epsilon} \frac{\partial T}{\partial X_l} - \frac{(C_{T2} - 1)}{C_{T1}} \frac{k}{\epsilon} \overline{-u_l\theta} \frac{\partial U_i}{\partial X_l}, \tag{5.24}$$

and in the k-ϵ-E model

$$-\overline{u_i\theta} = \frac{C_\mu k^2}{Pr_t \epsilon} \frac{\partial T}{\partial X_i}. \tag{5.25}$$

5.2 TURBULENT FLOW PREDICTIONS: THREE (TWO-DIMENSIONAL SEPARATED FLOWS)

5.2.1 Two-Dimensional Channel Expansion Flow

The fluid flowing through a sudden expansion in a pipe or a channel can easily separate from the wall. This separation phenomenon often occurs in engineering applications. For example, in a dump combustor, the recirculation zone that is created just behind the step can be used as a flameholder. In design of a nuclear heat extractor or a heat exchanger, turbulence created by a series of separations can promote the heat transfer efficiency. However, separation in a sudden expansion can inevitably increase the pressure loss. The backward facing step can be classified into three types.

1. two-dimensional symmetric step,
2. two-dimensional single step, and
3. axisymmetric annular step.

Table 6.1 lists the pertinent works on flows past a backward facing step. This table provides the range of expansion ratios and Reynolds numbers, classification of geometries, and flow types (laminar or turbulent). It is known from experiments of Abott and Kline (1962) that if the expansion ratio is equal or less than 1.5, turbulent flow is steady and separation is also symmetrical in a symmetric backward facing step. On the other hand if the expansion ratio is greater than 1.5, the flow may become unsteady and unsymmetrical. Further details of experimental works are reviewed by Chang (1984)[11]. In this section, the flow past a backward facing step, shown in Fig. 5.2 with expansion ratio of 1.5, is analyzed. The expansion ratio (h_2/h_1) is defined as the ratio of the height of the

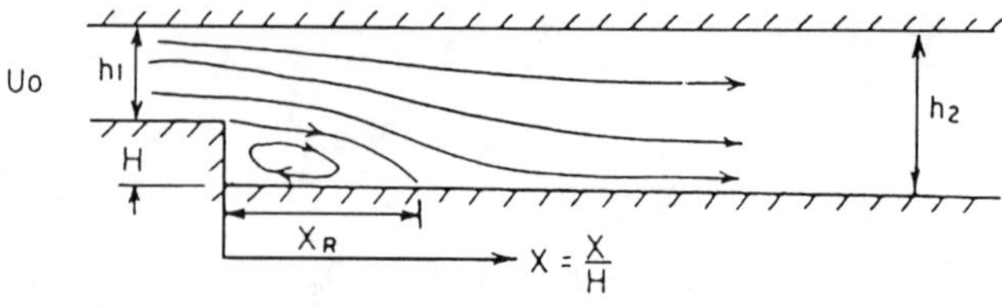

Figure 5.2 Two-dimensional backward facing step flow.

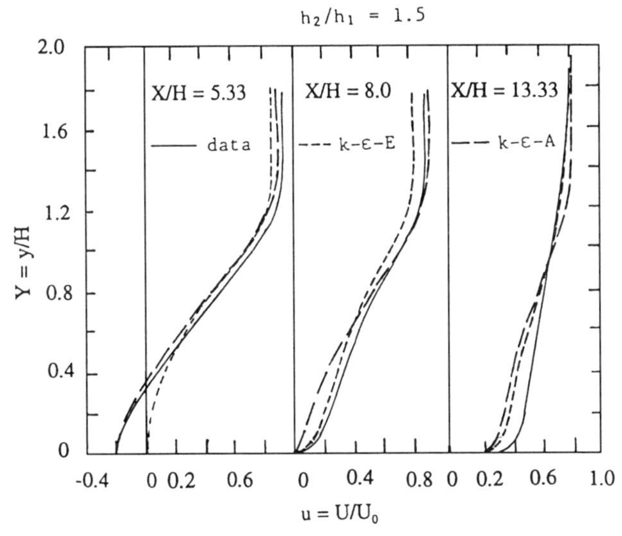

Figure 5.3 Velocity profiles in flow over a backward facing step. (Adapted from Kline et al. (1981).)

outlet of the channel (h_2) to the height of the inlet opening (h_1). In the present study, both the k-ϵ-E and the k-ϵ-A models are employed for prediction of turbulent quantities in the flow. Let us consider the prediction of turbulent flow in a channel with backward facing step, as shown in Fig. 5.2. The experiments of Kim, Kline, and Johnston (1978)[73], with $h_2/h_1 = 1.5$, were chosen as test cases for the 1980–1981 Stanford Conference on Complex Turbulent Flows, and this flow was calculated for the conference with a number of different methods (see Kline, Cantwell, and Lilley, 1981 [76]). Calculations shown in Fig. 5.3 with both the standard k-ϵ-E model and with models employing transport equations yielded reattachment lengths X_R on the order of 20 percent shorter than the measured value of $X_R/H = 7$. This is associated with rather poor predictions of the velocity profile in the recirculation flow.

Using the finite analytic (FA) method, Chang (1984)[11] obtained the results that are shown in Fig. 5.4. The prediction was done with

$$\frac{D\overline{u_i u_j}}{Dt} = D_{ij}, \tag{5.26}$$

or

$$-\overline{u_i u_j} = \frac{(C_2 - 1)}{C_1}\frac{k}{\epsilon}\left(P_{ij} - \frac{2}{3}\delta_{ij}\frac{P_k C_2}{C_2 - 1}\right) - \frac{2}{3}\delta_{ij}k\frac{C_1 - 1}{C_1}. \tag{5.27}$$

The prediction is only fair. In particular, the prediction behind the separation zone is not good.

Consider the k-ϵ-A model again with Rodi's (1972)[134] hypothesis,

$$\frac{D\overline{u_i u_j}}{Dt} - D_{ij} = \frac{\overline{u_i u_j}}{k}\left(\frac{Dk}{Dt} - D_k\right) = \frac{\overline{u_i u_j}}{k}\left(P_k - \epsilon\right). \tag{5.28}$$

Thus, we have the algebraic stress equation,

$$\frac{\overline{u_i u_j}}{k}\left(P_k - \epsilon\right) = P_{ij} - \frac{2}{3}\delta_{ij}\epsilon - C_1\frac{\epsilon}{k}\left(\overline{u_i u_j} - \frac{2}{3}\delta_{ij}k\right) - C_2\left(P_{ij} - \frac{2}{3}\delta_{ij}P_k\right). \tag{5.29}$$

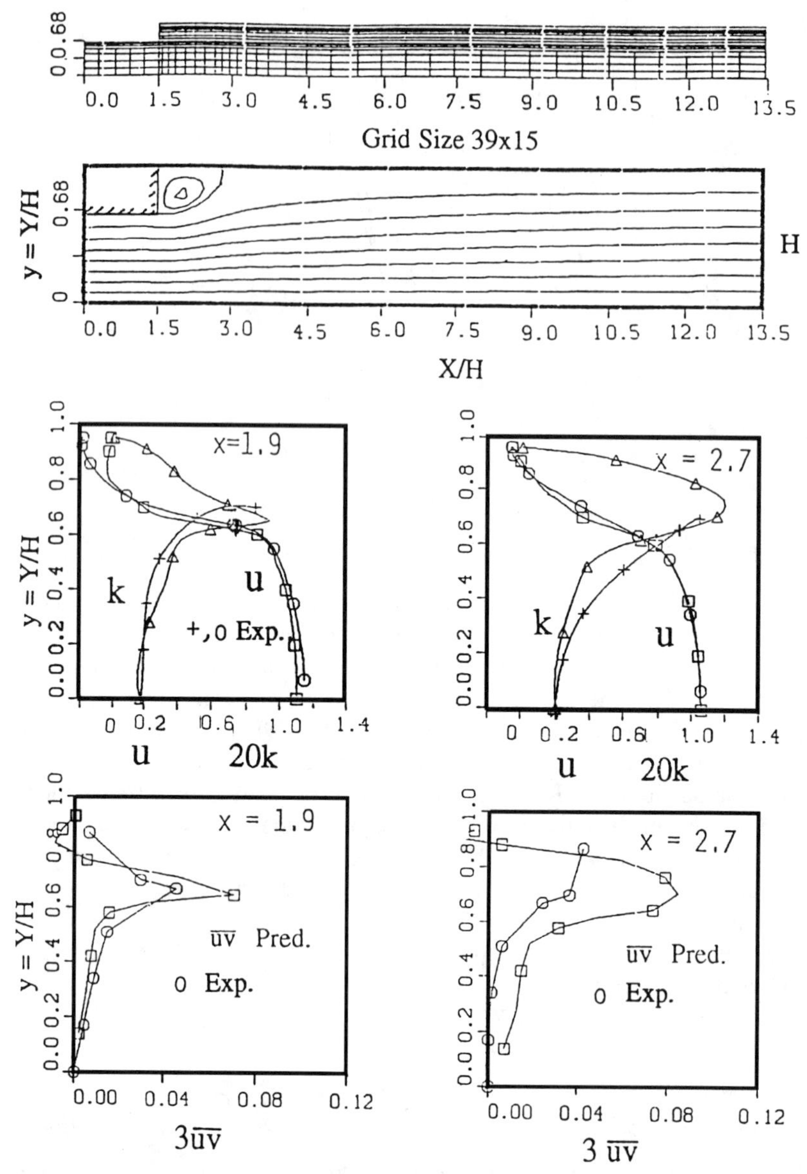

Figure 5.4 Predicted U, k, and $\overline{uv}$ profiles (k-ϵ-A model). (Adapted from Chang (1984).)

The k-ϵ-A model with Rodi's hypothesis is

$$-\overline{u_i u_j} = \frac{C_2 - 1}{C_1}\frac{k}{\epsilon}\left[\frac{P_{ij} - \frac{2}{3}\delta_{ij}P_k\frac{C_2}{C_2-1}}{1 + \frac{P_k/(\epsilon-1)}{C_1}}\right] - \frac{2}{3}\delta_{ij}k\left[\frac{C_1 - 1}{\frac{C_1+P_k}{\epsilon-1}}\right]. \tag{5.30}$$

Chang (1984)[11] repeated the calculations by using Rodi's hypothesis for the expansion channel flow. The improvements can be observed in Figs. 5.5 and 5.6.

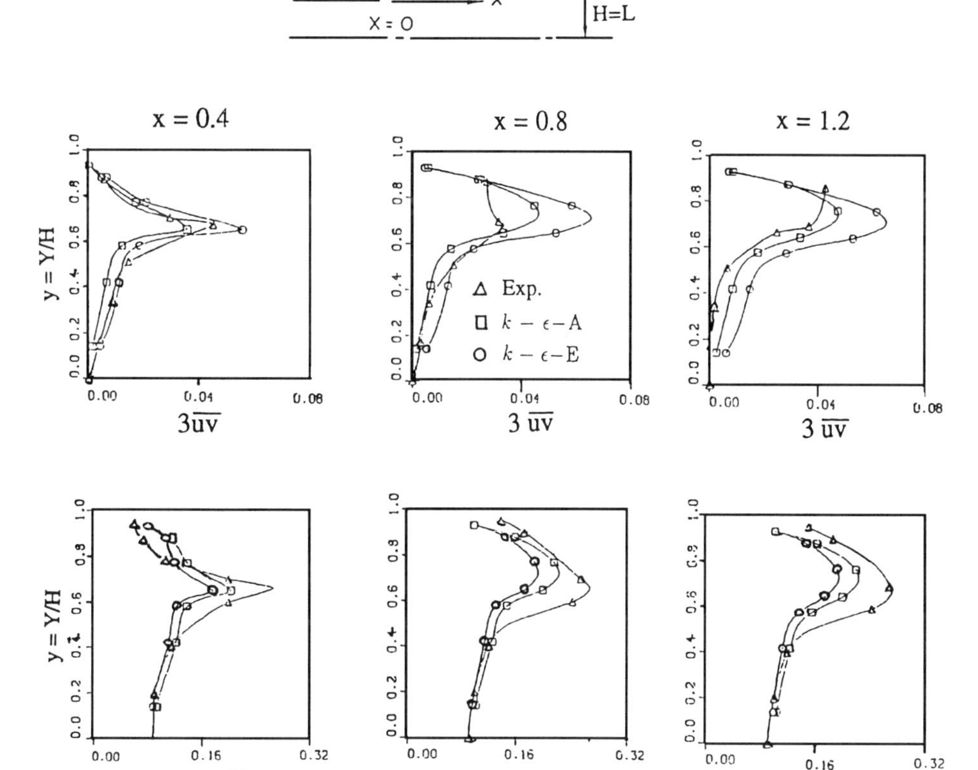

Figure 5.5 Predicted $\overline{uv}$ and $\overline{uu}$ profiles (k-ϵ-A model). (Adapted from Chang (1984).)

5.2.2 Axisymmetric Pipe Expansion Flow

Consider a pipe with sudden expansion as shown in Fig. 5.7. The prediction of Ha Minh and Chassaing (1979)[109] near separation is good, but there are some differences in the prediction of downstream behavior.

Chen, Yoon, and Yu (1983)[23] used the finite analytic numerical method to calculate a case of pipe expansion flow. They considered the geometry shown in Fig. 5.8, with $Re = 280{,}000$. Comparison between prediction of separation length and that of the experiments is satisfactory. The results in Figs. 5.9–5.13 were obtained by using the k-ϵ-E model. Prediction of the centerline decay is also satisfactory, as shown in Fig. 5.11. Note that the dimensionless value of ν_t is defined as

$$\nu_t = \frac{C_\mu}{\nu}\frac{k^2}{\epsilon}.$$

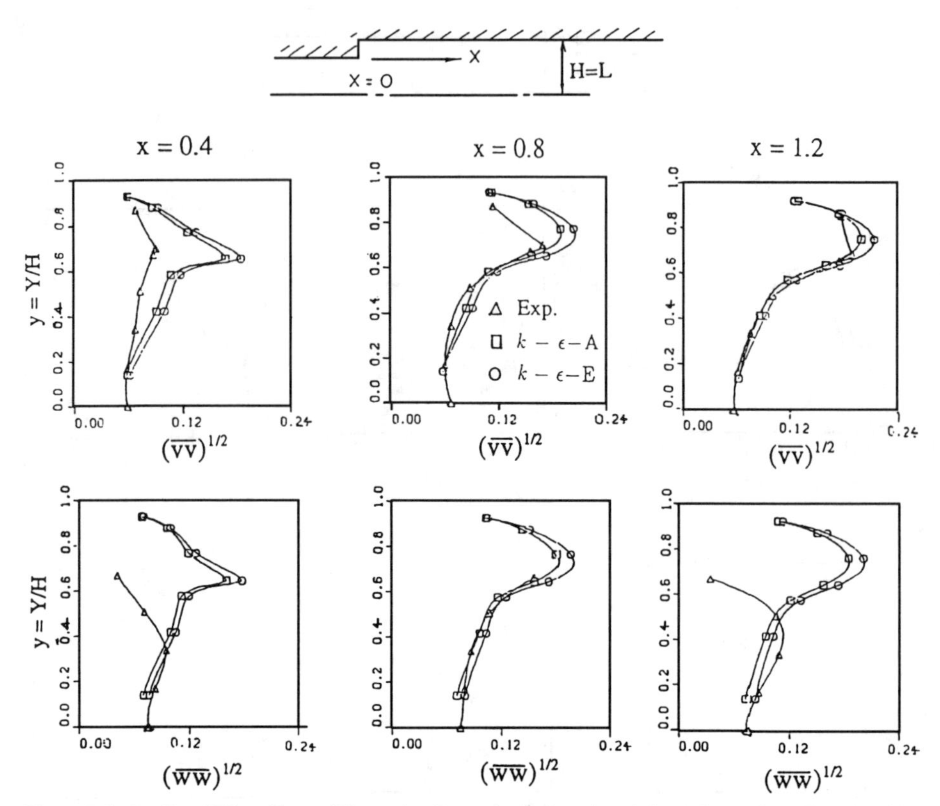

Figure 5.6 Predicted $\overline{uv}$ profiles at different locations of x/L (k-ϵ-A model). (Adapted from Chang (1984).)

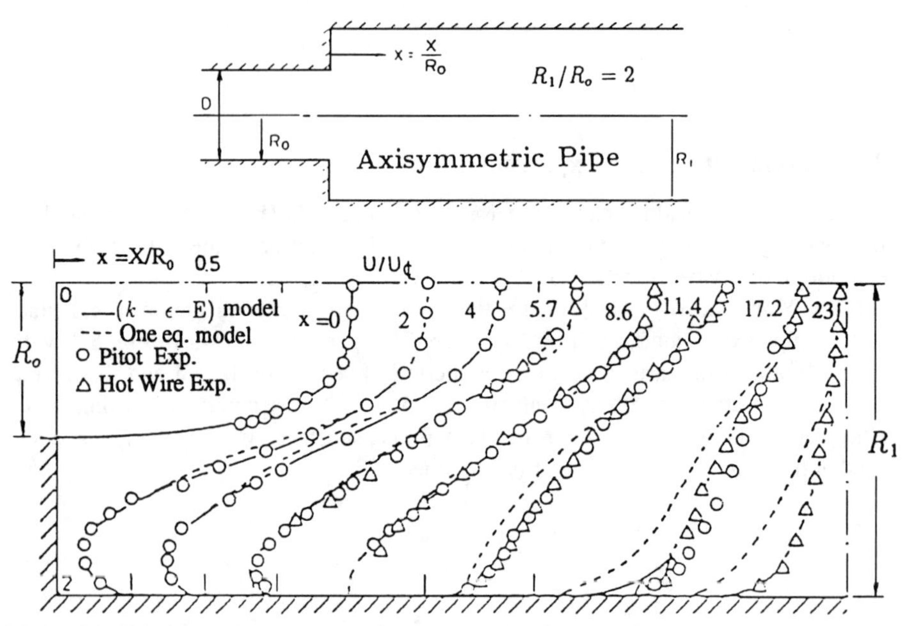

Figure 5.7 Velocity development in a sudden pipe expansion flow with $R_1/R_o = 2$ using (k-ϵ-E) model. (Adapted from Minh and Chassaing (1979).)

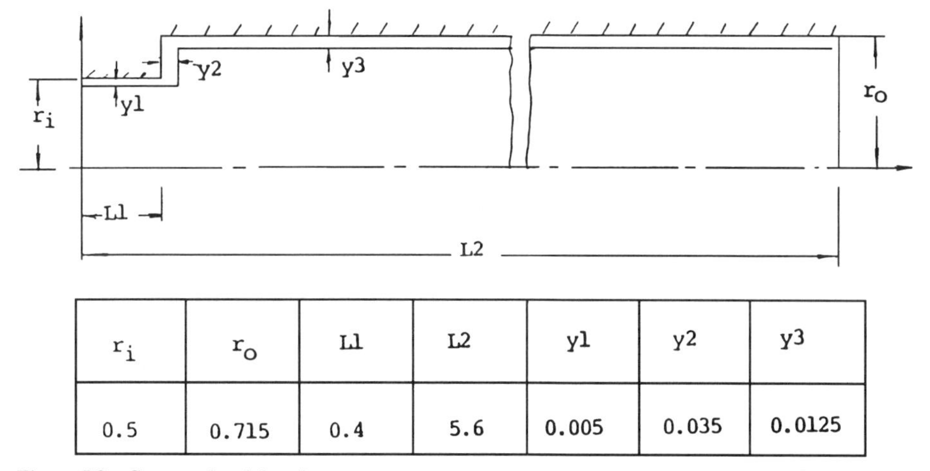

ri	ro	L1	L2	y1	y2	y3
0.5	0.715	0.4	5.6	0.005	0.035	0.0125

Figure 5.8 Computational domain.

From the prediction, it gives ν_t of the order of $10^2 \sim 10^3$. The other variables are defined as

$$
\begin{aligned}
r &= \frac{R}{D} \\
\overline{\acute{u}\acute{v}} &= \frac{\overline{uv}}{U_m^2} \\
x &= \frac{X}{D} \\
k &= \frac{\overline{u^2} + \overline{v^2} + \overline{w^2}}{2U_m^2} \\
P &= \frac{P - P_{\text{exit}}}{\rho U_m^2}.
\end{aligned}
$$

5.2.3 Flow Past a Square Obstacle in a Channel

Durst and Rastogi (1980)[44] calculated flow past a square obstacle in a channel, as shown in Fig. 5.14(a), and gave the velocity profiles, as shown in Fig. 5.14(b). It is observed that slightly incorrect velocity profiles were predicted at the end of recirculation (i.e., $x/H = 0.5$). It seems that there is not enough diffusion in the k-ϵ-E model.

5.2.4 Flow Past Pipe Orifice

Chen et al. (1983)[23] computed the flow past a pipe orifice using the k-ϵ-E model with the Jones and Launder low Reynolds number modification. There is no experimental data available for comparison with the prediction. The results of the calculations are shown in Figs. 5.15 and 5.16. The calculations were carried out by using the FA method with primitive variables and staggered grids. Note that there is a similiarity in contours for k,

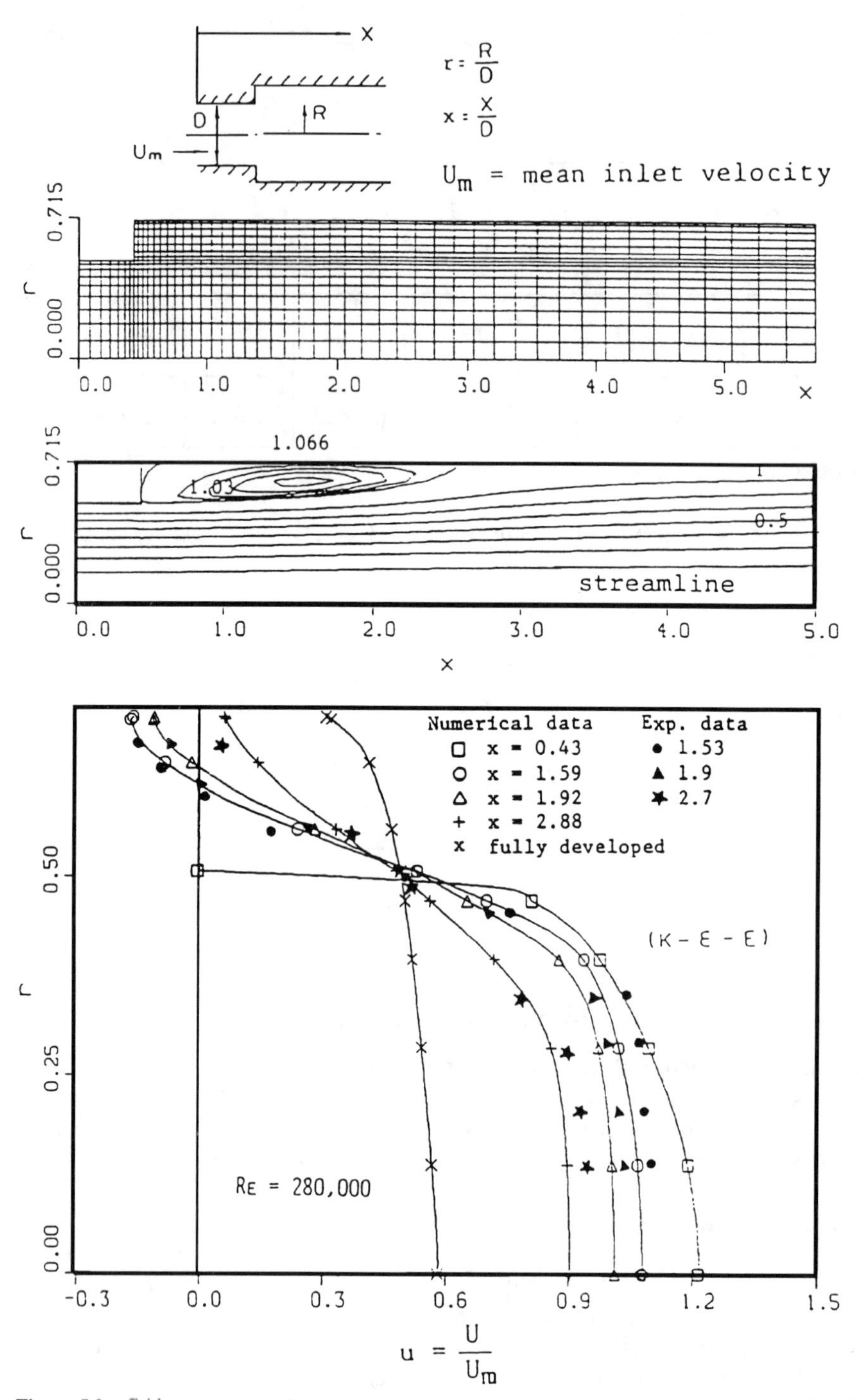

Figure 5.9 Grid system, streamline contours, and velocity profiles. (Adapted from Chen et al. (1983).)

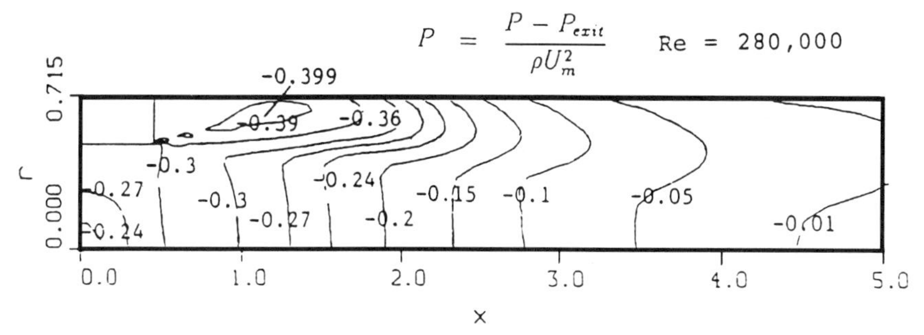

Figure 5.10 Pressure distribution. (Adapted from Chen et al. (1983).)

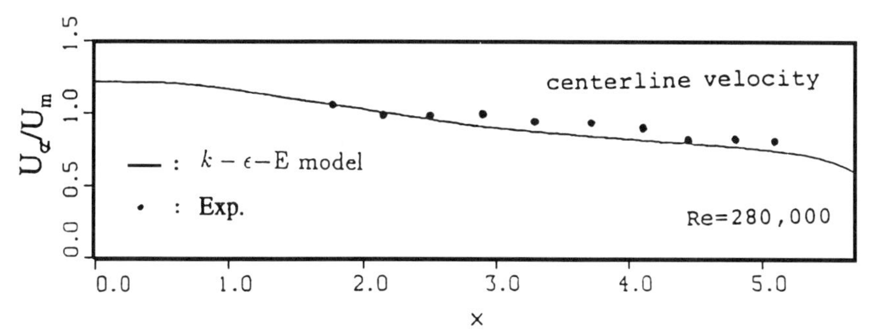

Figure 5.11 Decay of centerline velocity. (Adapted from Chen et al. (1983).)

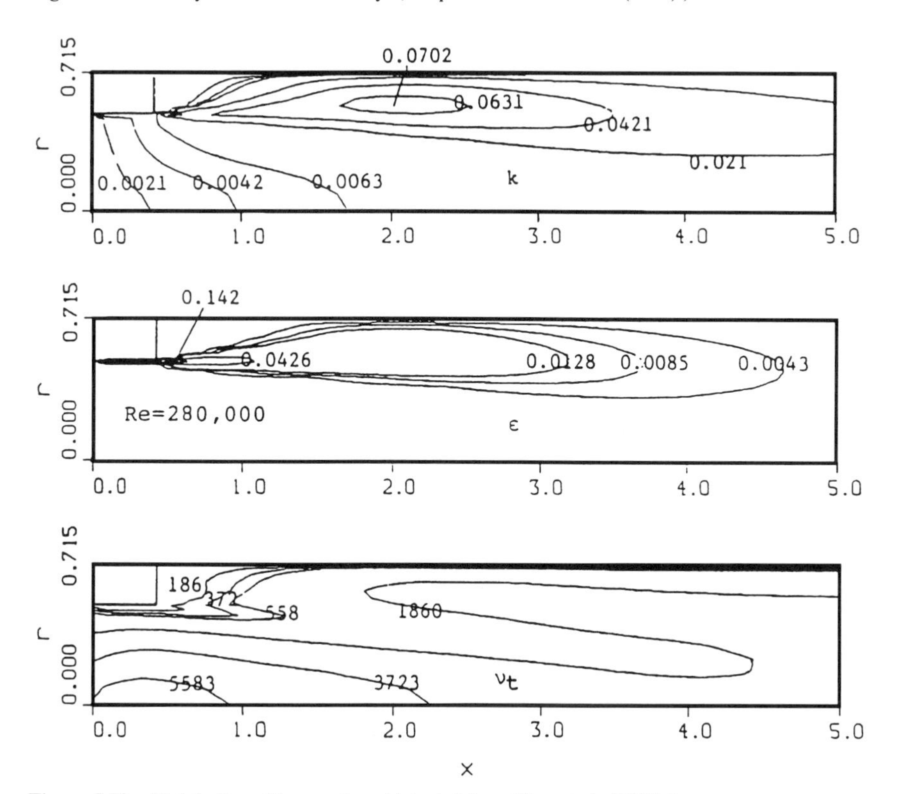

Figure 5.12 Distribution of k, ϵ, and ν_t. (Adapted from Chen et al. (1983).)

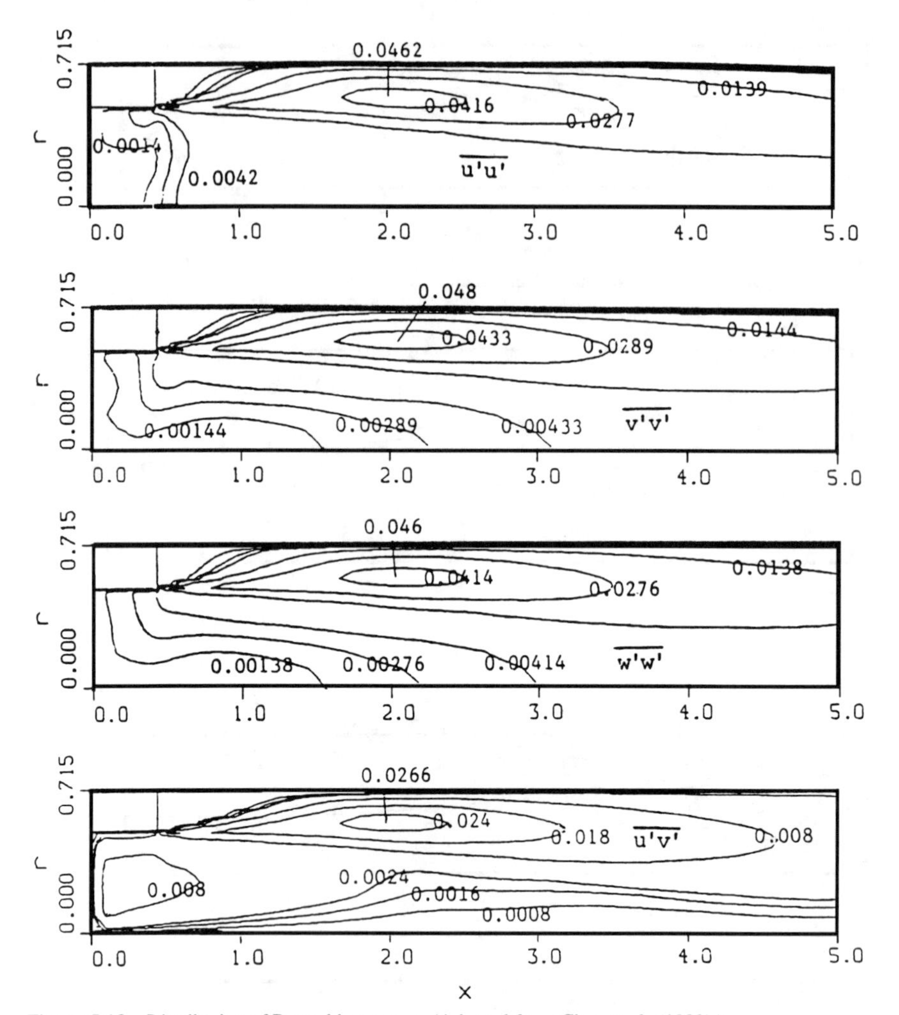

Figure 5.13 Distribution of Reynolds stresses. (Adapted from Chen et al. (1983).)

$\overline{u^2}$, $\overline{v^2}$, and $\overline{w^2}$, where

$$k = \frac{\overline{u^2} + \overline{v^2} + \overline{w^2}}{2U_m^2}$$

$$\overline{u_i' u_i'} = \frac{\overline{u_i^2}}{U_m^2}$$

$$\epsilon = \frac{\nu}{Re} \overline{\frac{\partial u_i}{\partial X_l} \frac{\partial u_i}{\partial X_l}}$$

$$Re = \frac{U_m D}{\nu}$$

$$\nu_t = \frac{C_\mu k^2}{\epsilon \nu},$$

ν_t for $Re = 2 \times 10^5$ is of the order of $10^2 \sim 10^3$.

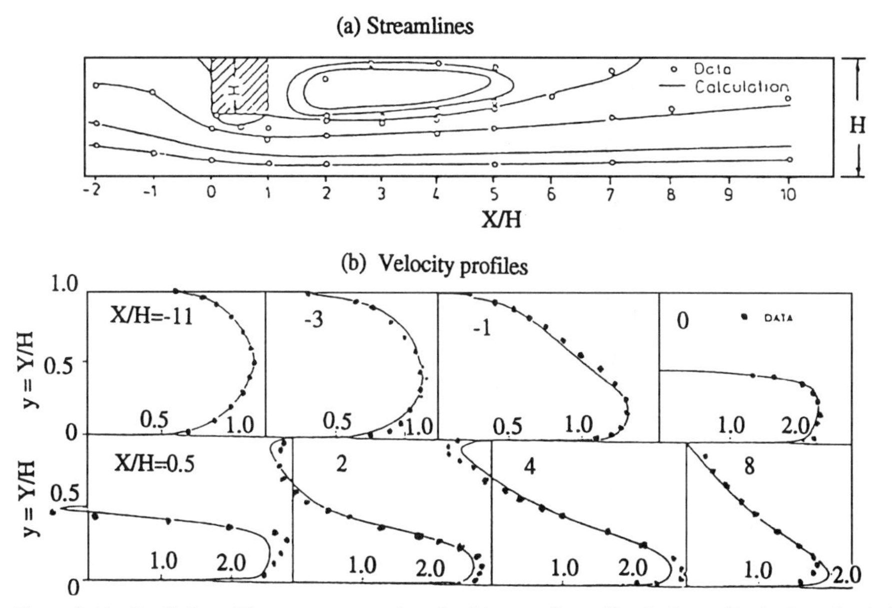

Figure 5.14 Prediction of flow past a square obstacle. (a) streamlines, (b) velocity profiles in the recirculating region, and (c) velocity profiles in the redevelopment region. (Adapted from Durst and Rastogi (1980).)

5.2.5 Flow Past Channel Cavity

The flow past channel cavity was measured by Girard and Curlet (1975)[52] and Roshko (1955)[141]. Ideriah (1977)[66] considered the two cases when the gap between the sliding plate and the cavity equals zero and when the gap has a finite length. Figure 5.17 shows the prediction of the two configurations.

The flow and the turbulence fields were predicted as shown in Fig. 5.18. The velocity vectors and the kinetic energy contours are shown in Fig. 5.17 for the two setups. Figure 5.18 shows the velocity profile at the cavity centerline, turbulence energy across midplane, and the pressure distribution along the cavity walls.

5.2.6 Flow Past a Pipe Cavity

Chen et al. (1983)[23] considered the pipe cavity as shown in Fig. 5.19. The two turbulent flows, one with $Re = 3 \times 10^5$ and the other with $Re = 6 \times 10^4$, were found to be almost independent of the Reynolds number. Plots of turbulent transport properties are given in Fig. 5.20. Plots of temperature and heat transfer distribution are given in Figs. 5.21 and 5.22, where Nu is the Nusselt number, which relates the heat transfer that is due to convection and is given by $Nu = hD/k$, and the amount of heat energy transferred is given by $Q = Ah(T_i - T_w)$. Here, h is the convective heat transfer coefficient, k is the thermal conductivity, D is the pipe diameter, A is the surface area, and the subscripts i and w indicate inlet and wall conditions, respectively.

5.2.7 Flows in Rectangular Cavity with k-ϵ-A and k-ϵ-E Models

Chen and Chang (1985)[14] obtained the FA numerical solutions with the k-ϵ-E and k-ϵ-A models for turbulent internal recirculating flows in a rectangular cavity, as shown in

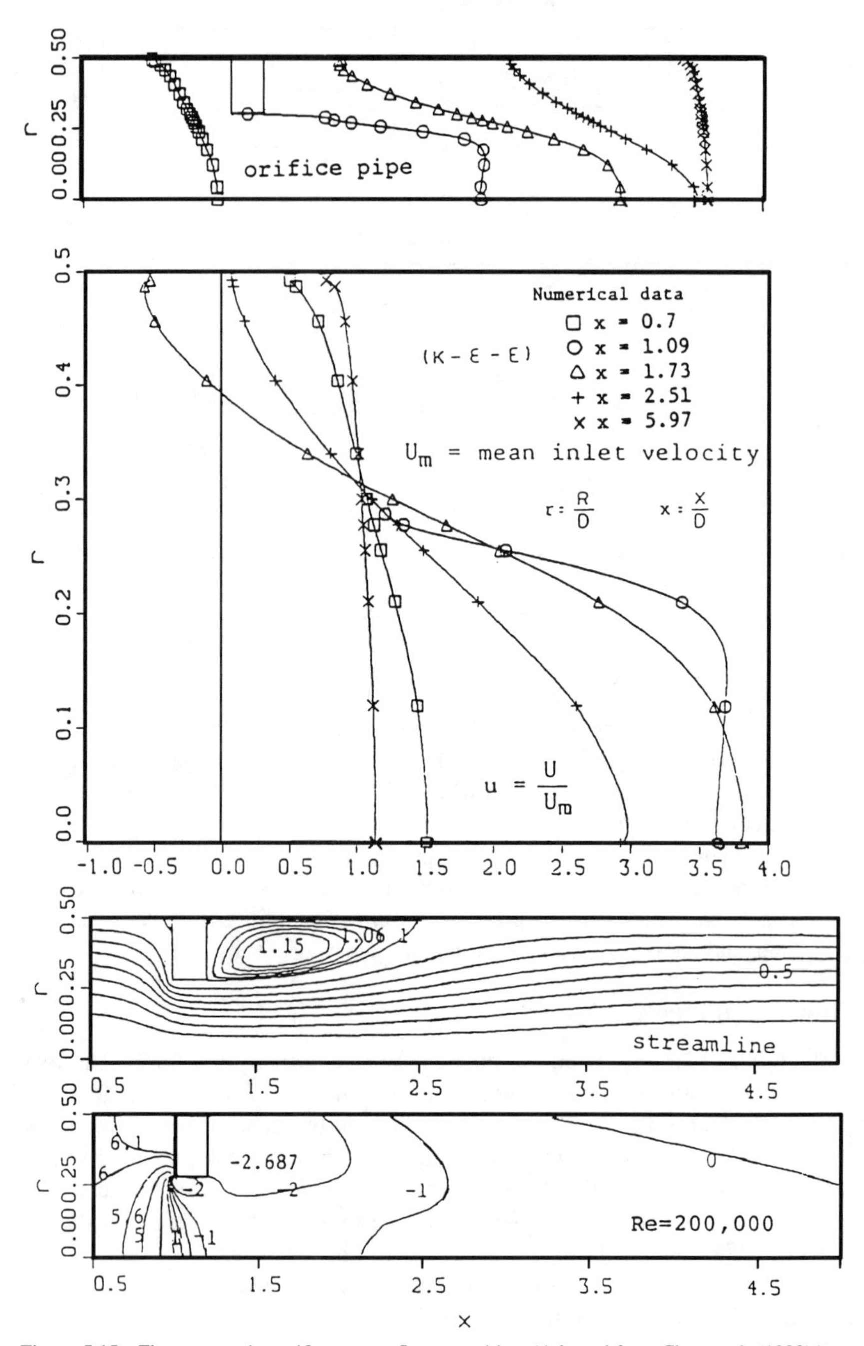

Figure 5.15 Flow past a pipe orifice: mean-flow quantities. (Adapted from Chen et al. (1983).)

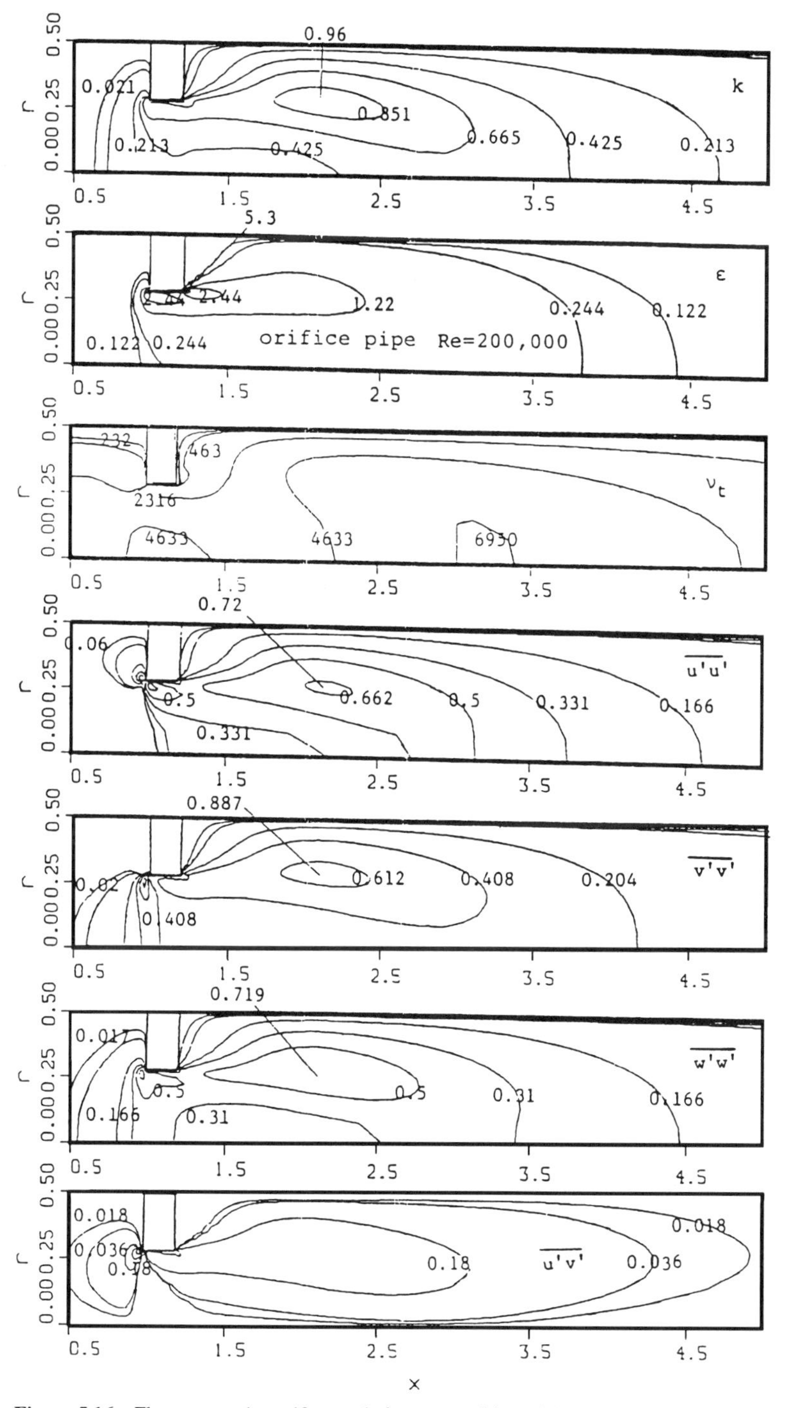

Figure 5.16 Flow past a pipe orifice: turbulence quantities. (Adapted from Chen et al. (1983).)

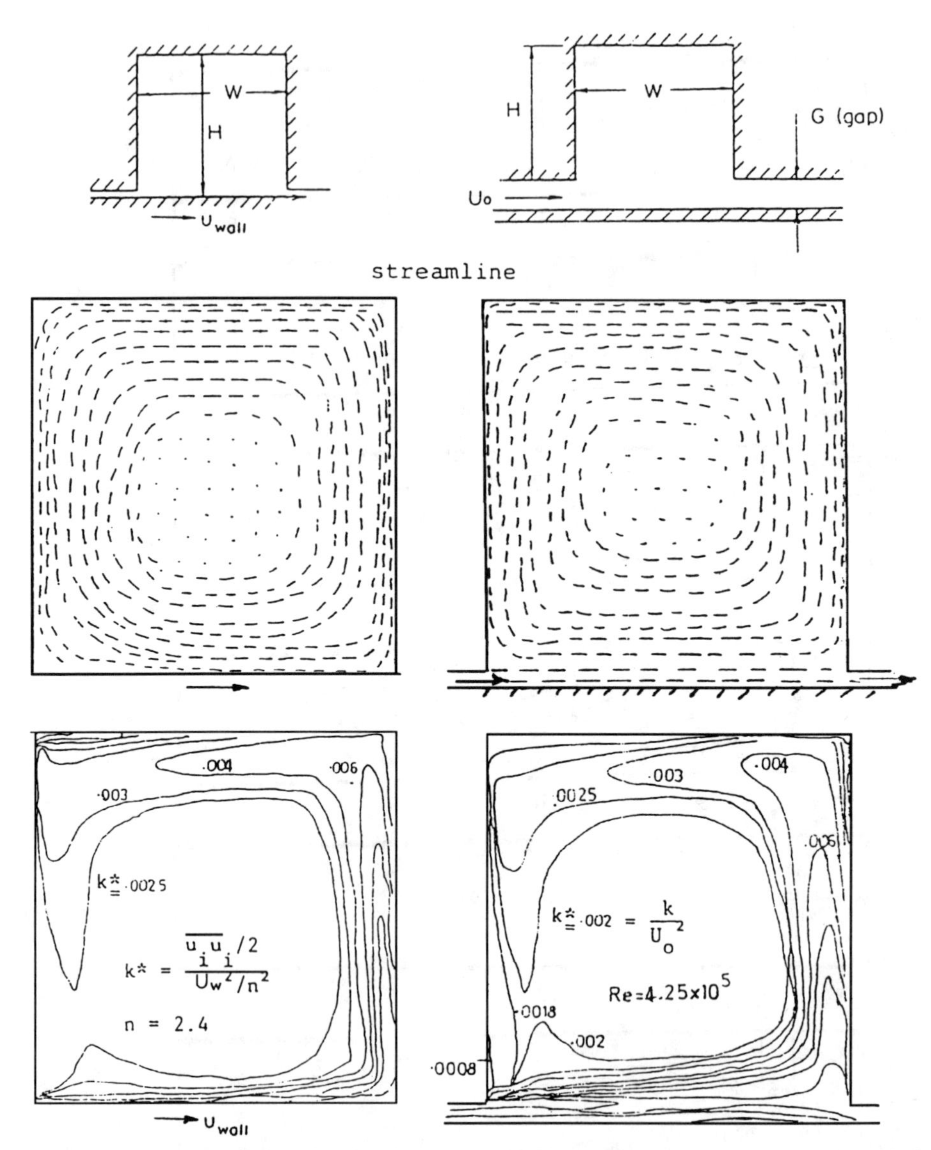

Figure 5.17 Flow past a channel cavity: Velocity and kinetic energy predictions. (Adapted from Iderih (1977).)

Fig. 5.23a. Predictions are made for a square cavity with a Reynolds number of 4.8×10^5 and a rectangular cavity with an aspect ratio of three and a Reynolds number of 2.0×10^5. The predictions were modeled with the k-ϵ-E and k-ϵ-A turbulence models. The time averaged Navier–Stokes equations for incompressible fluid with constant viscosity are

- the continuity equation,

$$\frac{\partial U_i}{\partial X_i} = 0, \tag{5.31}$$

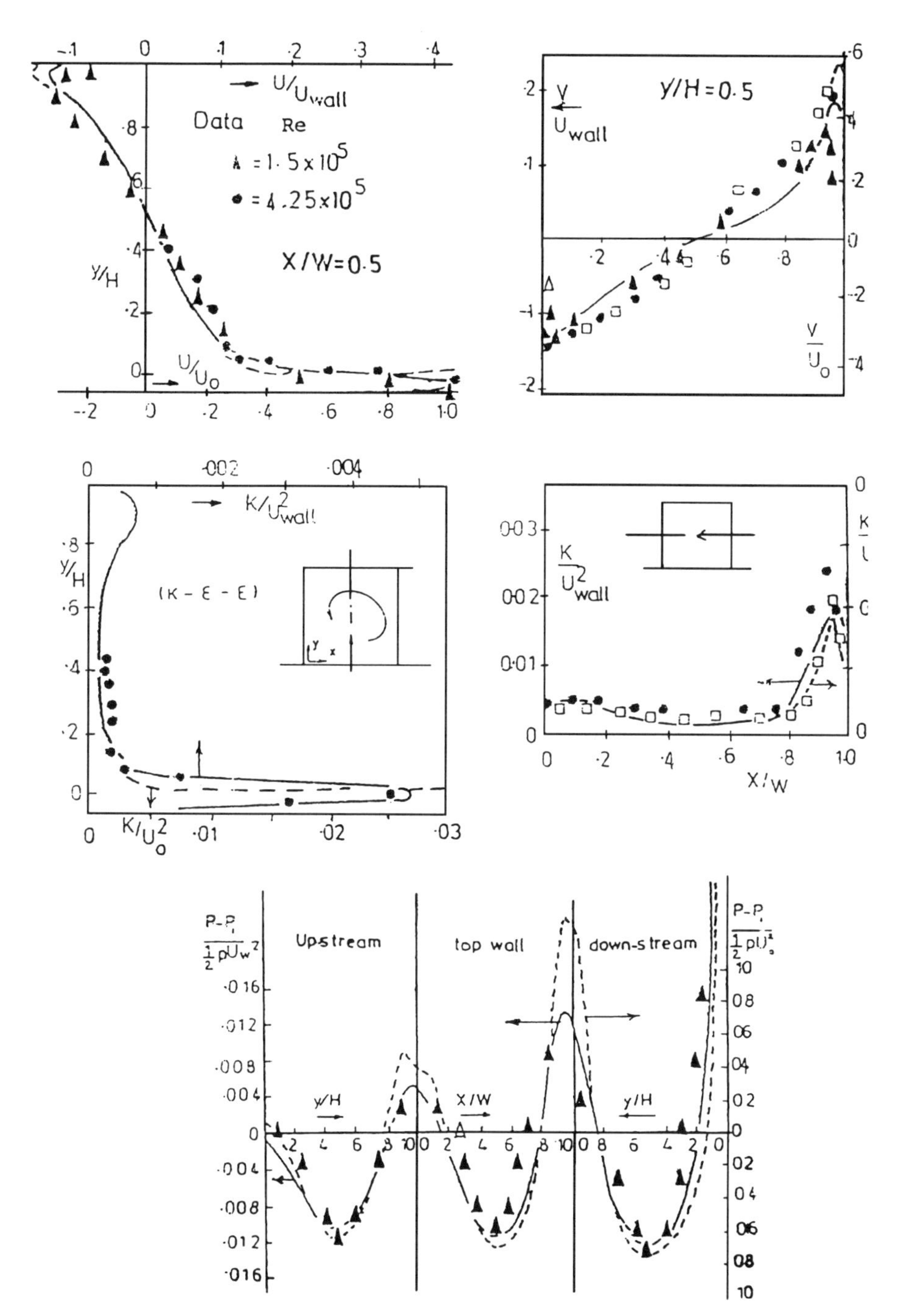

Figure 5.18 Flow past a channel cavity: Flow and turbulence fields predictions. (Adapted from Iderih (1977).)

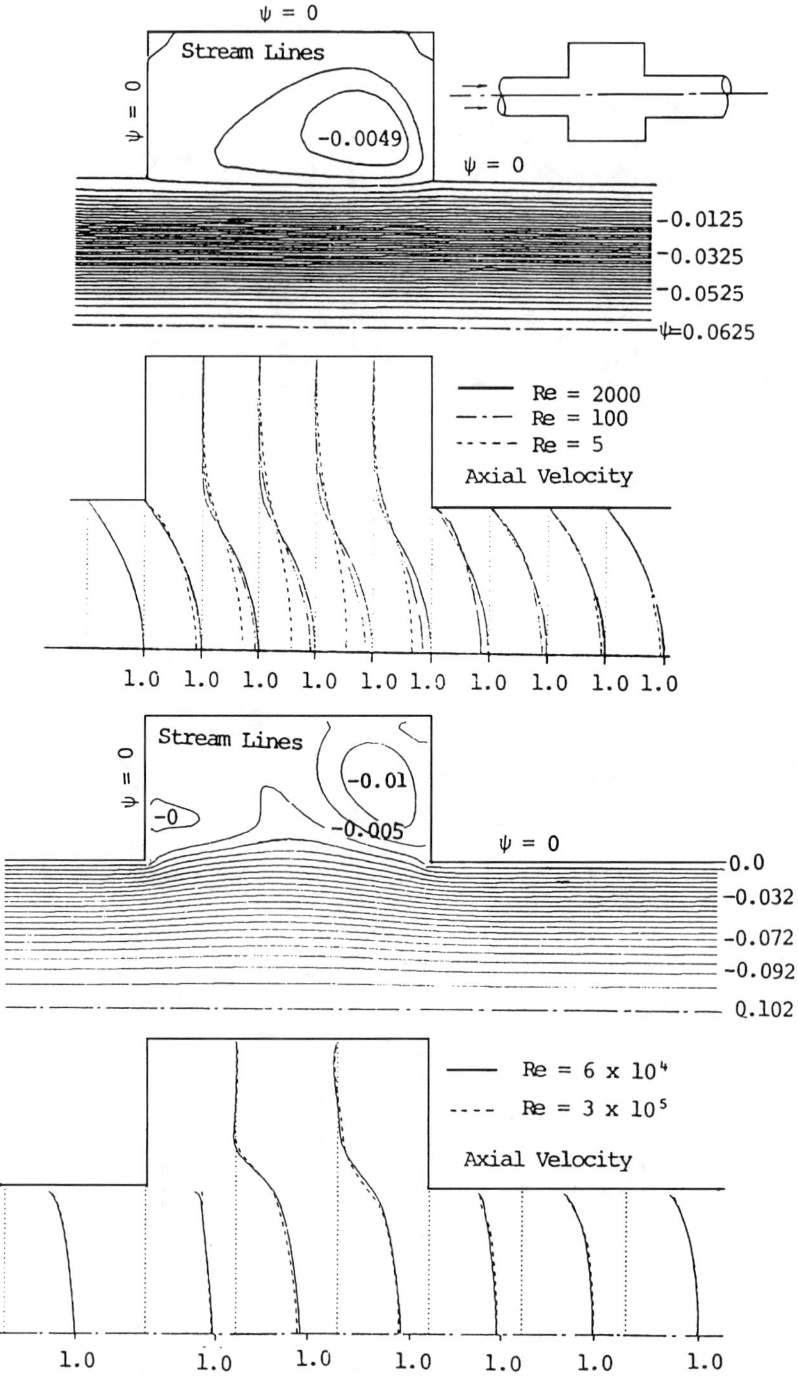

Figure 5.19 Stream function and velocity distribution. (Adapted from Chen et al. (1983).)

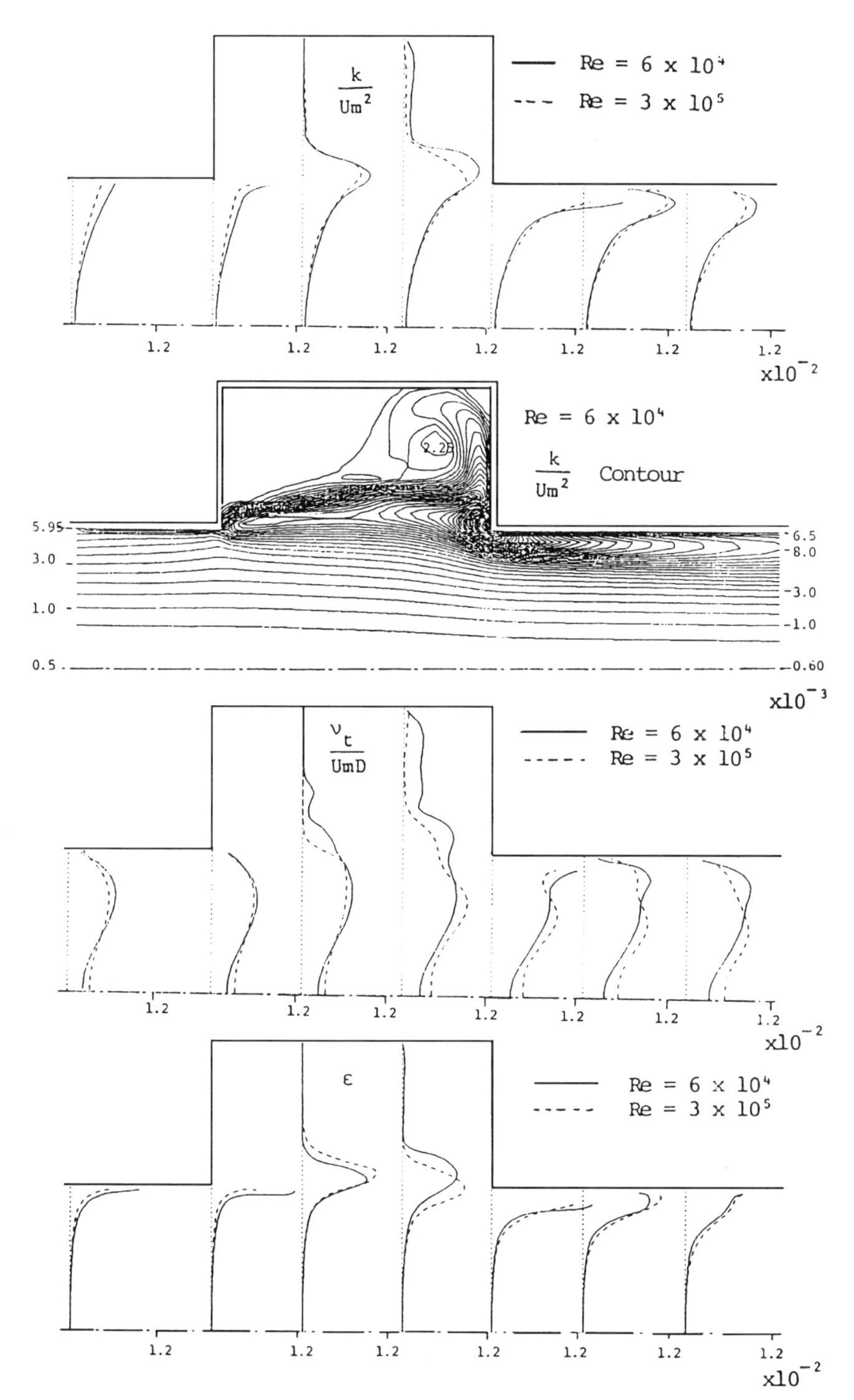

Figure 5.20 Turbulent transport properties distribution. (Adapted from Chen et al. (1983).)

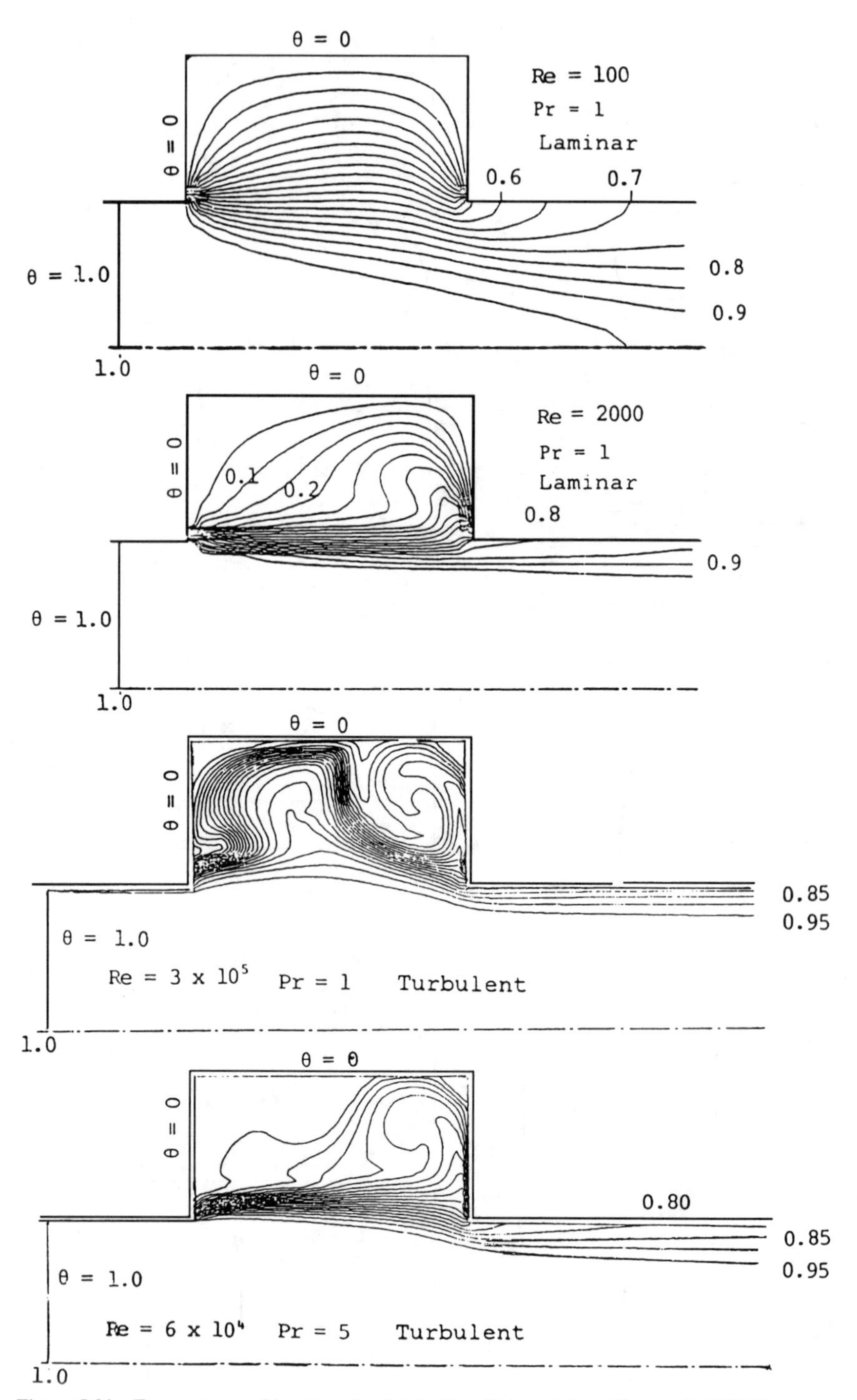

Figure 5.21 Temperature and heat transfer distribution. (Adapted from Chen et al. (1983).)

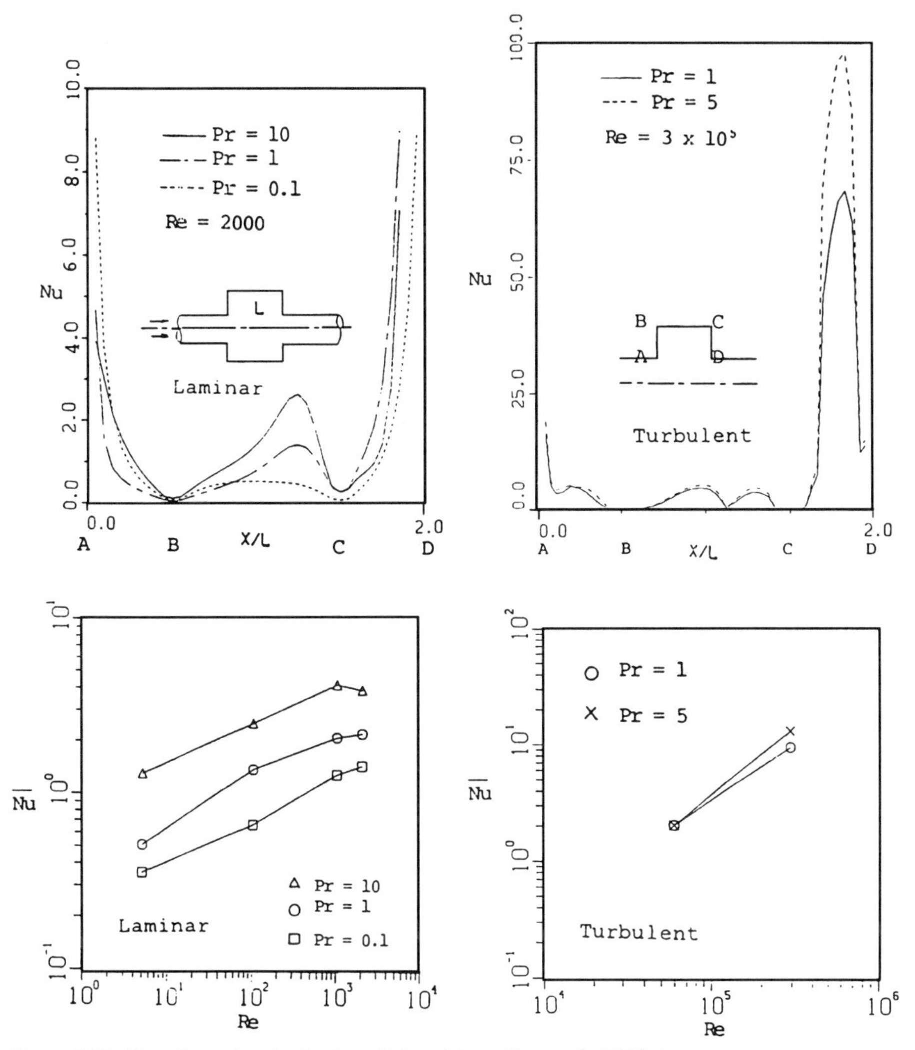

Figure 5.22 Nusselt number distribution. (Adapted from Chen et al. (1983).)

- the momentum equations,

$$\frac{\partial U_i}{\partial t} + U_l \frac{\partial U_i}{\partial X_l} = -\frac{1}{\rho}\frac{\partial p}{\partial X_i} + \nu_l \frac{\partial^2 U_i}{\partial X_l \partial X_l} - \frac{\partial \overline{u_i u_l}}{\partial X_l} \tag{5.32}$$

- for the Reynolds stresses, the algebraic stress (k-ϵ-A) model gives

$$\overline{u_i u_j} = k \left[\frac{\frac{2}{3}(C_1 - 1)\epsilon\delta_{ij} + (1 - C_2) P_{ij} + \frac{2}{3} C_2 P_k \delta_{ij}}{C_1 \epsilon + P_k - \epsilon} \right], \tag{5.33}$$

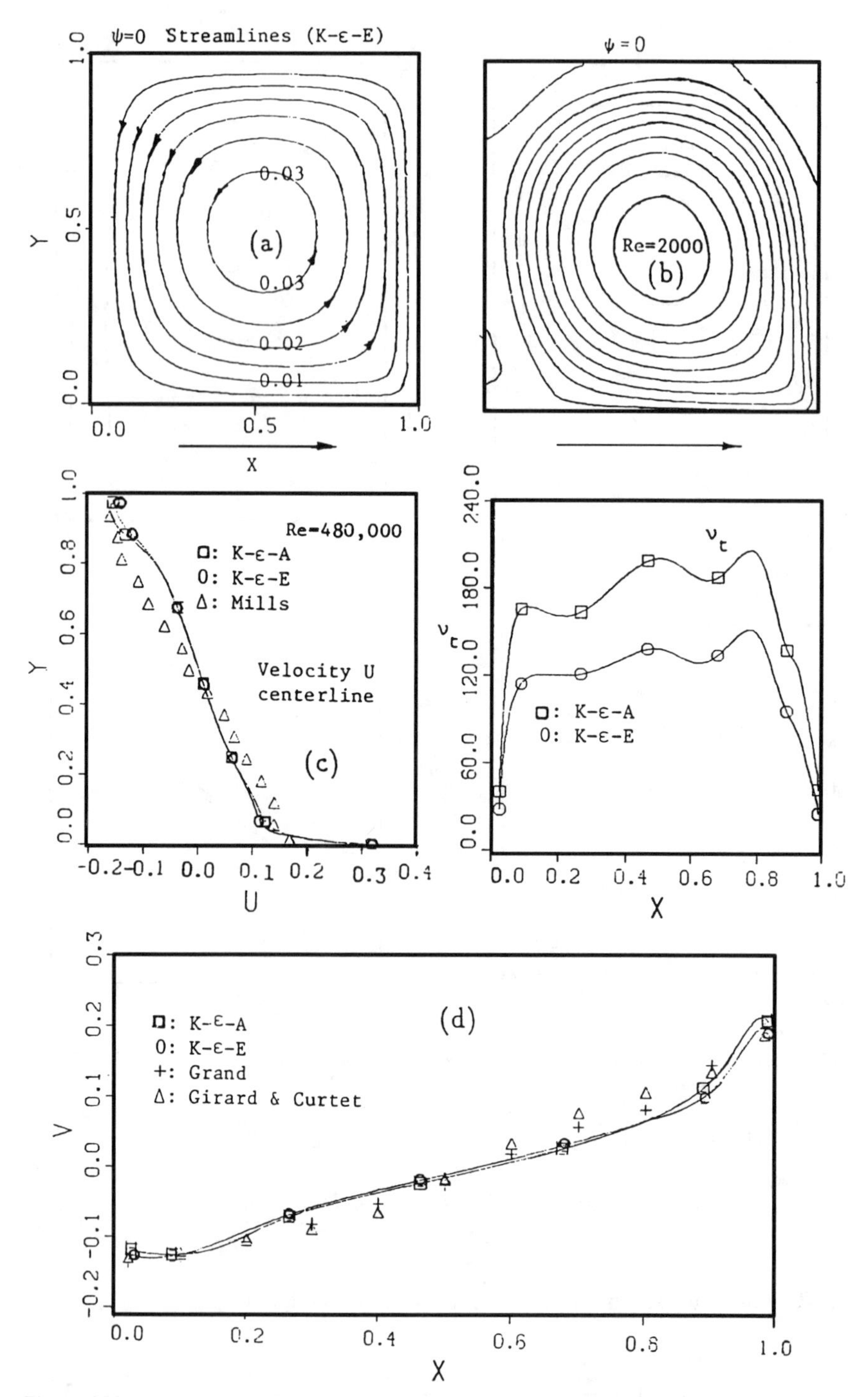

Figure 5.23 Prediction of square cavity flow. (Adapted from Chen and Chang (1985).)

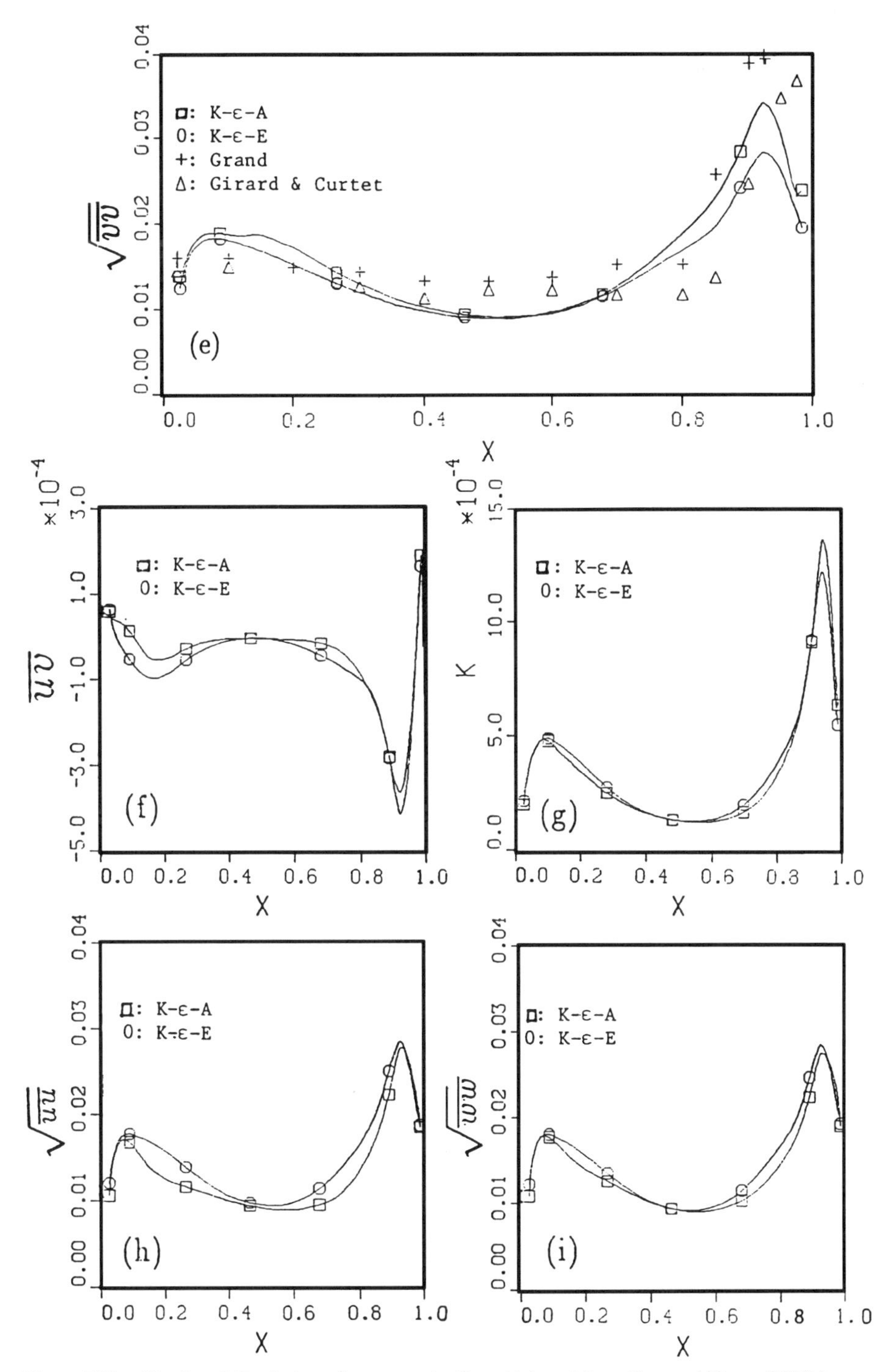

Figure 5.23 (Continued) Prediction of square cavity flow. (Adapted from Chen and Chang (1985).)

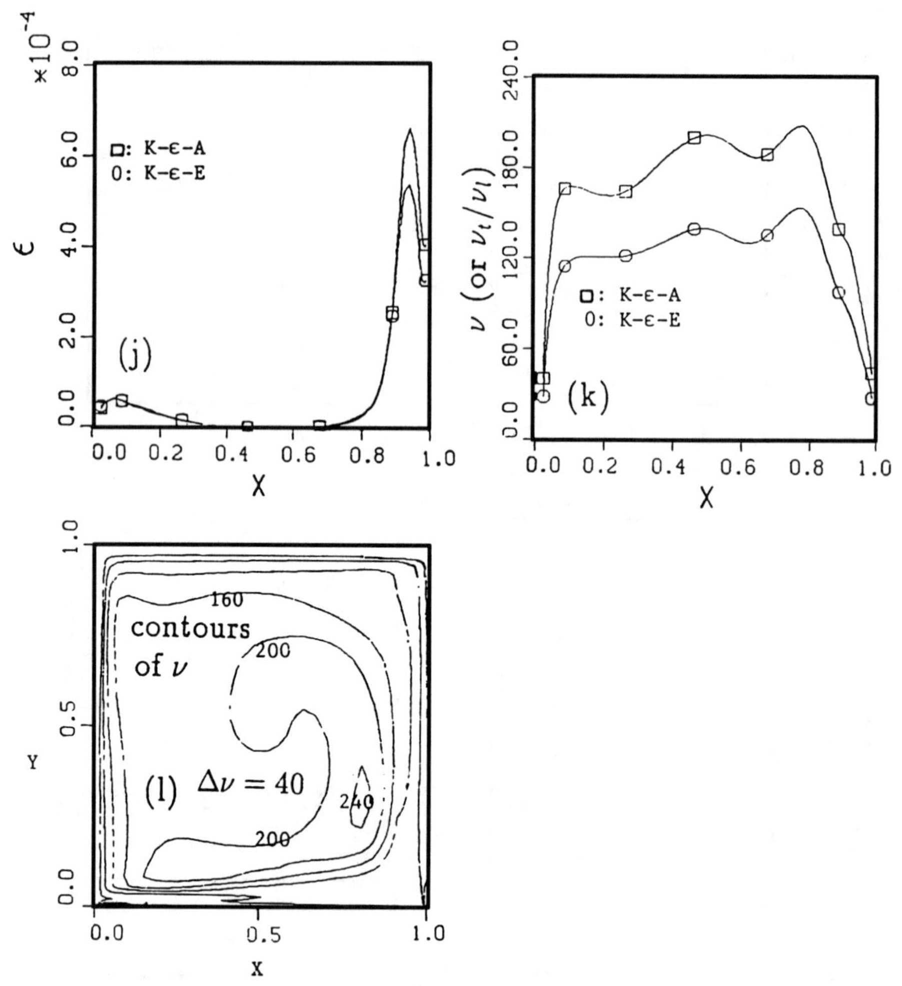

Figure 5.23 (Continued) Prediction of square cavity flow. (Adapted from Chen and Chang (1985).)

where

$$P_{ij} = -\left(\overline{u_i u_m}\frac{\partial U_i}{\partial X_m} + \overline{u_j u_m}\frac{\partial U_i}{\partial X_m}\right) \tag{5.34}$$

$$P_k = -\overline{u_n u_m}\frac{\partial U_n}{\partial X_m} \tag{5.35}$$

and the eddy viscosity (k-ϵ-E) model gives

$$-\overline{u_i u_j} = \nu_t\left(\frac{\partial U_i}{\partial X_j} + \frac{\partial U_j}{\partial X_i}\right) - \frac{2}{3}\delta_{ij}k, \tag{5.36}$$

where

$$\nu_t = C_\mu \frac{k^2}{\epsilon}.$$

- The kinetic energy equation is

$$\frac{Dk}{Dt} = \frac{\partial}{\partial X_l}\left(C_k \frac{k^2}{\epsilon}\frac{\partial k}{\partial Xl} + \nu_l \frac{\partial k}{\partial Xl}\right) - \overline{u_i u_l}\frac{\partial U_i}{\partial X_l} - \epsilon. \tag{5.37}$$

- The ϵ equation is

$$\frac{D\epsilon}{Dt} = \frac{\partial}{\partial X_l}\left(C_\epsilon \frac{k^2}{\epsilon}\frac{\partial \epsilon}{\partial Xl} + \nu_l \frac{\partial \epsilon}{\partial Xl}\right) - C_{\epsilon 1}\left(\frac{\epsilon}{k}\right)\overline{u_i u_l}\frac{\partial U_i}{\partial X_l} - C_{\epsilon 2}\left(\frac{\epsilon}{k}\right)\epsilon. \tag{5.38}$$

The various constants are $C_1 = 2.80$, $C_2 = 0.40$, $C_{\epsilon 1} = 1.43$, $C_{\epsilon 2} = 1.92$, $C_k = 0.09$, $C_\epsilon = 0.07$, and $C_\mu = 0.09$.

In order to obtain a unique solution for cavity flow, boundary conditions must be specified to make this problem well posed. In this study, the wall function approach is used to specify the boundary conditions. On the stationary boundary, the following equations are used

$$U_p^+ = \frac{U_p}{U_\tau} = \frac{1}{\kappa}\ln\left(Ey_p^+\right) \tag{5.39}$$

$$k_p^+ = \frac{1}{\sqrt{C_\mu}} \tag{5.40}$$

$$\epsilon_p^+ = \frac{1}{\kappa y_p^+}. \tag{5.41}$$

The subscript p stands for the values evaluated at one nodal point away from the wall. κ and E are the von Kármán and log-law constants and are 0.41 and 9.0, respectively. The above-indicated equations are valid for $y_p^+ > 10$. When $y_p^+ < 10$, $U_p^+ = y_p^+$, $k_p^+ = 0.1(y_p^+)^2$, and $\epsilon_p^+ = 0.2$ are used. Here

$$y_p^+ = \frac{y_p U_\tau}{\nu_l}$$

and

$$U_\tau = \sqrt{\nu_l \frac{\partial U}{\partial n}\bigg|_{\text{wall}}}.$$

U_τ is the friction velocity, and ν_l is the molecular viscosity. Because the bottom wall moves with a velocity U_w, the velocity wall function for this boundary should be

$$U_p = U_w - \frac{U_\tau}{\kappa}\ln\left(Ey_p^+\right). \tag{5.42}$$

Turbulent flow in a square cavity. FA solutions obtained for flow in a square cavity with a Reynolds number of 4.8×10^5 are given below. The cavity width w and the sliding wall velocity U_w are used to normalize the results and data.

Figure 5.23(a) is the predicted results of streamlines with an equal interval of $\Delta\psi = 0.005$. The first streamline with $\Delta\psi = 0.005$ is approximately square because of the influence of the square boundary geometry, whereas the inner streamline with $\psi = 0.03$ is almost a circle. Contrary to the laminar flow solution obtained by Chen et al. (1981)[17], shown in Fig. 5.23(b), no separation bubbles near the corners are detected for turbulent flow. The turbulent vortex center ($x = 0.51$, $y = 0.5$; coordinates are normalized by the cavity width w) is nearly coincident with the geometric center of the cavity. In turbulent flow, the stronger mixing is provided by the turbulent eddy motion. Thus, the momentum diffusion is much easier to diffuse the fluid toward cavity corners. Consequently, there is no separation.

Figure 5.23(c) is a plot of the mean-velocity U on the vertical centerline, whereas Fig. 5.23(d) shows the mean-velocity V on the horizontal centerline. The line with $\circ$ is the prediction of the k-ϵ-E model and that with $\square$ it is the prediction of the k-ϵ-A model. The experimental data are denoted by $+$ and Δ. In general, the prediction agrees very well with the experimental data. The prediction of U and V by the k-ϵ-A model is almost identical to those of the k-ϵ-E model. In the central region, both models slightly underpredict the magnitude of the mean velocity. It is noted that Grand's (1975)[53] data are closer to the predicted results than the data of Girard and Curlet (1975)[52] in the central region. From Figs. 5.23(c) and (d), it is observed that the velocity distributions of U and V in the central region are approximately linear. This implies that the fluid is rotating around a centerlike solid body.

Figure 5.23(e) shows the distribution of the Reynolds normal stress $\overline{vv}$ on the horizontal centerline. It is seen that the stress is largest near the right wall where the fluid just leaves the driven wall and turns upward. The stress is also predicted to be larger ($\overline{vv}^{1/2}$ is about 0.02) on the left wall than that at the center, but it is only about two thirds as large as that at the right wall ($\overline{vv}^{1/2}$ is about 0.03). The reason that larger stress occurs near the wall is because the mean-velocity gradient near the wall region is larger, as shown in Fig. 5.23(d). Thus, it is expected to produce more turbulence near the wall. In general, the k-ϵ-A model predicts larger $\overline{vv}$ ($\overline{vv}^{1/2}|_{\max} = 0.035$ than the k-ϵ-E model ($\overline{vv}^{1/2}|_{\max} = 0.028$). Both experimental data of Grand (1975)[53] and Girard and Curlet (1975)[52] showed that only near the right wall are there larger $\overline{vv}$ values ($\overline{vv}^{1/2}|_{\max} = 0.04$) and that $\overline{vv}$ values are almost the same in the rest region ($\overline{vv}^{1/2} = 0.015$). The maximum Reynolds stress $\overline{vv}$ of both experiments is larger than both numerical predictions. Both numerical results predict the location of $\overline{vv}|_{\max}$ at $X = 0.9$, which agrees with the measurement of Grand [53], whereas the measurement of Girard and Curlet [52] is closer to the wall at about $x = 0.95$.

Figures 5.23(f)–(j) give the shear stress $\overline{uv}$, the turbulent kinetic energy k, the Reynolds stresses $\overline{uu}$ and $\overline{ww}$, and the rate of dissipation of turbulent kinetic energy ϵ, on the horizontal centerline, respectively. Because there are no available experimental data, only numerical results are reported. It is seen that turbulence is stronger near the wall, especially near the upstream wall than near the central region. Physically, turbulence is created by the driving wall at the bottom and convects upward and rotates counterclockwise. As the flow circulates, the turbulent kinetic energy is dissipated and diffused.

Figures 5.23(k) and (l) give the dimensionless viscosity ν (or ν_t/ν_l) on the horizontal centerline and the contours of ν in the cavity with an increment of $\Delta\nu = 40$, respectively. There is an obvious difference between the predicted values of ν by the two models as shown in Fig. 5.23(k). This is because a small change in k or ϵ will greatly influence the value of ν. Inspite of this deviation, the flow patterns predicted by these two models are almost the same.

Turbulent flow in rectangular cavity. Figure 5.24 shows the predictions of the rectangular cavity flow with an aspect ratio of 3 and a Reynolds number of 2×10^5. Figure 5.24(a) shows the streamlines. Only one large vortex is found. The vortex center is located at $X = 0.57$, $Y = 1.24$, which is in good agreement with the experimental result of $X = 0.6$, $Y = 1.4$ given by Normandin (1980)[118].

Chen et al. (1981)[17] solved for a rectangular cavity with an aspect ratio of 3 for Reynolds number of 400 as shown in Figure 5.24(b), and showed that there were three primary vortices with two separation bubbles on the two upper corners in laminar flow. The difference is due to the strong mixing provided by the turbulent eddy stresses. For a larger aspect ratio, turbulent flow near the upper wall may become laminarized. In this situation, the flow may have laminar-like patterns, and multiple cells may occur.

Figure 5.24(c) also shows the result of the mean-velocity U on the vertical centerline. It can be seen that, except near the driving wall, the velocity is relatively small and is of the order of 0.01 in the upper region. The flow exhibits a boundary layerlike flow near the driving wall with thickness of approximately $Y_\delta = 0.1$. Figure 5.24(d) gives the prediction of the mean-velocity V on the horizontal centerline, and this result is compared with Normandin's experimental data (1980)[118]. In general, the comparison is good except that the predicted magnitudes near the wall are slightly smaller than the experimental data. Figure 5.24(e) compares the predicted and measured Reynolds stress $\overline{vv}$. The experimental data ($\overline{vv}^{1/2} = 0.03$) are substantially higher than the predicted values ($\overline{vv}^{1/2} = 0.02$). Figures 5.24(f)–(j) show the shear stress $\overline{uv}$, the turbulent kinetic energy k, the Reynolds stresses $\overline{uu}$ and $\overline{ww}$, and the dissipation function ϵ on the horizontal line, respectively. The prediction of the two models is found to be approximately the same.

5.2.8 Comparison of the k-ϵ Model and RSM

Ushijima et al. (1985)[180] compared the prediction of the k-ϵ model and two RSMs for flows in rectangular plenum, as shown in Fig. 5.25. The agreement of these results is examined, and the properties of these models are compared. The main results are summarized as follows.

1. Concerning the mean-velocity distributions, although a few differences exist, the results of the three models agree with experimental values.
2. It can be found that nonisotropy of normal Reynolds-stress ($\overline{u'^2}$ and $\overline{v'^2}$) distribution is quite well simulated by the two RSMs, but not adequately by the k-ϵ model. Shear

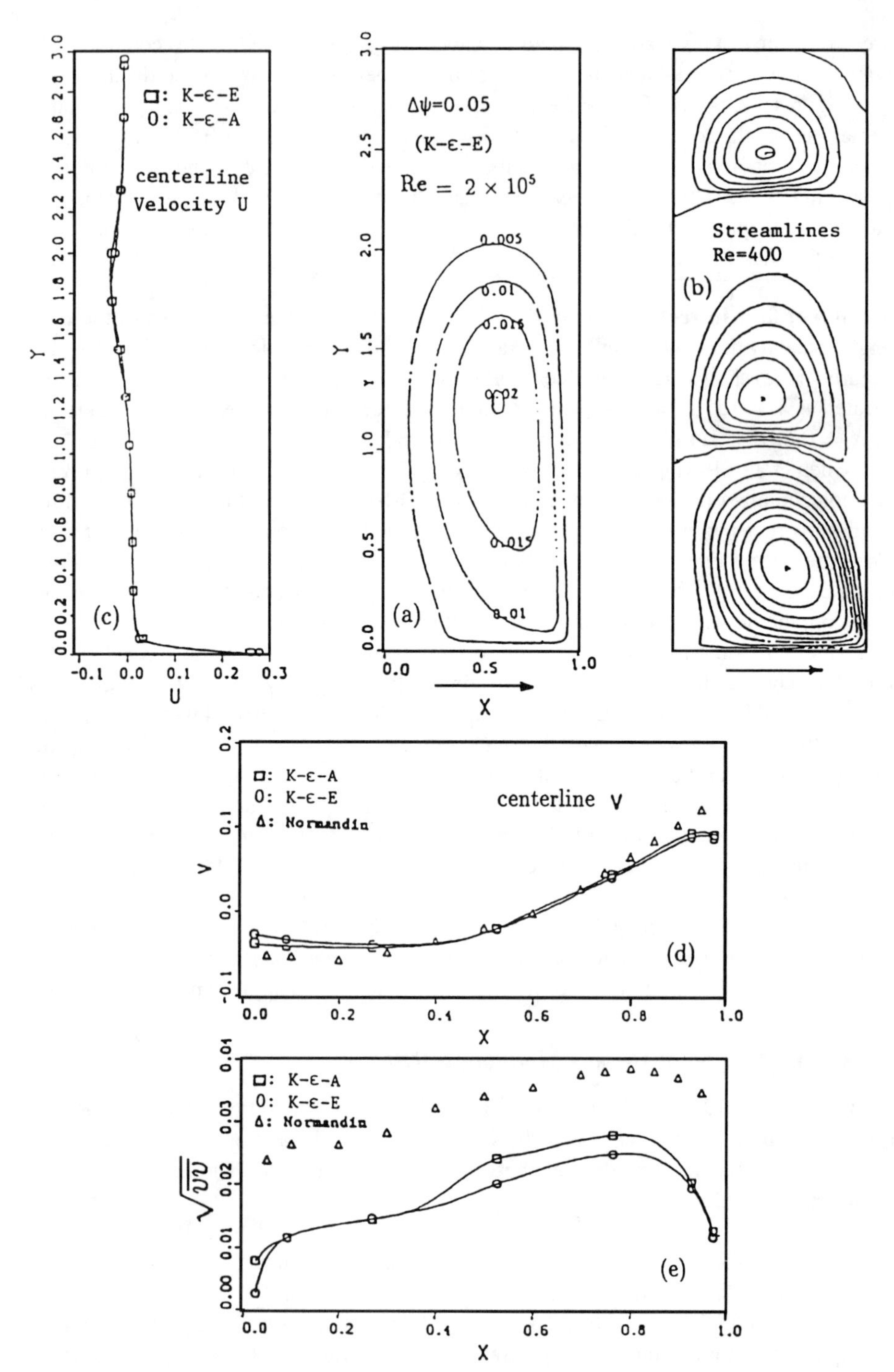

Figure 5.24 Prediction of rectangular cavity flow. (Adapted from Chen and Chang (1985).)

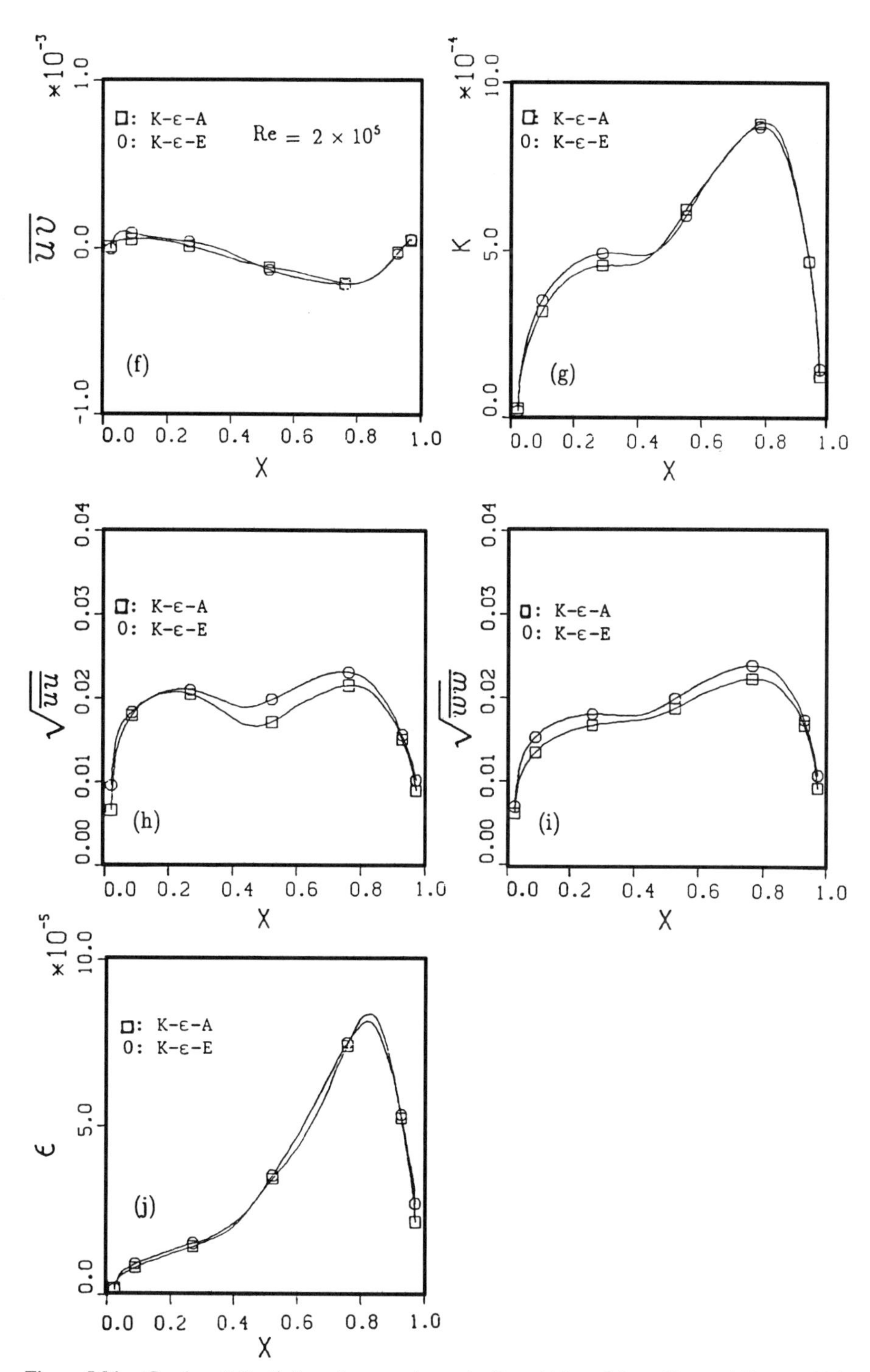

Figure 5.24 (Continued) Prediction of rectangular cavity flow. (Adapted from Chen and Chang (1985).)

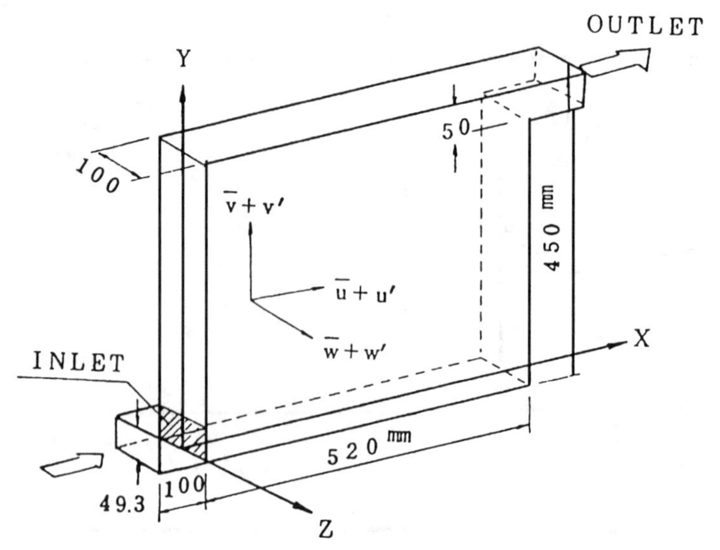

Figure 5.25 Rectangular plenum. (Adapted from Ushijima et al. (1985).)

Reynolds-stress ($\overline{uv}$) distributions of the three models have less difference and agree well with experiments.

3. The balances of the various terms of the Reynolds stress equations are examined. When comparing the results obtained by analyses and those of previous experiments, both distributions show qualitative agreement.

The governing equations are

- the continuity equation,

$$\frac{\partial U_m}{\partial X_m} = 0, \tag{5.43}$$

- the momentum equation,

$$\frac{\partial U_i}{\partial t} = -U_m \frac{\partial U_i}{\partial X_m} + F_i - \frac{1}{\rho}\frac{\partial P}{\partial X_i} + \frac{\partial}{\partial X_m}\left(\nu \frac{\partial U_i}{\partial X_m} - \overline{u_i u_m}\right), \tag{5.44}$$

and

- for the Reynolds stress equations, let

$$P_{ij} = -\left(\overline{u_i u_m}\frac{\partial U_i}{\partial X_m} + \overline{u_j u_m}\frac{\partial U_j}{\partial X_m}\right) \tag{5.45}$$

$$P = -\overline{u_m u_n}\frac{\partial U_m}{\partial X_n} \tag{5.46}$$

$$D_{ij} = -\left(\overline{u_i u_m}\frac{\partial U_m}{\partial X_j} + \overline{u_j u_m}\frac{\partial U_m}{\partial X_i}\right). \tag{5.47}$$

For the Launder et al. (1975)[88] Reynolds Stress Model (LSRM), we have

$$\frac{\partial \overline{u_i u_j}}{\partial t} = -U_m \frac{\partial \overline{u_i u_j}}{\partial X_m} + P_{ij} - \frac{2}{3}\delta_{ij}\epsilon - C_1 \frac{\epsilon}{k}\left(\overline{u_i u_j} - \frac{2}{3}\delta_{ij}k\right)$$
$$-\left[\frac{C_2+8}{11}\left(P_{ij} - \frac{2}{3}\delta_{ij}P\right) - \frac{30C_2-2}{55}k\left(\frac{\partial U_i}{\partial X_j} + \frac{\partial U_j}{\partial X_i}\right)\right.$$
$$\left. - \frac{8C_2-2}{11}\left(D_{ij} - \frac{2}{3}\delta_{ij}P\right)\right]$$
$$- C_s \frac{\partial}{\partial X_m}\frac{k}{\epsilon}\left(\overline{u_i u_j}\frac{\partial \overline{u_i u_m}}{\partial X_i} + \overline{u_j u_i}\frac{\partial \overline{u_m u_i}}{\partial X_i} + \overline{u_m u_i}\frac{\partial \overline{u_i u_j}}{\partial X_i}\right), \quad (5.48)$$

whereas the Standard Reynolds Stress Model (Eq. 2.17 in Section 2.3) gives

$$\frac{\partial \overline{u_i u_j}}{\partial t} = -U_m \frac{\partial \overline{u_i u_j}}{\partial X_m} + P_{ij} - \frac{2}{3}\delta_{ij}\epsilon - C_1' \frac{\epsilon}{k}\left(\overline{u_i u_j} - \frac{2}{3}\delta_{ij}k\right)$$
$$- C_2'\left(P_{ij} - \frac{2}{3}\delta_{ij}P\right) + \frac{\partial}{\partial X_m}\left(C_k \frac{k^2}{\epsilon}\frac{\partial \overline{u_i u_j}}{\partial X_m}\right), \quad (5.49)$$

and the k-ϵM (or k-ϵ-E) model gives

$$\overline{u_i u_j} = -\nu_t\left(\frac{\partial U_i}{\partial X_j} + \frac{\partial U_j}{\partial X_i}\right) + \frac{2}{3}\delta_{ij}k, \quad (5.50)$$

where

$$\nu_t = C_\nu \frac{k^2}{\epsilon}.$$

- The turbulent kinetic energy equation for the LRSM is

$$\frac{\partial k}{\partial t} = -U_m \frac{\partial k}{\partial X_m} + P - \epsilon + \frac{\partial}{\partial X_m}\left[C_s \frac{k}{\epsilon}\left(\overline{u_i u_n}\frac{\partial \overline{u_i u_m}}{\partial X_n} + \overline{u_m u_n}\frac{\partial k}{\partial X_n}\right)\right]. \quad (5.51)$$

For the SRSM and k-ϵM the kinetic energy equation is

$$\frac{\partial k}{\partial t} = -U_m \frac{\partial k}{\partial X_m} + P - \epsilon + \frac{\partial}{\partial X_m}\left(C_k \frac{k^2}{\epsilon}\frac{\partial k}{\partial X_m}\right). \quad (5.52)$$

- The dissipation equation for the various models is for the LRSM model,

$$\frac{\partial \epsilon}{\partial t} = -U_m \frac{\partial \epsilon}{\partial X_m} + \frac{\partial}{\partial X_m}\left(C_\epsilon \frac{k}{\epsilon}\overline{u_m u_n}\frac{\partial \epsilon}{\partial X_m}\right) - C_{\epsilon 1}\frac{\epsilon}{k}P - C_{\epsilon 2}\frac{\epsilon^2}{k}, \quad (5.53)$$

and
for the SRSM and k-ϵM,

$$\frac{\partial \epsilon}{\partial t} = -U_m \frac{\partial \epsilon}{\partial X_m} + \frac{\partial}{\partial X_m}\left(C_\epsilon' \frac{k}{\epsilon}\frac{\partial \epsilon}{\partial X_m}\right) - C_{\epsilon 1}'\frac{\epsilon}{k}P - C_{\epsilon 2}'\frac{\epsilon^2}{k}. \quad (5.54)$$

The constants for the models are

LRSM: $C_1 = 1.50$, $C_2 = 0.40$, $C_s = 0.11$, $C_{\epsilon 1} = 1.44$, $C_{\epsilon 2} = 1.90$, $C_\epsilon = 0.15$
SRSM: $C_1' = 2.80$, $C_2' = 0.50$, $C_k = 0.09$, $C_{\epsilon 1}' = 1.44$, $C_{\epsilon 2}' = 1.92$, $C_\epsilon' = 0.07$
k-ϵM: $C_\nu = 0.09$, $C_k = 0.09$, $C_{\epsilon 1}' = 1.44$, $C_{\epsilon 2}' = 1.92$, $C_\epsilon' = 0.07$.

The inlet boundary conditions are shown in Fig. 5.26 assuming $\epsilon = P$. The wall boundary conditions are

- for velocity, (tangential to the wall, U_t, with normal direction X_n)

$$\frac{\partial U_t}{\partial X_n} = \frac{U_*^2}{\nu + \nu_t}$$

$$\nu_t = C_w \left(C_\nu \frac{k^2}{\epsilon} \right) = C_w \epsilon X_n U_*,$$

  where $C_w = 3.0$,
- In the region $X = 0$ to $X = 250$ mm

$$\frac{U}{U_*} = -7.8 + 5.67 \ln \frac{U_* Y}{\mu}$$

  where U_* is the friction velocity
- for the turbulent properties, in the region $X = 0$ to $X = 250$ mm

$$k = a_1 U_*^2$$

$$\epsilon = a_2 \left(\frac{U_*^3}{\kappa X_n} \right),$$

  where $\kappa = 0.4$, $\overline{u^2} = a_3 U_*^2$, $\overline{v^2} = a_4 U_*^2$, $\overline{w^2} = a_5 U_*^2$, $-\overline{uv} = a_6 U_*^2$, $-\overline{vw} = a_7 U_*^2$, $-\overline{wu} = a_8 U_*^2$.

The gradient of turbulence quantities at the wall are assumed to be zero in other regions. The coefficients are given as

$a_1 = 3.60$, $a_2 = 1.00$, $a_3 = 6.20$, $a_4 = 0.53$,
$a_5 = 0.53$, $a_6 = \pm 0.46$, $a_7 = 0$, $a_8 = 0$.

Numerical solutions are obtained for two dimensional case by assuming $w = 0$ and $\partial/\partial z = 0$. The results of the comparison are given in Figs. 5.27–5.30. The results show that the LRSM, is somewhat better than SRSM or k-ϵ-E models.

5.3 TURBULENT FLOW PAST DISC TYPE VALVES

The steady and unsteady turbulent flow past a fully open disc type valve in an axisymmetric tube was studied by Chen, Yu, and Chandran (1987, 1988)[24, 25]. The Reynolds stresses that appear in the N-S equations are closed by a two-equation k-ϵ turbulence model. A wall function method is applied to determine the wall boundary conditions. Velocity, stream function, pressure distribution, Reynolds stresses, and turbulent kinetic

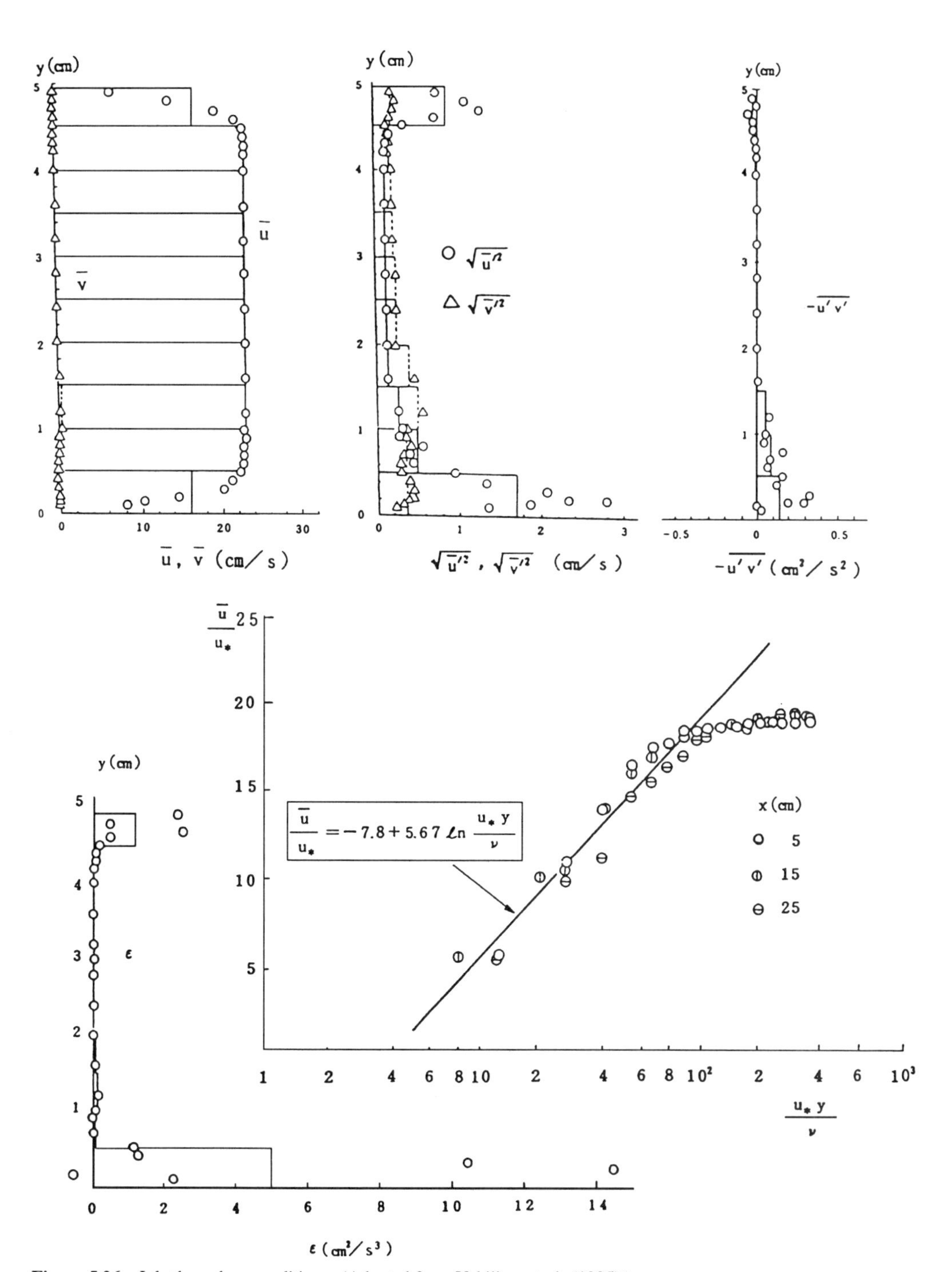

Figure 5.26 Inlet boundary conditions. (Adapted from Ushijima et al. (1985).)

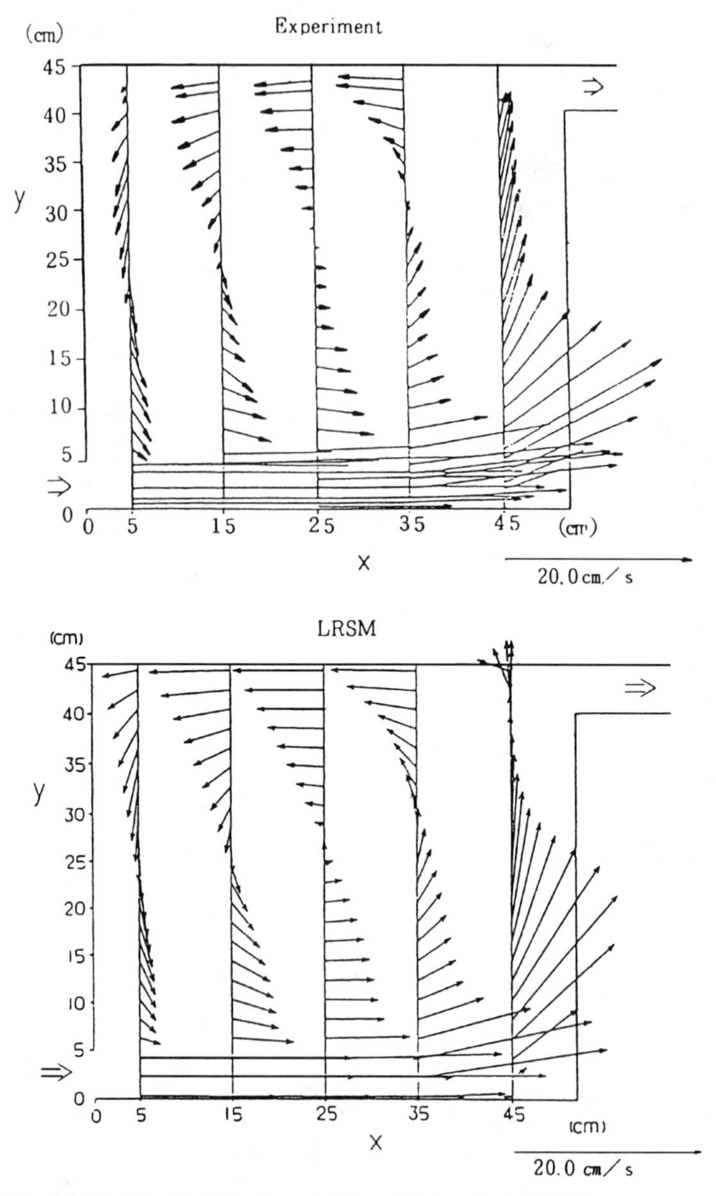

Figure 5.27 Velocity vector. (Adapted from Ushijima et al. (1985).)

energy and its dissipation rate are obtained for Reynolds number in the range of 10,000–200,000. Figure 5.31 shows the computational domain. The problem under consideration consists of a constant diameter pipe with a valve orifice ring and a disc valve. The fluid enters from the left and exits to the right. This disc valve system can be thought to simulate the caged disc heart valve similar to the Kay–Shiley valve in the fully open position. All of the dimensions in Fig. 5.32 are normalized by the diameter D of the pipe. The

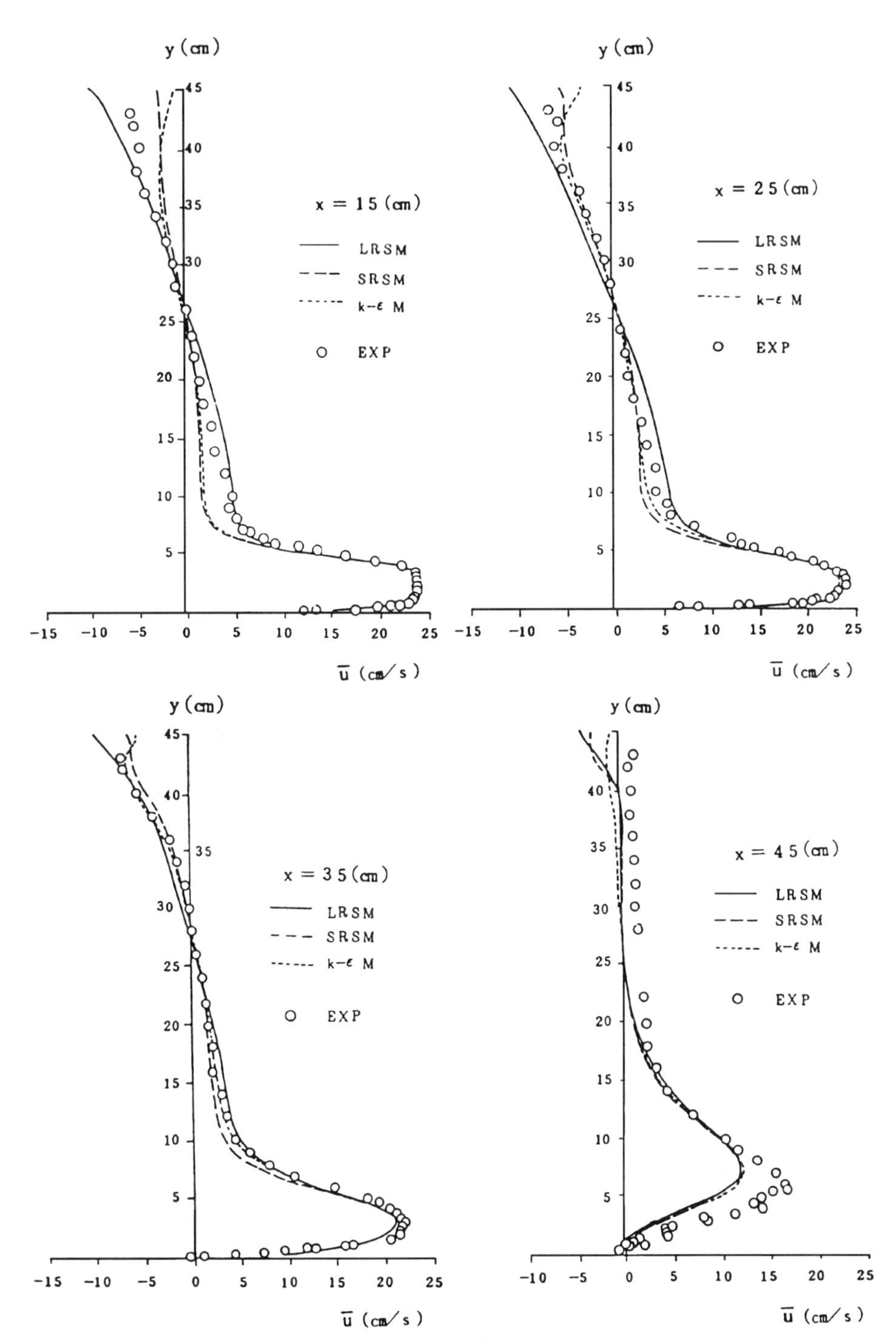

Figure 5.28 *U* profile. (Adapted from Ushijima et al. (1985).)

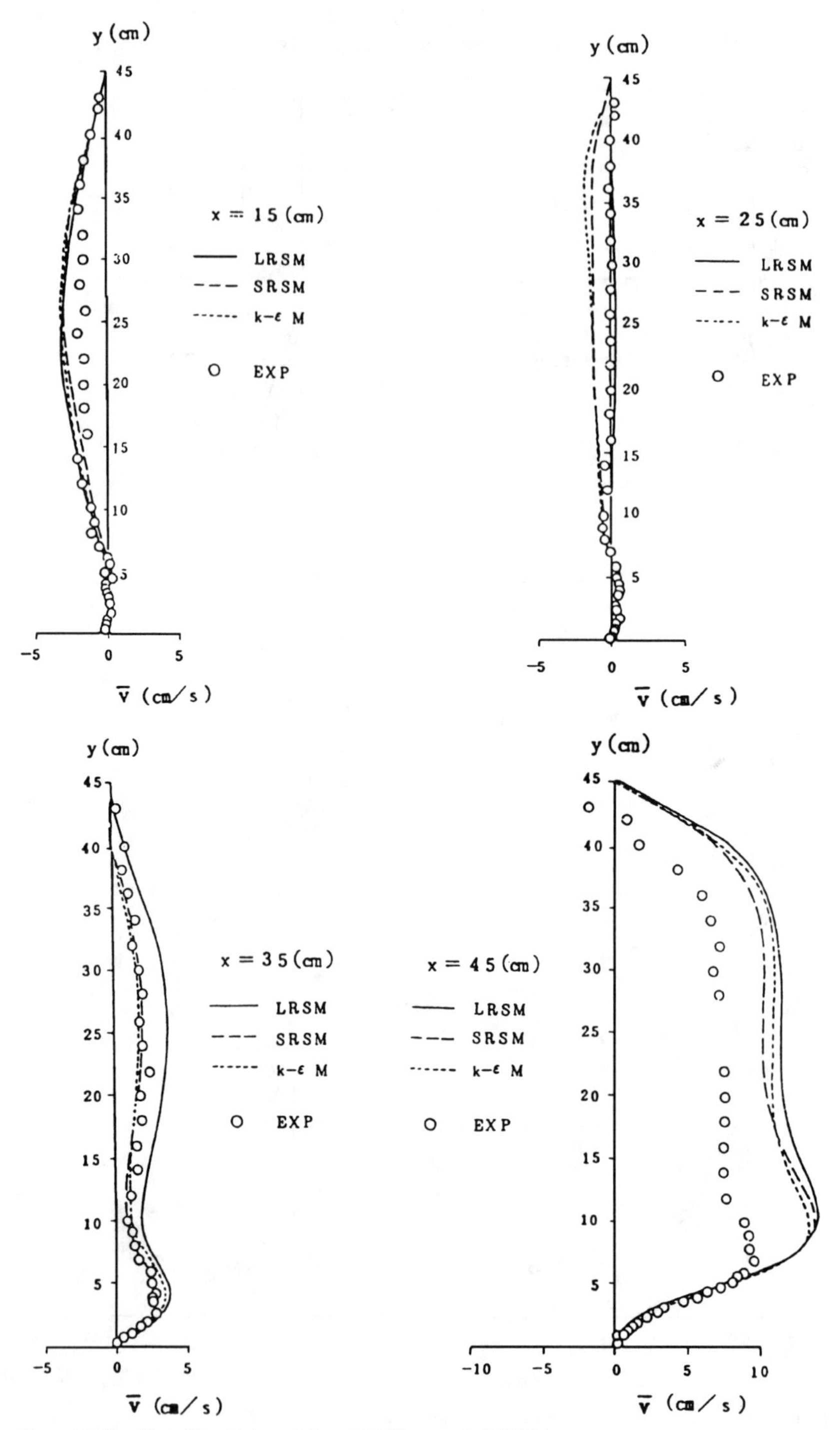

Figure 5.29 *V* profile. (Adapted from Ushijima et al. (1985).)

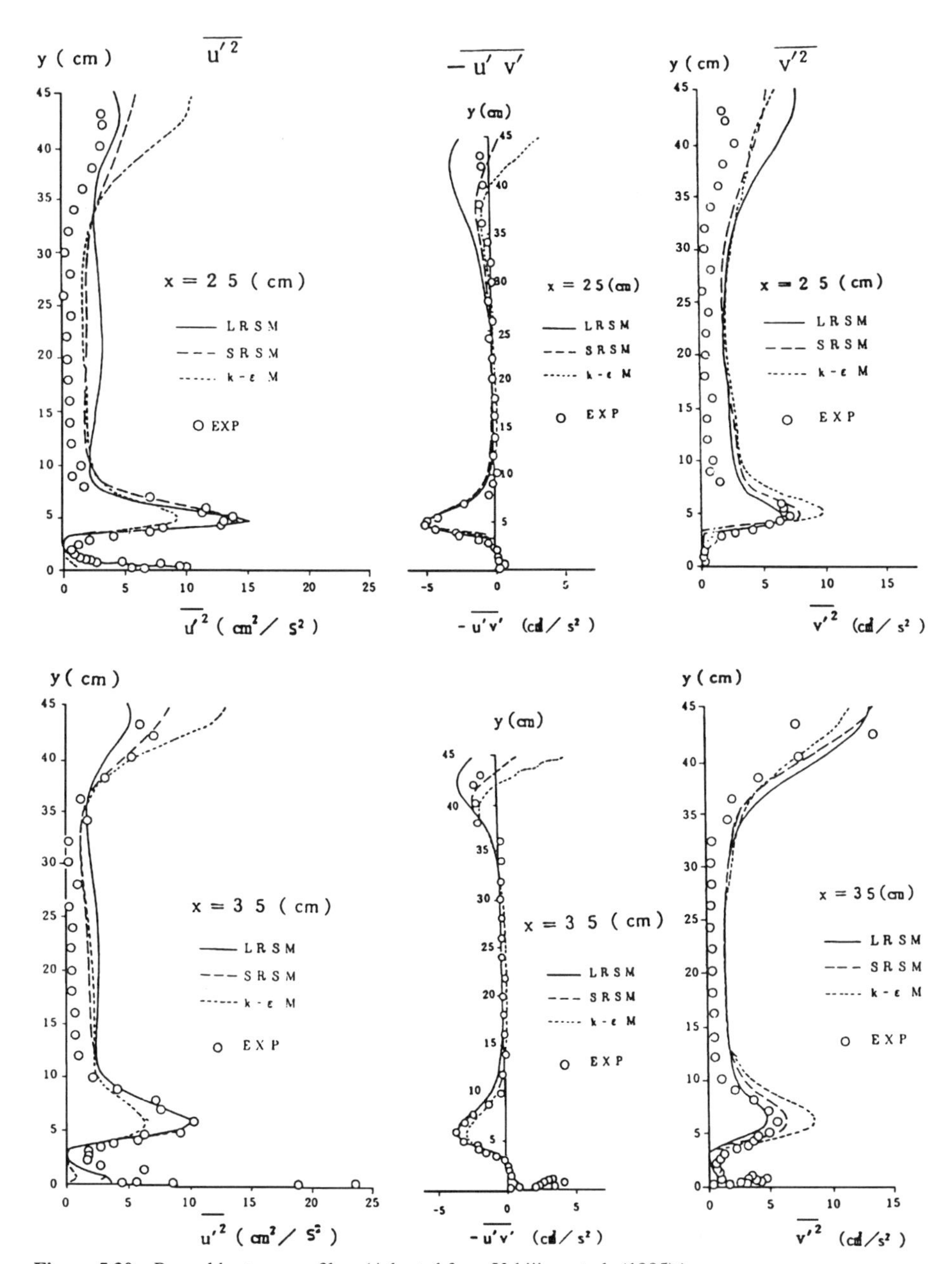

Figure 5.30 Reynolds-stress profiles. (Adapted from Ushijima et al. (1985).)

valve is located at (1/7) D downstream from the orifice ring. The thickness and the inner diameter of the orifice are (5/28) D and (4/7) D, respectively. The thickness and the diameter of the disc valve are (1/14) D and (5/7) D, respectively. The length of the inlet pipe is $3D$. The total length of the computational domain is 10 D. The modified Jones and Launder (1972)[70] low-Reynolds-number k-ϵ turbulence model is adopted

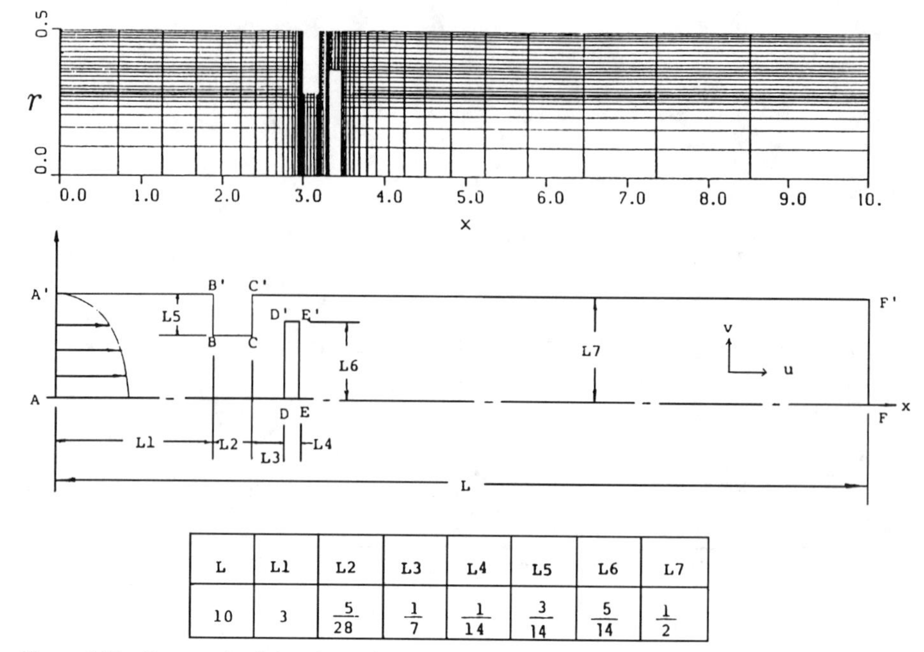

L	L1	L2	L3	L4	L5	L6	L7
10	3	$\frac{5}{28}$	$\frac{1}{7}$	$\frac{1}{14}$	$\frac{3}{14}$	$\frac{5}{14}$	$\frac{1}{2}$

Figure 5.31 Computational domain. (Adapted from Chen et al. (1987, 1988).)

here. The governing equations are as follows:

- The continuity equation,

$$\frac{\partial U}{\partial x} + \frac{\partial r V}{r \partial r} = 0. \tag{5.55}$$

- The momentum equations are

$$\frac{DU}{Dt} = -\frac{\partial P}{\partial x} + \frac{1}{Re}\left[\frac{\partial^2 V}{\partial x^2} + \frac{\partial}{r \partial r}\left(r \frac{\partial U}{\partial r}\right)\right] - \left(\frac{\partial \overline{u^2}}{\partial x} + \frac{\partial \overline{uv}}{\partial r} + \frac{\overline{uv}}{r}\right) \tag{5.56}$$

$$\frac{DV}{Dt} = -\frac{\partial P}{\partial r} + \frac{1}{Re}\left[\frac{\partial^2 V}{\partial x^2} + \frac{\partial}{r \partial r}\left(r \frac{\partial V}{\partial r}\right) + \frac{V}{r^2}\right] - \left(\frac{\partial \overline{uv}}{\partial x} + \frac{\partial \overline{v^2}}{\partial r} + \frac{\overline{v^2}}{r} - \frac{\overline{w^2}}{r}\right). \tag{5.57}$$

- The kinetic energy equation is

$$\frac{Dk}{Dt} = \frac{1}{Re}\left\{\frac{\partial}{r \partial r}\left[r\left(1 + \frac{\nu_t}{\sigma_k}\right)\frac{\partial k}{\partial r}\right] + \left[\left(1 + \frac{\nu_t}{\sigma_k}\right)\frac{\partial k}{\partial x}\right]\right\} + G - \epsilon. \tag{5.58}$$

- The ϵ equation is

$$\frac{D\epsilon}{Dt} = \frac{1}{Re}\left\{\frac{\partial}{r \partial r}\left[r\left(1 + \frac{\nu_t}{\sigma_k}\right)\frac{\partial \epsilon}{\partial r}\right] + \left[\left(1 + \frac{\nu_t}{\sigma_k}\right)\frac{\partial \epsilon}{\partial x}\right]\right\}$$
$$+ (C_{\epsilon 1} G - C_{\epsilon 2}\epsilon)\frac{\epsilon}{k} - 2.0 \frac{\nu_t}{Re^2}\left(\frac{\partial^2 U_i}{\partial x_j \partial x_l}\right)^2, \tag{5.59}$$

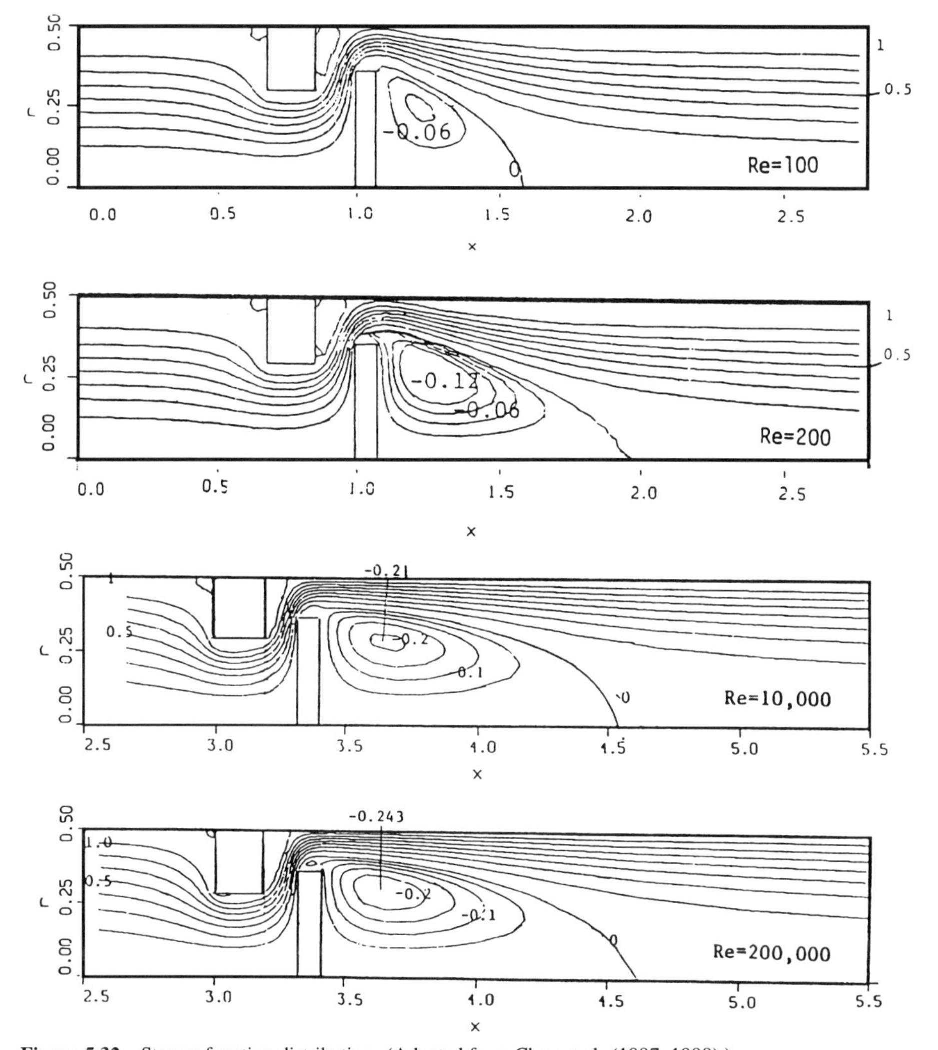

Figure 5.32 Stream function distribution. (Adapted from Chen et al. (1987, 1988).)

where

$$G = \frac{\nu_t}{Re}\left\{2\left[\left(\frac{\partial U}{\partial x}\right)^2 + \left(\frac{\partial V}{\partial r}\right)^2 + \left(\frac{V}{r}\right)^2\right] + \left(\frac{\partial U}{\partial r} + \frac{\partial U}{\partial r}\right)^2\right\} \tag{5.60}$$

and

$$C_\mu = C_{\mu h} \exp\left(\frac{-2.5}{1 + Rt/50}\right),$$

where

$$C_{\epsilon 1} = 1.44,\ C_{\epsilon 2} = C_{\epsilon 2h}[1 - 0.3\exp(-Rt^2)].$$

The five empirical constants in the above governing equation are $C_{\mu h} = 0.09$, $C_{\epsilon 1} = 1.44$, $C_{\epsilon 2h} = 1.92$, $\sigma_k = 1.00$, $\sigma_\epsilon = 1.33$, and $Rt = Rek^2/\epsilon$. The relevant equations for Reynolds stresses are

$$\overline{u^2} = -2\frac{\nu_t}{Re}\frac{\partial U}{\partial x} + \frac{2}{3}k \tag{5.61}$$

$$\overline{v^2} = -2\frac{\nu_t}{Re}\frac{\partial V}{\partial r} + \frac{2}{3}k \tag{5.62}$$

$$\overline{w^2} = -2\frac{\nu_t}{Re}\frac{\partial V}{\partial r} + \frac{2}{3}k \tag{5.63}$$

$$\overline{uv} = -\frac{\nu_t}{Re}\left(\frac{\partial V}{\partial x} + \frac{\partial U}{\partial r}\right), \tag{5.64}$$

where

$$\nu_t = ReC_\mu\frac{k^2}{\epsilon}.$$

For steady flow, the inlet and exit velocity functions are assumed to be fully developed turbulent velocity profiles as follows

$$U_i(r) = U_o(r) = (2r)^{1/n}\frac{(n+1)(2n+1)}{2n^2}, \tag{5.65}$$

where $n = 6.6$ for $Re = 10{,}000$, and $n = 7.0$ for $Re = 200{,}000$. The inlet and outlet boundary conditions for k and ϵ are specified as

$$k_i = 0.001,$$

$$\epsilon_i = \frac{0.1(k_i)^{3/2}}{\kappa(0.5 - r)},$$

$$\frac{\partial k_o}{\partial x} = 0,$$

and

$$\frac{\partial \epsilon_o}{\partial x} = 0,$$

where $\kappa(=0.435)$ is the von Kármán constant. In an attempt to reduce the computational time, the wall function method is chosen to approximate the wall boundary conditions. The basic idea of the wall function approach is that instead of solving the governing equations through the near wall region, a semianalytic solution is applied for this region. In this approach, the first computational nodal point is placed at a distance y_{p1} away from the wall. The wall boundary conditions are specified as

$$U_w^+ = \tfrac{1}{\kappa}\ln\left(9y_{p1}^+\right),\ V_w = 0,\ k_w = \tfrac{|-\overline{uv}|}{\sqrt{C_w}},\ \epsilon_w = \tfrac{|-\overline{uv}|U_\tau}{\kappa y_{p1}} : 150 > y_{p1}^+ > 12$$

$$U_w^+ = y_{p1}^+,\ V_w = 0,\ k_w = \tfrac{|-\overline{uv}|}{\sqrt{C_w}},\ \epsilon_w = ReU_\tau^2|-\overline{uv}| : y_{p1}^+ < 12,$$

where $y^+ = yU_\tau Re$, $U^+ = U/U_\tau$, and U_τ is the friction velocity. Near the wall, if $y_{p1}^+ > 12$, $|\overline{uv}|$ is assumed to be U_τ^2; if $y_{p1}^+ < 12$, then $|\overline{uv}| = U_\tau^2 Y_{pl}^+/12$ is used to approximate the value of $|\overline{uv}|$ in the sublayer region. Furthermore, because the flow is symmetric about the centerline axis, the radial velocity and the radial derivative of the axial velocity and other turbulent transport properties are zero.

The results are shown in Figs. 5.32–5.40. Details of the discussion of the results are found in Chen et al. (1987, 1988)[24, 25].

5.4 THIRD-ORDER CLOSURE MODEL

To predict the diffusion process of the Reynolds stresses in reattaching shear flow, the transport model for the triple velocity products has been developed by Amano and Goel (1984)[3] and tested for the computation of the flow in a channel with a backward facing step. On comparison of the results of $\overline{uuv}$, $\overline{uvv}$, and $\overline{vvv}$ with those obtained by using the existing algebraic correlations, it was shown that the third-order closure model improved the prediction of the triple-velocity products.

5.4.1 The Third-Order Closure Model

The transport equation for the kinematic Reynolds-stress $\overline{u_i u_j}$ can be written as

$$(U_k \overline{u_i u_j})_{,k} = -(\overline{u_j u_k} U_{i,k} + \overline{u_i u_k} U_{j,k}) - 2\nu \overline{u_{i,k} u_{j,k}} + \overline{\left(\frac{p}{\rho}\right)(u_{i,j} + u_{j,i})}$$
$$- [\overline{u_i u_j u_k} - \nu(\overline{u_i u_j})_{,k} + \overline{\left(\frac{p}{\rho}\right)}(\delta_{jk} u_i + \delta_{ik} u_j)]_{,k}. \tag{5.66}$$

The transport equations of the triple-velocity products are

$$U_l \frac{\partial}{\partial x_l}(\overline{u_i u_j u_k}) = -\left(\overline{u_i u_j u_l} \frac{\partial U_k}{\partial x_l} + \overline{u_j u_k u_l} \frac{\partial U_i}{\partial x_l} + \overline{u_k u_i u_l} \frac{\partial U_j}{\partial x_l}\right)$$
$$+ \left(\overline{u_i u_j} \overline{\frac{\partial u_k u_l}{\partial x_l}} + \overline{u_j u_k} \overline{\frac{\partial u_i u_l}{\partial x_l}} + \overline{u_k u_i} \overline{\frac{\partial u_j u_l}{\partial x_l}}\right) - \frac{\partial}{\partial x_l}(\overline{u_i u_j u_k u_l})$$
$$- \frac{1}{\rho}\left(\overline{u_i u_j \frac{\partial p}{\partial x_k}} + \overline{u_j u_k \frac{\partial p}{\partial x_i}} + \overline{u_k u_i \frac{\partial p}{\partial x_j}}\right), \tag{5.67}$$

where

- the terms

$$\left(\overline{u_i u_j u_l} \frac{\partial U_k}{\partial x_l} + \overline{u_j u_k u_l} \frac{\partial U_i}{\partial x_l} + \overline{u_k u_i u_l} \frac{\partial U_j}{\partial x_l}\right)$$
$$+ \left(\overline{u_i u_j} \overline{\frac{\partial u_k u_l}{\partial x_l}} + \overline{u_j u_k} \overline{\frac{\partial u_i u_l}{\partial x_l}} + \overline{u_k u_i} \overline{\frac{\partial u_j u_l}{\partial x_l}}\right)$$

  represent the generations that are due to the mean-strain rate and turbulent stresses, respectively, and
- the terms

$$-\frac{\partial}{\partial x_l}(\overline{u_i u_j u_k u_l}) - \frac{1}{\rho}\left(\overline{u_i u_j \frac{\partial p}{\partial x_k}} + \overline{u_j u_k \frac{\partial p}{\partial x_i}} + \overline{u_k u_i \frac{\partial p}{\partial x_j}}\right)$$

  both represent the diffusion of the triple-velocity products.

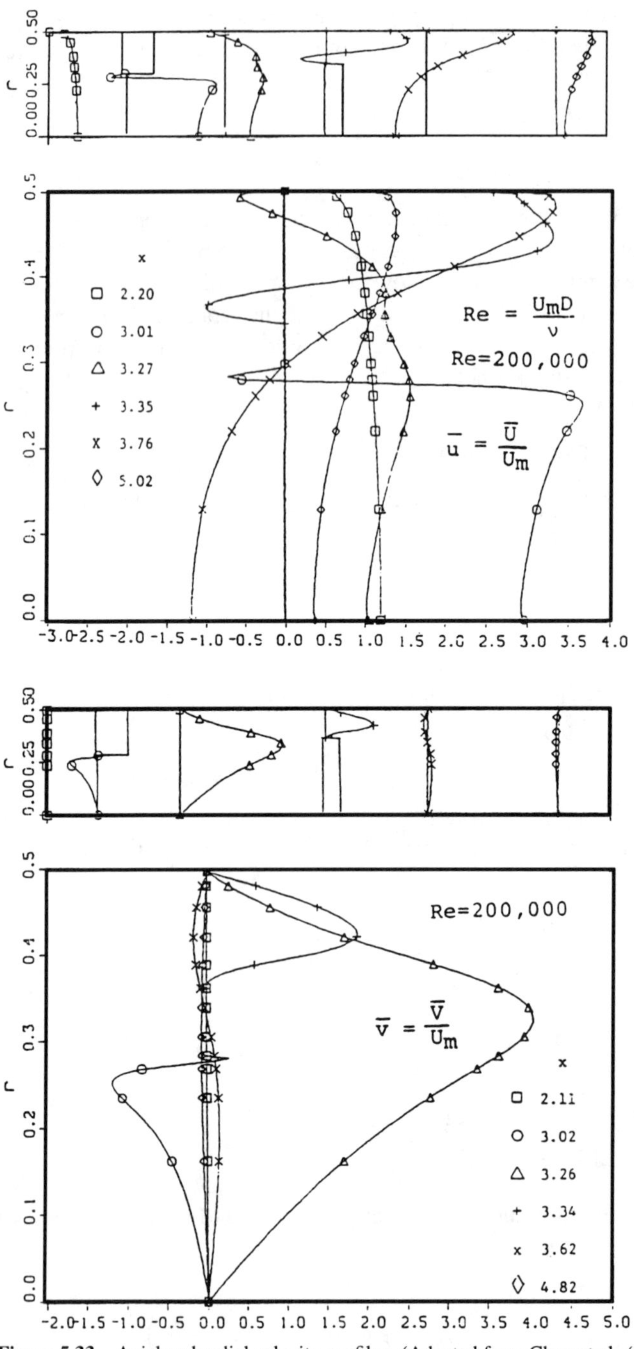

Figure 5.33 Axial and radial velocity profiles. (Adapted from Chen et al. (1987, 1988).)

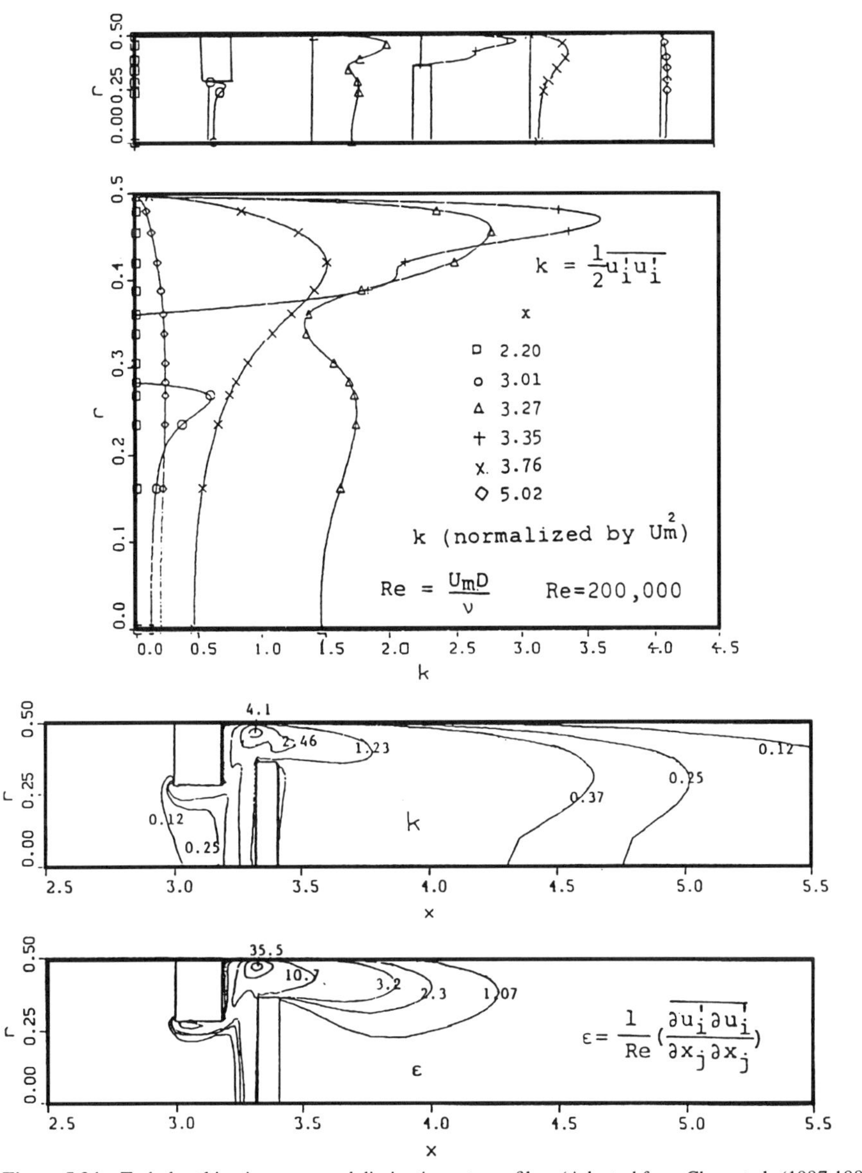

Figure 5.34 Turbulent kinetic energy and dissipation rate profiles. (Adapted from Chen et al. (1987,1988).)

In Eq. 5.67, the terms associated with the molecular viscosity are neglected. Amano and Goel modeled Eq. 5.67 as

$$U_l \frac{\partial}{\partial x_l}(\overline{u_i u_j u_k}) = -\left(\overline{u_i u_j u_l}\frac{\partial U_k}{\partial x_l} + \overline{u_j u_k u_l}\frac{\partial U_i}{\partial x_l} + \overline{u_k u_i u_l}\frac{\partial U_j}{\partial x_l}\right)$$

$$-\left(\overline{u_k u_l}\frac{\partial \overline{u_i u_j}}{\partial x_l} + \overline{u_j u_l}\frac{\partial \overline{u_i u_k}}{\partial x_l} + \overline{u_i u_l}\frac{\partial \overline{u_j u_k}}{\partial x_l}\right) - \Gamma(\overline{u_i u_j u_k}) \quad (5.68)$$

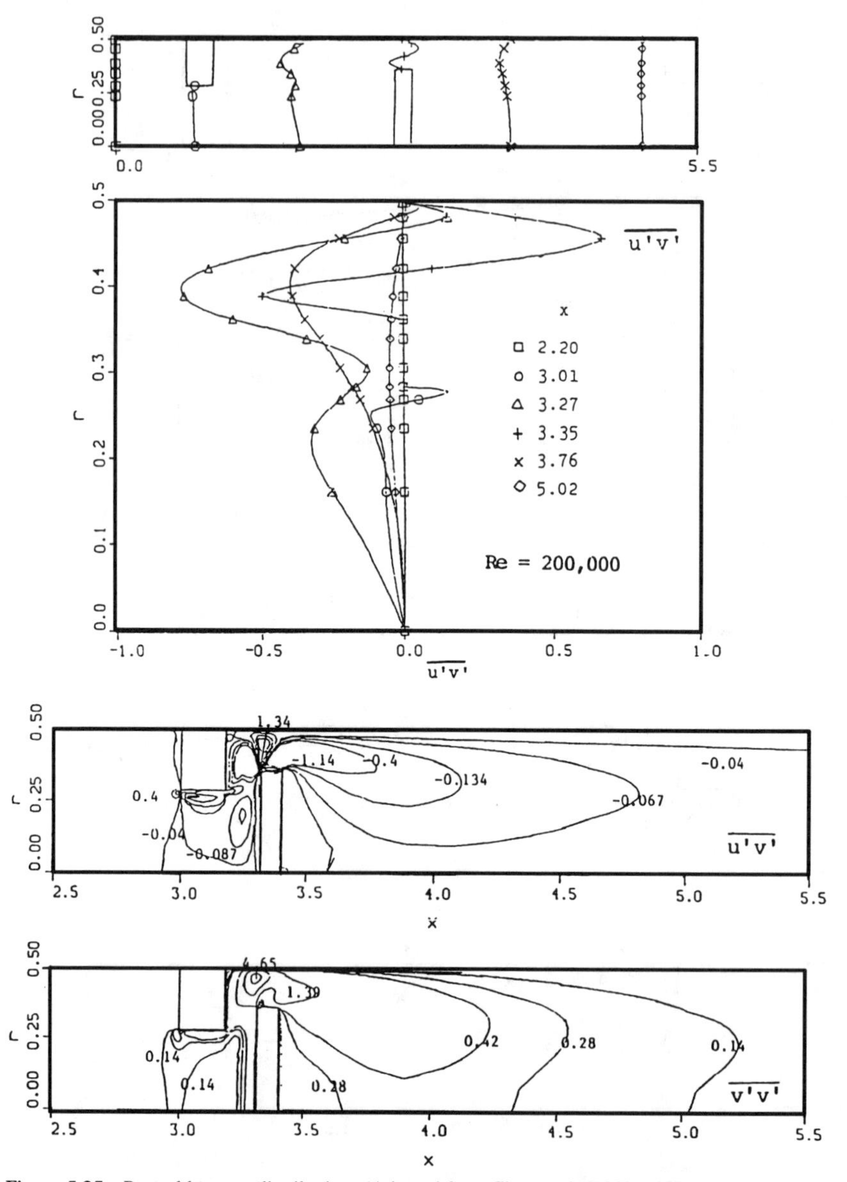

Figure 5.35 Reynolds-stress distribution. (Adapted from Chen et al. (1987, 1988).)

The function Γ was proposed as $\Gamma = C_\gamma(\epsilon/k)$, where C_γ is an empirical constant. The value of C_γ has been recommended to be 5.8 for the best agreement with several experimental data.

The models used for comparison with the transport equation are given as follows:

- The model of Daly and Harlow (1970)[37] is

$$\overline{u_i u_j u_k} = -0.25\tau\overline{u_k u_l}(\overline{u_i u_j})_{,l}. \tag{5.69}$$

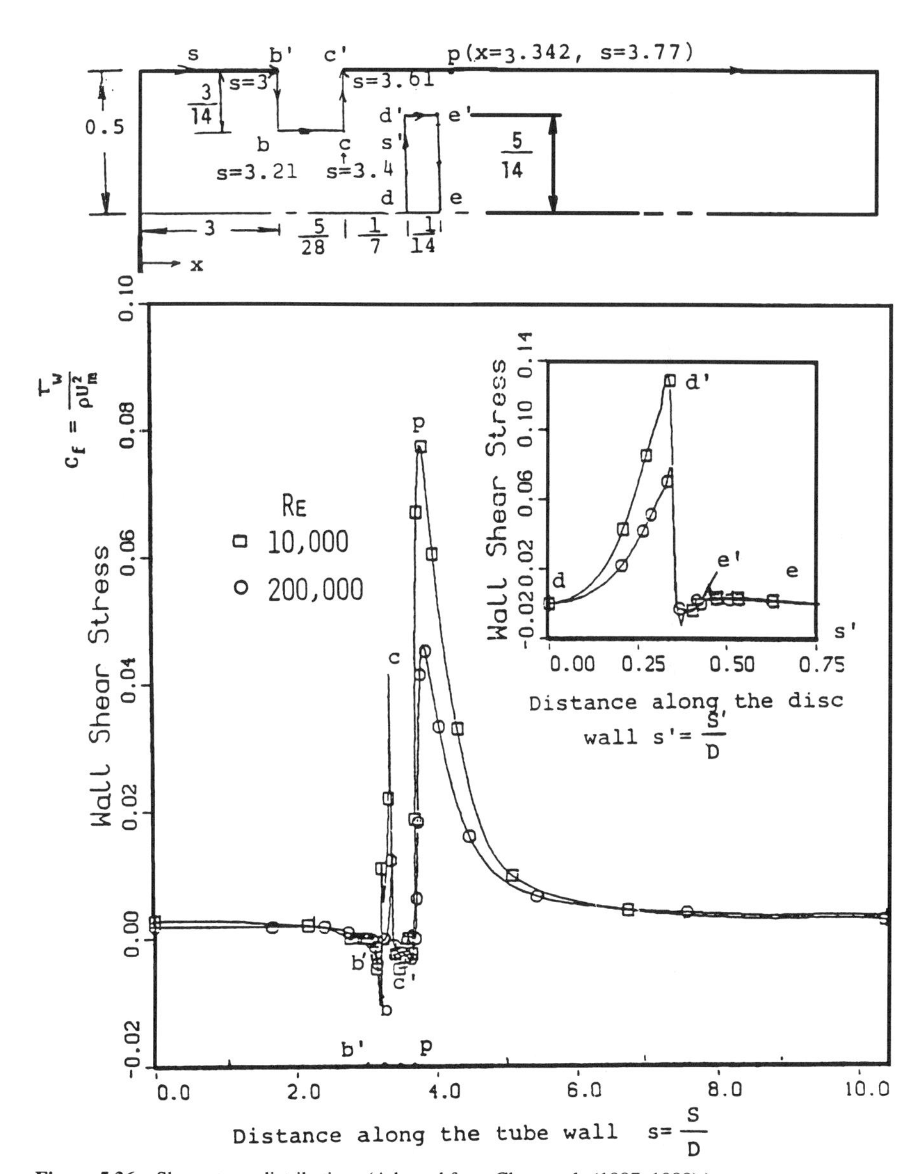

Figure 5.36 Shear stress distribution. (Adapted from Chen et al. (1987, 1988).)

- The model of Hanjalic and Launder (1972)[55] is

$$\overline{u_i u_j u_k} = -0.11\tau[\overline{u_l u_j}\,(\overline{u_i u_k})_{,l} + \overline{u_l u_i}(\overline{u_j u_k})_{,l} + \overline{u_l u_k}(\overline{u_i u_j})_{,l}]. \tag{5.70}$$

- The model of Shir (1973)[155] is

$$\overline{u_i u_j u_k} = -0.04\tau k(u_i u_j)_{,k}. \tag{5.71}$$

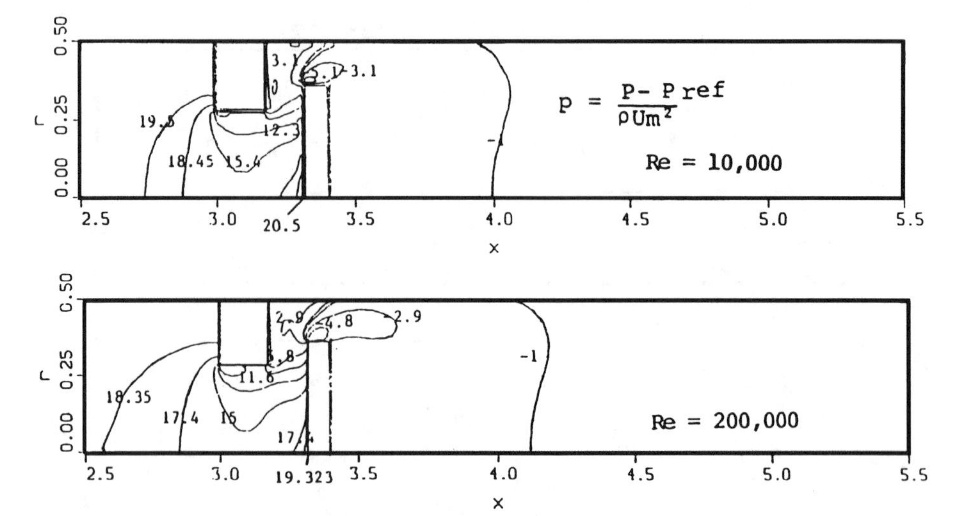

Figure 5.37 Pressure contours. (Adapted from Chen et al. (1987, 1988).)

- The model of Cormack et al. (1978)[34] is

$$\begin{aligned}\overline{u_i u_j u_k} = {} & 4\tau k[2\alpha_1(\delta_{ij}\delta_{kl} + \delta_{ik}\delta_{jl} + \delta_{kj}\delta_{il})k_{,l} + \alpha_2(a_{ik,l} + a_{ij,k} + a_{kj,i})] \\ & + 2\tau[2\alpha_3(\delta_{ij}a_{kl} + \delta_{ik}a_{jl} + \delta_{kj}a_{il})k_{,l} + \alpha_4(a_{ik}a_{jl,l} + a_{ij}a_{kl,l} + a_{kj}a_{il,l}],\end{aligned}$$

where

$$a_{ij} = \overline{u_i u_j} - \frac{2}{3}\delta_{ij}k,$$

and where the coefficients α_i are given in Table 5.1.

5.4.2 Flow Past Backward Facing Step

For comparison, the flow past a backward facing step with a step height H, shown in Fig. 5.38, is considered. At the channel inlet, with inlet width, Y_o, and inlet velocity, U_{in}, the triple-velocity products are evaluated by using the model of Daly and Harlow (1970)[37] (Eq. 5.69). The outlet is located at 50 H downstream from the step where the continuative boundary condition is used; that is, the normal gradient of the triple-velocity products to the boundary is zero. Although the values of $\overline{u_i u_j u_k}$ at the wall are zero, these are evaluated by using an algebraic model at the nodes next to the wall. Because the turbulence kinetic energy can be better defined with *the law of the wall* than with the Reynolds stresses in the near-wall region, the model of Shir (Eq. 5.71), which has fewer

Table 5.1 Values of the coefficients α_i

α_1	α_2	α_3	α_4
-8.14×10^{-3}	-1.72×10^{-2}	-4.80×10^{-2}	-1.02×10^{-1}

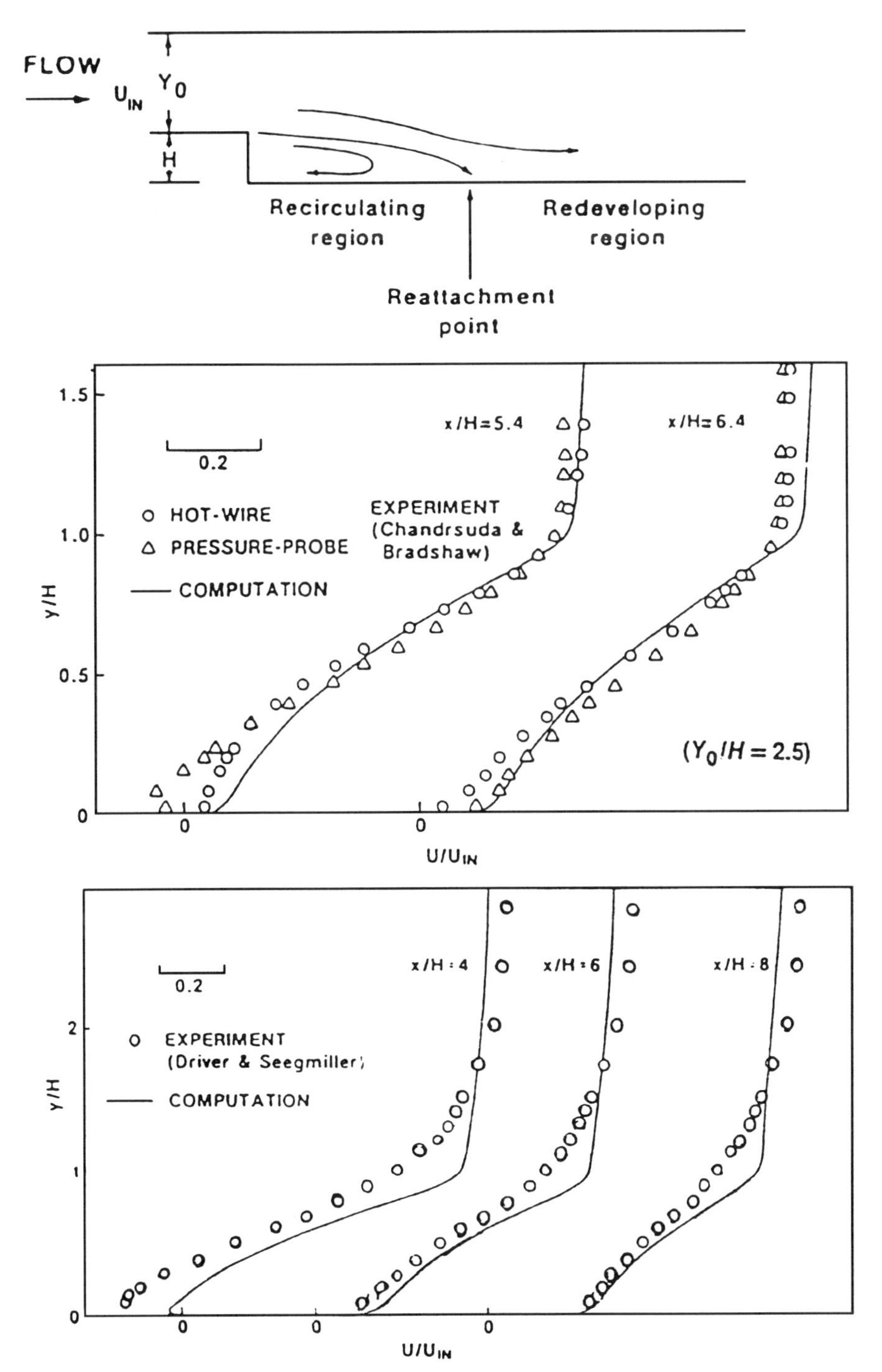

Figure 5.38 U velocity profiles. (Adapted from Amano and Goel (1984).)

terms with the Reynolds stresses than the other models, was employed along with the so-called *near-wall Reynolds stresses*. The near-wall Reynolds stresses are given as

$$\overline{u_i u_j} = C_{ij}k - (1 - \delta_{ij})\frac{y}{\rho}\frac{dp}{dx}, \tag{5.72}$$

where x is the streamwise coordinate, and y is the normal coordinate to the wall. The coefficients are determined on the basis of several experimental data as $C_{11} = 1.21$, $C_{22} = 0.24$, $C_{12} = 0.24$.

5.4.3 Results

Calculations were made for Reynolds number (which is based on inlet velocity U_{in} and step height H) of 32,000. Figures 5.38–5.41 show the mean-velocity profiles, Reynolds-stress profiles, and triple-velocity profiles as obtained by the numerical method and the experimental method at $Y_o/H = 2.5$ and 8.0. The experimental data are taken from Chandrsuda and Bradshaw (1980)[10] and Driver and Seegmiller (1985)[41].

Amano and Goel (1984)[3] concluded that

1. the diffusion coefficient of the turbulence energy for the computation of reattaching shear layers should be smaller than the commonly used value for free-mixing layers;
2. the diffusion coefficient for the dissipation rate equation was found to be larger for the computation of reattaching layers than for free-mixing layers;
3. the production and dissipation rates of the turbulence energy predominate in the region upstream of the reattachment point, but they decay rapidly downstream from it. In contrast, the convection and diffusion rates do not change appreciably in the streamwise direction; and
4. the low-Reynolds-number-transport model for the third moment of the turbulence velocity fluctuations improves the predictions of triple-velocity correlations as compared with algebraic models.

5.5 THREE-DIMENSIONAL FLOWS

5.5.1 Remarks and Examples

There are many flows and heat transfer problems that require three-dimensional treatment. Indeed sometimes it is easier to create three-dimensional flows than two-dimensional flows. However, three-dimensional computation at present, although it can be done, is still costly and time consuming. Most of the work done in three-dimensional computing utilizes the k-ϵ-(E or A) model. Also, the type of flow considered is such that one flow direction dominates or that the flow can approximately be solved by a partially parabolized equation (semielliptic problem).

In three-dimensional calculations, the grid cannot be refined with most of the present computer capacity to obtain grid-independent solutions. In this respect, it demands further research in numerical methods.

On the other hand, there are many secondary flows in a three-dimensional category. An example of such flow is a rectangular duct flow (see Fig. 5.42). One thus questions

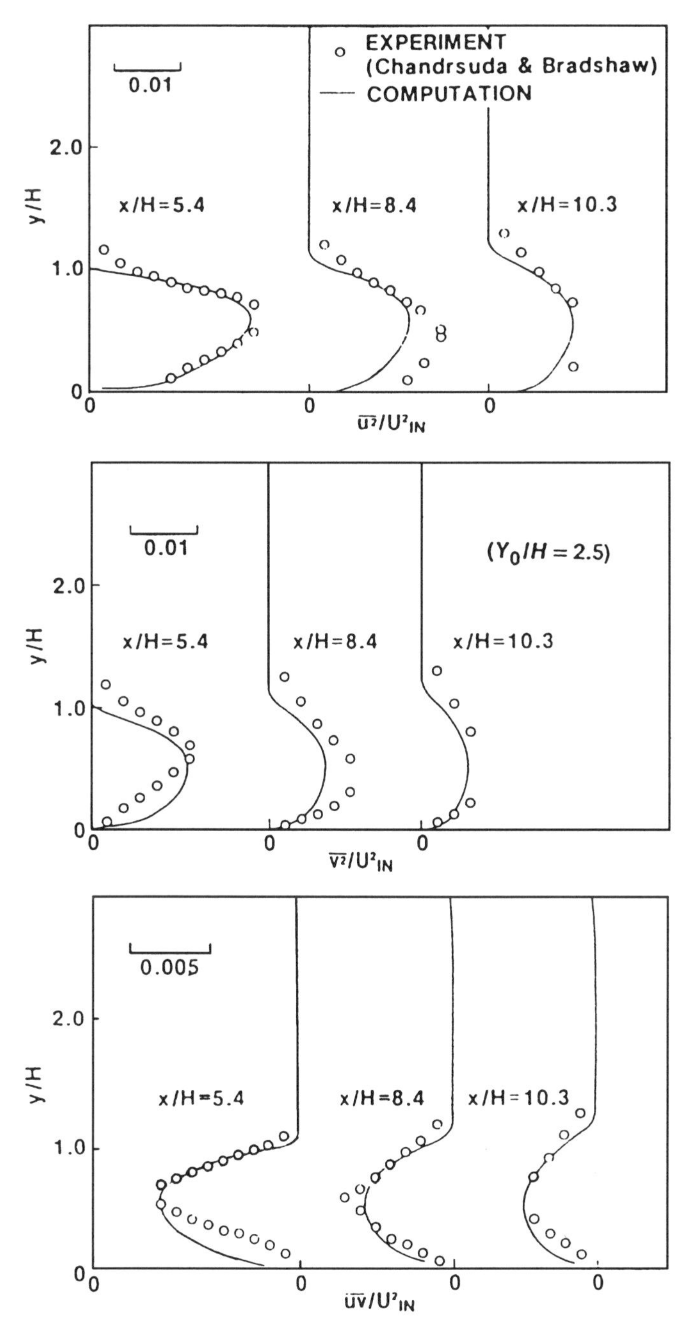

Figure 5.39 Reynolds stress profiles. (Adapted from Amano and Goel (1984).)

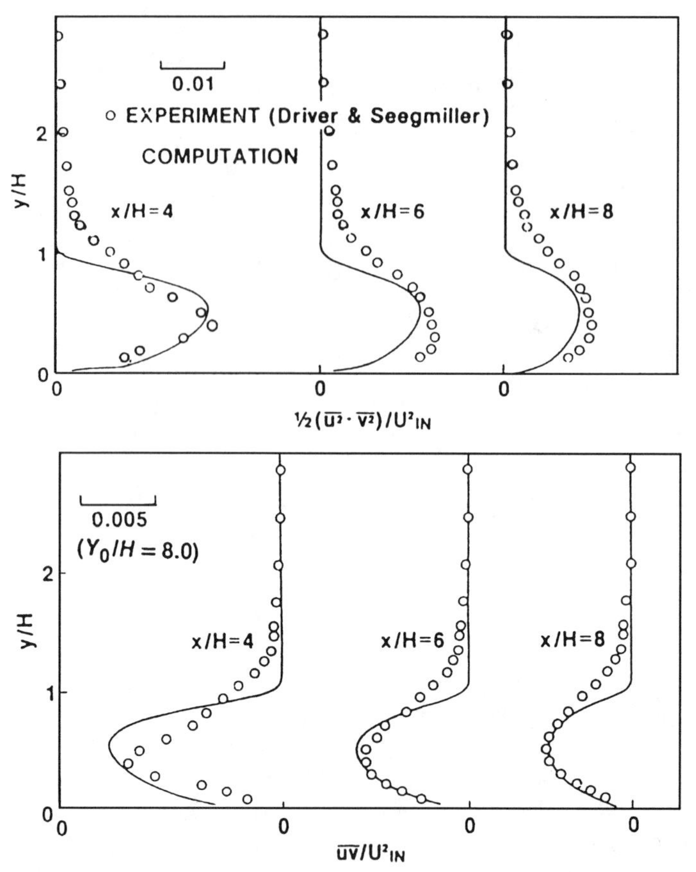

Figure 5.40 Reynolds stress profile.

the ability of the k-ϵ model to predict such flows. This issue will be examined in the following sections.

Some examples of three-dimensional flows are shown in Figs. 5.44 and 5.45.

- Rectangular duct flow, including secondary flow
- Plane strain in channel flow
- Open channel flow
- Curved channel pipe flow
- Jet in cross-flow
- Tank type plenum flow

5.5.2 Turbulence-Driven Secondary Flows

In order to examine which turbulence model is capable of predicting turbulence-driven secondary flow, we consider a fully developed rectangular duct flow as shown in Fig. 5.46. Thus, as per the fully developed condition, we have

$$\frac{\partial \text{velocity}}{\partial Z} = 0.$$

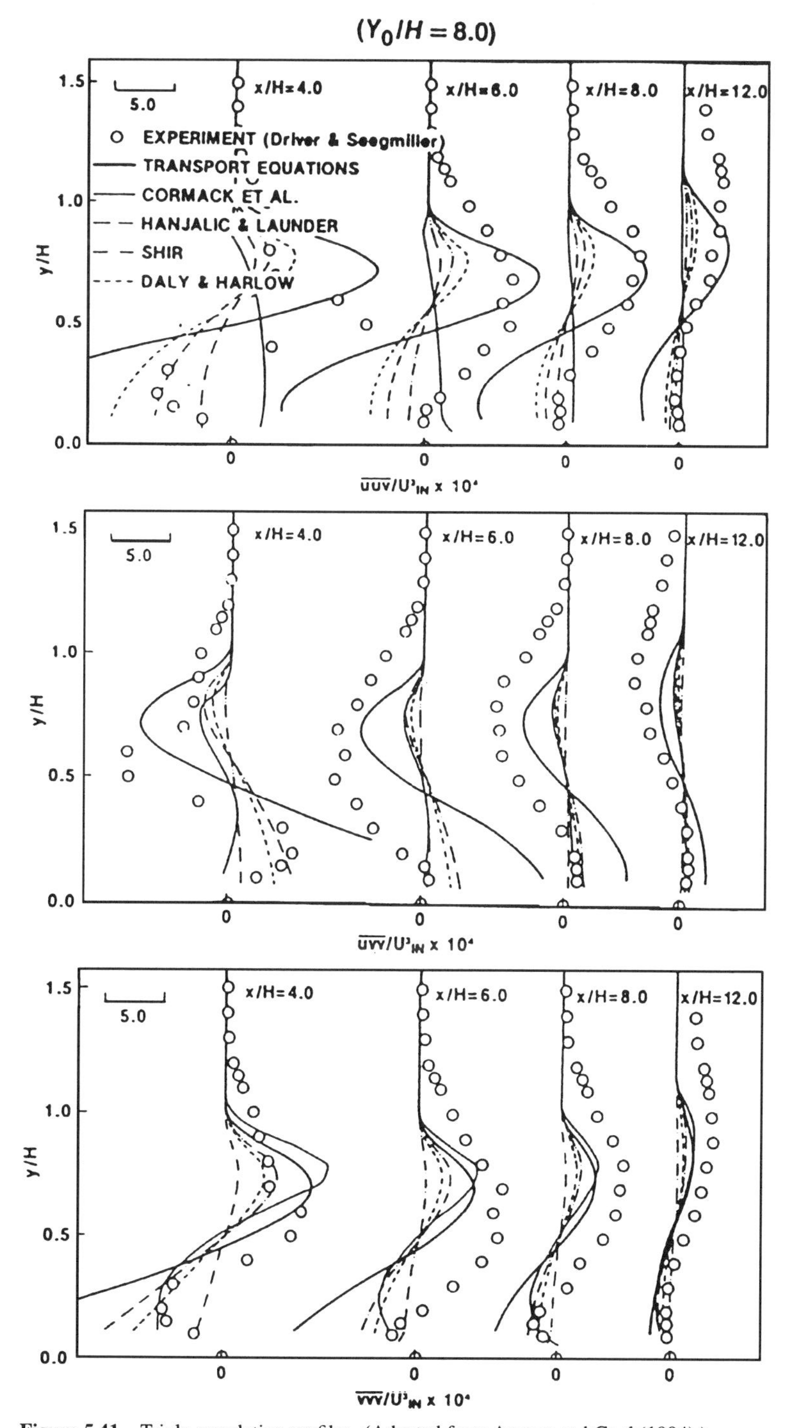

Figure 5.41 Triple correlation profiles. (Adapted from Amano and Goel (1984).)

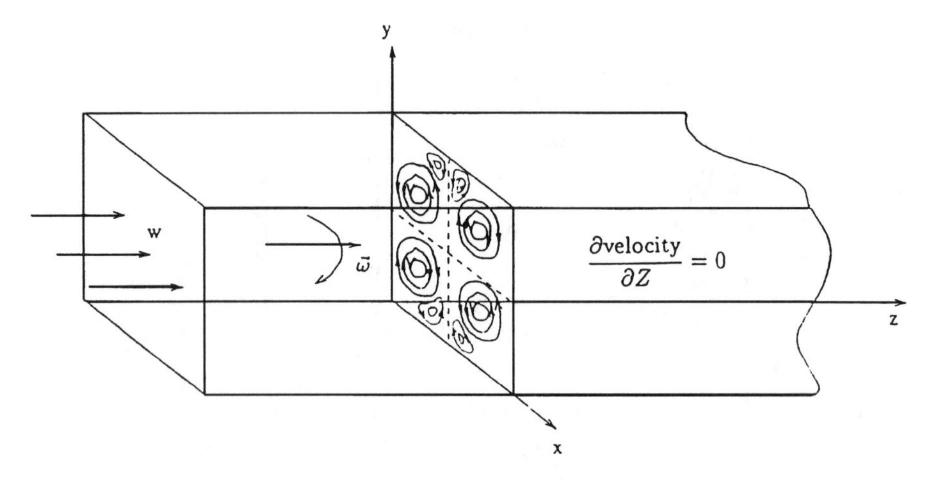

Rectangular Channel Flow

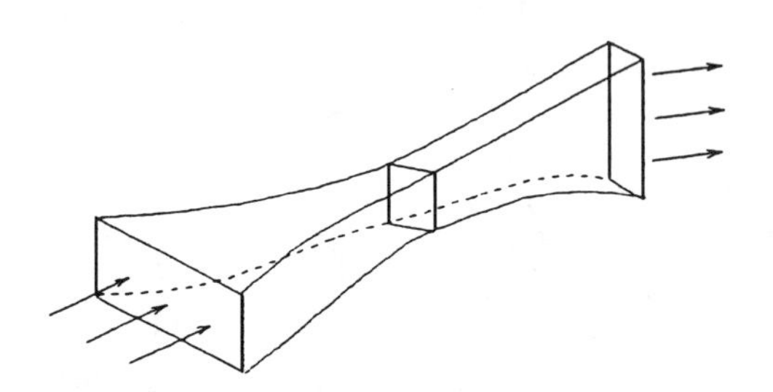

Plane Strain in Channel Flow

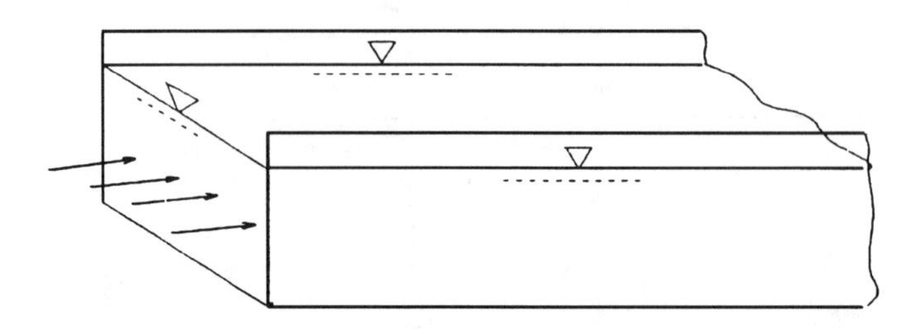

Open Channel Flow

Figure 5.42 Channel flows.

If there is secondary flow, then,

$$U = U(X, Y) \neq 0$$
$$V = V(X, Y) \neq 0$$
$$W = W(X, Y) \neq 0.$$

Hence, the governing equations are as follows:

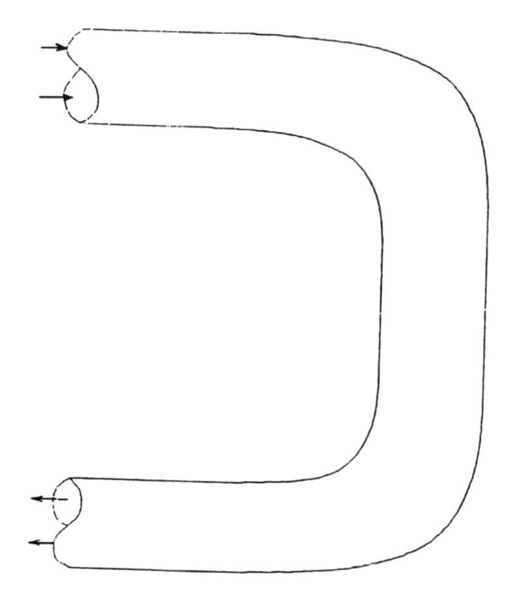

Figure 5.43 Curved channel pipe flow.

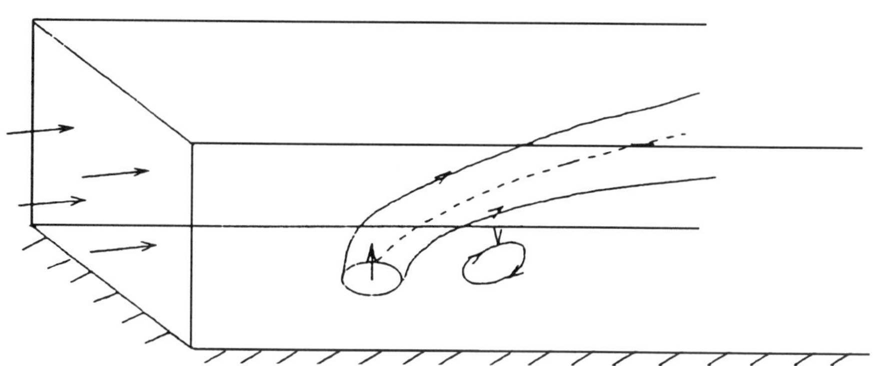

Figure 5.44 Jet in cross-flow.

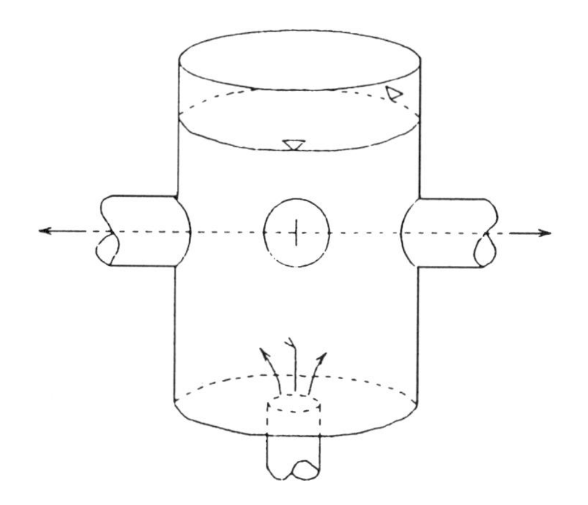

Figure 5.45 Tank type plenum flow.

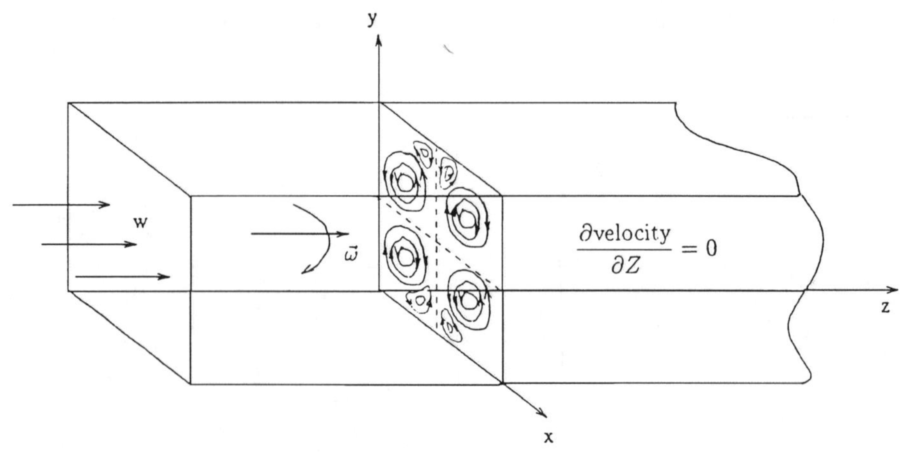

Figure 5.46 Fully developed rectangular duct flow.

- The continuity equation is

$$\frac{\partial U}{\partial X} + \frac{\partial V}{\partial Y} = 0. \tag{5.73}$$

- The momentum equations become

$$\frac{DU_i}{Dt} = -\frac{1}{\rho}\frac{\partial P}{\partial X_i} + \nu \nabla^2 U_i + \frac{\partial \tau_{ij}^t}{\partial X_j}, \tag{5.74}$$

  where

$$\tau_{ij}^t = -\overline{u_i u_j},$$

  and

$$\boldsymbol{U} = (U_1, U_2, U_3) = (U, V, W)$$

Taking the curl of Eq. 5.74, we have

$$\frac{D\omega_k}{Dt} = \omega_l \frac{\partial U_k}{\partial X_l} + \epsilon_{klm} \frac{\partial^2 \tau_{nm}^t}{\partial X_l \partial X_n} + \nu \frac{\partial^2 \omega_k}{\partial X_l \partial X_l}, \tag{5.75}$$

where

$$\boldsymbol{\omega} = (\omega_1, \omega_2, \omega_3) = \nabla \times \boldsymbol{U},$$

and ϵ_{klm} is the alternating tensor term. Depending on the value of k, l, and m, ϵ_{klm} can attain the value of -1, 0, and 1; that is,

$$\begin{aligned} \epsilon_{klm} &= 1 && k, l, \text{ and } m \text{ in clockwise order,} \\ &= -1 && k, l, \text{ and } m \text{ in counterclockwise order, and} \\ &= 0 && \text{otherwise.} \end{aligned}$$

Thus, we have $\omega_3 = \omega_z$ given by

$$\omega_z = \frac{\partial U}{\partial Y} - \frac{\partial V}{\partial X} = \nabla^2 \psi,$$

where from Eq. 5.73, we have

$$U = \frac{\partial \psi}{\partial Y}$$

and

$$V = -\frac{\partial \psi}{\partial X}.$$

Thus, ω_z must be nonzero if the secondary flow is to exist (because $U \neq 0$ and $V \neq 0$). Let us examine the ω_z equation from Eq. 5.75:

$$\frac{D\omega_z}{Dt} = \omega_x \frac{\partial W}{\partial X} + \omega_Y \frac{\partial W}{\partial Y} + \nu \nabla^2 \omega_z + \frac{\partial^2 \left(\tau^t_{yy} - \tau^t_{XX}\right)}{\partial X \partial Y} + \frac{\partial^2 \tau^t_{XY}}{\partial X^2} - \frac{\partial^2 \tau^t_{YX}}{\partial Y^2} \tag{5.76}$$

While the first two terms on the right in Eq. 5.76 are the vortex stretching terms, the last three terms in Eq. 5.76 are the source of ω_z that is due to turbulence. Hence,

$$\frac{\partial^2 \left(\tau^t_{YY} - \tau^t_{XX}\right)}{\partial X \partial Y} + \frac{\partial^2 \tau^t_{XY}}{\partial X^2} - \frac{\partial^2 \tau^t_{YX}}{\partial Y^2}$$

must not be zero if the secondary flow is to develop.

If $\tau^t_{XY} \neq 0$, then there exists a secondary flow. However, initially if the flow is $W(x, y, 0) \neq 0$, $U = V = 0$, and $\tau_{XY} = 0$, then the only condition that will generate a secondary flow is

$$\frac{\partial^2 \left(\tau^t_{YY} - \tau^t_{XX}\right)}{\partial X \partial Y}.$$

Hence, any turbulent flow model that gives $\tau^t_{yy} = \tau^t_{xx}$ will not produce turbulence-driven secondary flow. For example, with the Boussinesq eddy viscosity model with $(U \sim V \sim 0)$ we have

$$-\overline{u_i u_j} = \nu_t \left(\frac{\partial U_i}{\partial X_j} + \frac{\partial U_j}{\partial X_i}\right) - \frac{2}{3}\delta_{ij} k$$

$$\Rightarrow \tau^t_{yy} = \tau^t_{xx}.$$

Therefore, the Boussinesq eddy viscosity model or the k-ϵ-E model will not produce turbulence-driven secondary flow. However, the algebraic stress model (ASM) and the differential stress model (RSM) will produce the turbulence-driven secondary flows. This illustrates another difference between the turbulence models.

In the flows such as curved flows or buoyant driven flows, the secondary flow may still be predicted by the k-ϵ-E model because the source of the secondary flow is not due to turbulence but is due to centrifugal and buoyant forces.

5.5.3 Governing Equations

The governing equations for three dimensional flow are as follows:

- The continuity equation is

$$\frac{\partial U}{\partial X} + \frac{\partial V}{\partial Y} + \frac{\partial W}{\partial Z} = 0. \tag{5.77}$$

- The momentum equation is

$$\frac{DU_i}{Dt} = -\frac{1}{\rho}\frac{\partial p}{\partial X_i} + \nu\frac{\partial^2 U_i}{\partial X_j \partial X_j} - \frac{\partial \overline{u_i u_j}}{\partial X_j}. \tag{5.78}$$

- The thermal energy equation is

$$\frac{DT}{Dt} = \Phi + \alpha\frac{\partial^2 T}{\partial X_i \partial X_i} - \frac{\partial \overline{u_i \theta}}{\partial X_i}, \tag{5.79}$$

where Φ is the frictional heating term and is given by

$$\Phi = \frac{\mu}{\rho C_p}\left(\frac{\partial U_i}{\partial X_j} + \frac{\partial U_j}{\partial X_i}\right)\left(\frac{\partial U_i}{\partial X_j}\right).$$

- The k equation is

$$\frac{Dk}{Dt} = \frac{\partial}{\partial X_l}\left[\left(C_k\frac{k^2}{\epsilon} + \nu\right)\frac{\partial k}{\partial X_l}\right] - \overline{u_i u_j}\frac{\partial U_i}{\partial X_j} - \epsilon, \tag{5.80}$$

- and the ϵ equation,

$$\frac{D\epsilon}{Dt} = \frac{\partial}{\partial X_l}\left[\left(C_\epsilon\frac{k^2}{\epsilon} + \nu\right)\frac{\partial \epsilon}{\partial X_l}\right] - C_{\epsilon 1}\frac{\epsilon}{k}\overline{u_i u_j}\frac{\partial U_i}{\partial X_j} - C_{\epsilon 2}\frac{\epsilon^2}{k}. \tag{5.81}$$

The algebraic $\overline{u_i u_j}$ equation gives

$$(1 - C_2)P_{ij} + (C_1 - 1)\frac{23}{\delta_{ij}}\epsilon - C_1\frac{\epsilon}{k}\overline{u_i u_j} + \frac{2}{3}C_2\delta_{ij}P_k = 0,$$

which yields the expression for the Reynolds stresses as

$$-\overline{u_i u_j} = \frac{(C_2 - 1)}{C_1}\frac{k}{\epsilon}P_{ij} - \frac{2}{3}\frac{C_2}{C_1}\delta_{ij}\frac{k}{\epsilon}P_k + \frac{(1 - C_1)}{C_1}\frac{2}{3}\delta_{ij}k. \tag{5.82}$$

Similarly, the algebraic $\overline{u_i \theta}$ equation yields an expression for the Reynolds heat fluxes as

$$-\overline{u_i \theta} = \frac{1}{C_{T1}}\frac{k}{\epsilon}\overline{u_i u_l}\frac{\partial T}{\partial X_l} - \frac{(C_{T2} - 1)}{C_{T1}}\frac{k}{\epsilon}\overline{u_l \theta}\frac{\partial U_i}{\partial X_l}, \tag{5.83}$$

where

$$P_{ij} = -\left(\overline{u_i u_l}\frac{\partial U_j}{\partial X_l} + \overline{u_j u_l}\frac{\partial U_i}{\partial X_l}\right),$$

and

$$P_k = -\overline{u_n u_m}\frac{\partial U_n}{\partial X_m}.$$

As mentioned earlier, the k-ϵ-A model may predict turbulence-driven secondary flows.

5.6 TURBULENCE FLOW PREDICTIONS: FOUR (THREE-DIMENSIONAL FLOWS)

5.6.1 Straight Duct Flow

In straight, noncircular duct flows or open-channel flows, secondary motions are known to exist. The secondary velocity components are only 2 to 5 percent of the mean longitudinal velocity. These secondary flows are driven by turbulent stresses. Here we discuss the predictions that are based on the RSM and ASM models.

From Reece's work (1977)[126] of flow through a square duct, we find that the RSM model and the k-ϵ-A model perform about the same. Figure 5.47 shows the computational domain and presents the streamwise velocity contours on the left and the secondary velocity profiles on the right, respectively, as obtained by the RSM, the ASM, and experimentally. In Fig. 5.47,

$$U_\tau = u^* = \sqrt{\frac{\tau_w}{\rho}}.$$

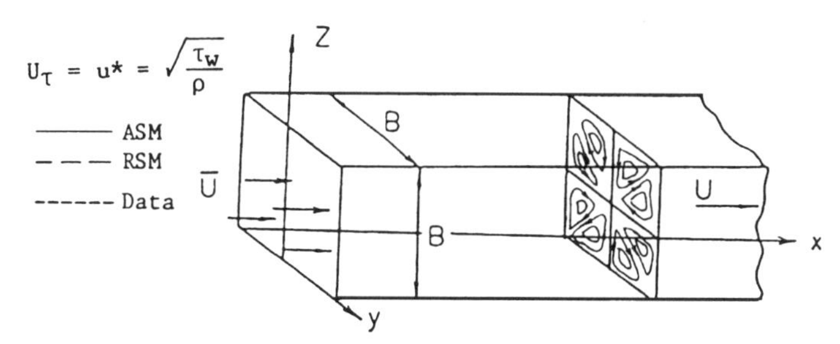

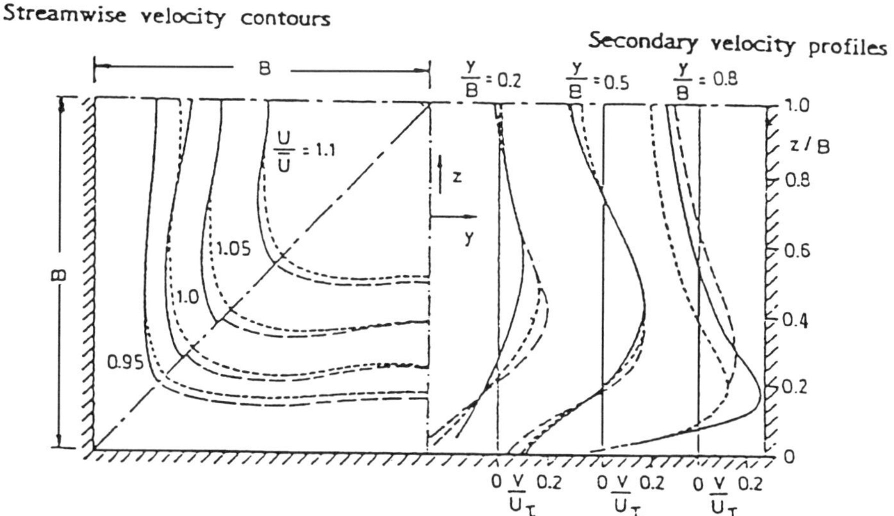

Figure 5.47 Mean-velocity profiles in square duct flow. (Adapted from Reece (1977).)

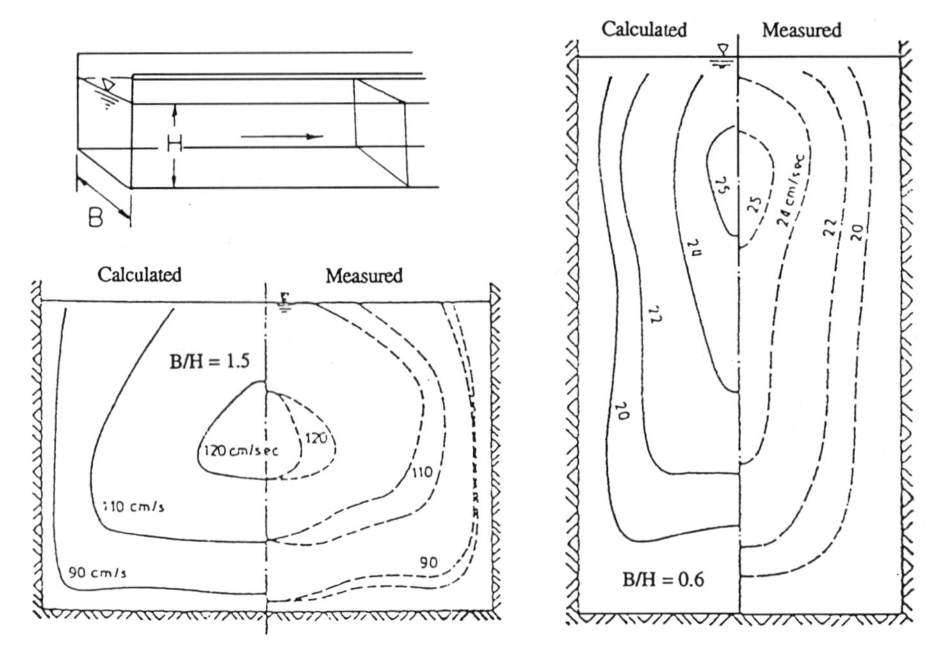

Figure 5.48 Mean velocity profiles in open channel flow. (Adapted from Naot and Rodi (1982).)

5.6.2 Open-Channel Flow

Naot and Rodi (1982)[113] predicted the flow in rectangular open channel. The setup is shown in Fig. 5.48, where $B/H =$ width : depth ratio. The k-ϵ-A model with near-wall pressure-strain (PS) correction was found to predict a general secondary flow pattern fairly well, considering the small magnitude of secondary flow.

5.6.3 Flow in Curved Pipes

Flow in a curved pipe was examined by Patankar, Pratap, and Spalding (1975)[120]. The predicted profile by the k-ϵ-A model compares well with the experiments of Rowe (1966)[146]. This is partly due to a strong pressure gradient that is responsible for the secondary flow. Figure 5.49 shows the setup and compares the nondimensionalized velocity contours at different locations in the pipe curvature, as obtained by the experiments and the numerical method.

5.6.4 Flow in a Curved Open Channel

The flow in a curved open channel was considered by Leschziner and Rodi (1979)[94] with the k-ϵ-A model. The result compares fairly well with Rosovskii's (1957)[142] experiment. Figure 5.50 shows the flow configuration and depth-averaged longitudinal velocity profiles. Figure 5.50 also shows the water-surface super-elevation at outer and inner banks and gives the typical calculated secondary velocity field at $\theta = 102°$.

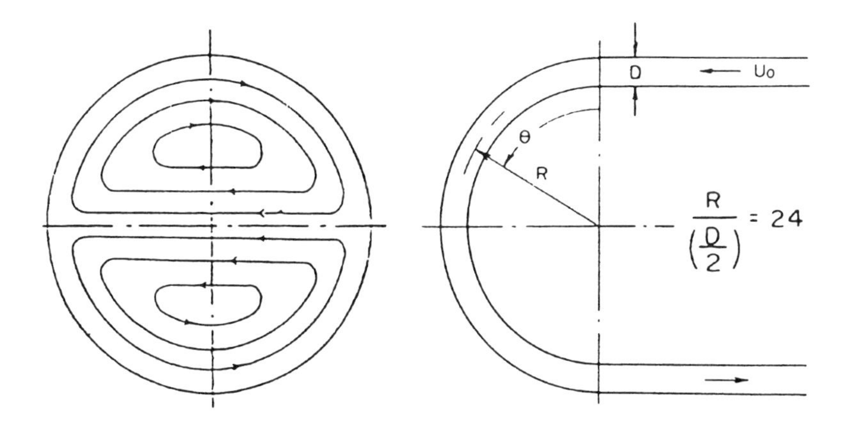

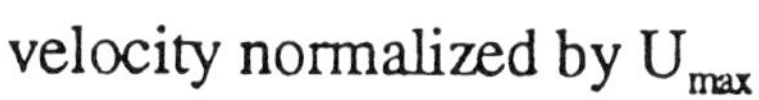

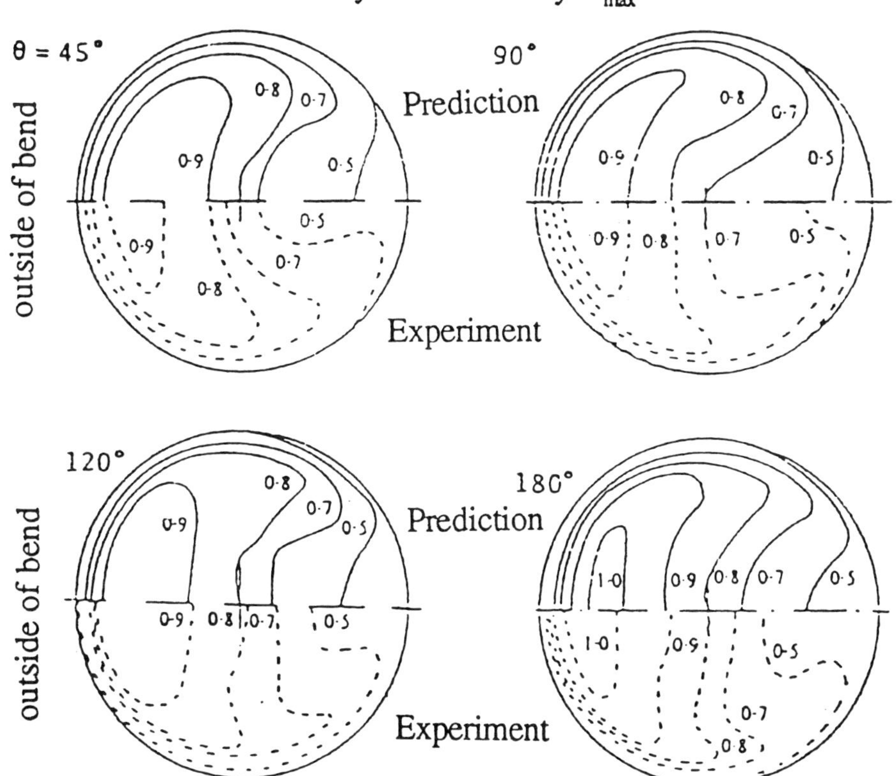

Figure 5.49 Velocity distribution in curved pipe flow. (Adapted from Patanber et al. (1975).)

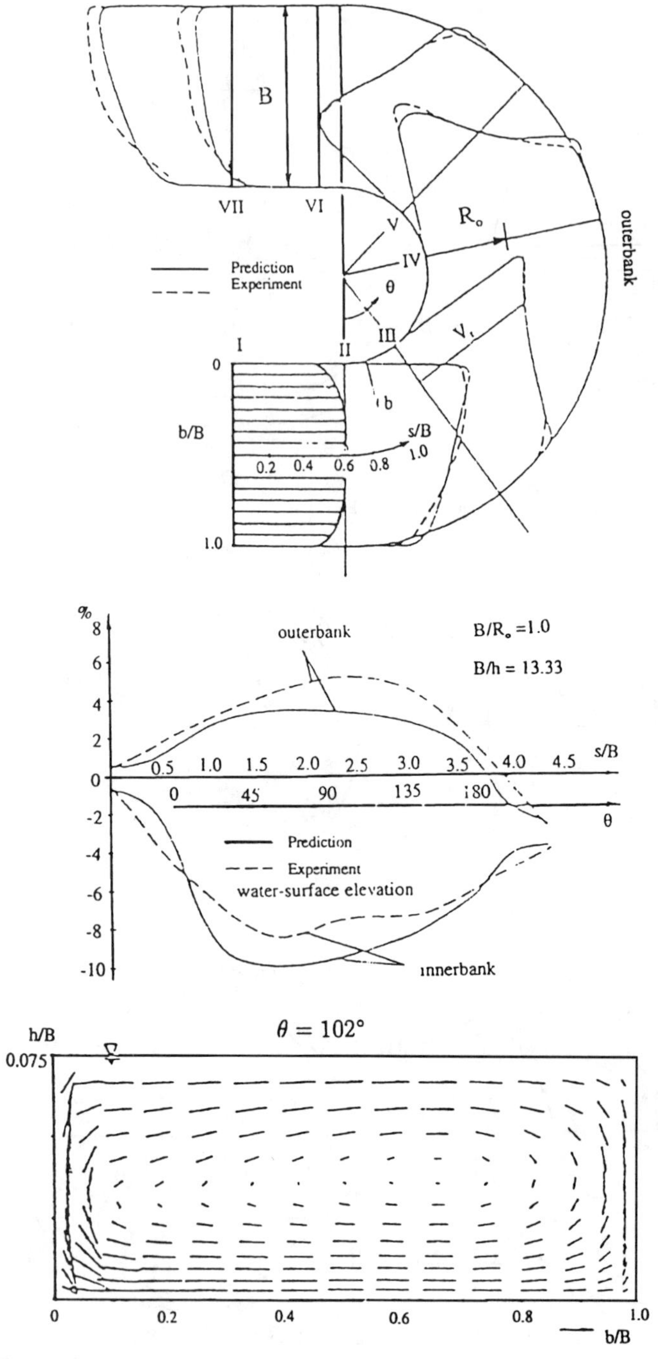

Figure 5.50 Velocity profile and surface-elevation in curved open channel. (Adapted from Leschziner and Rodi (1979).)

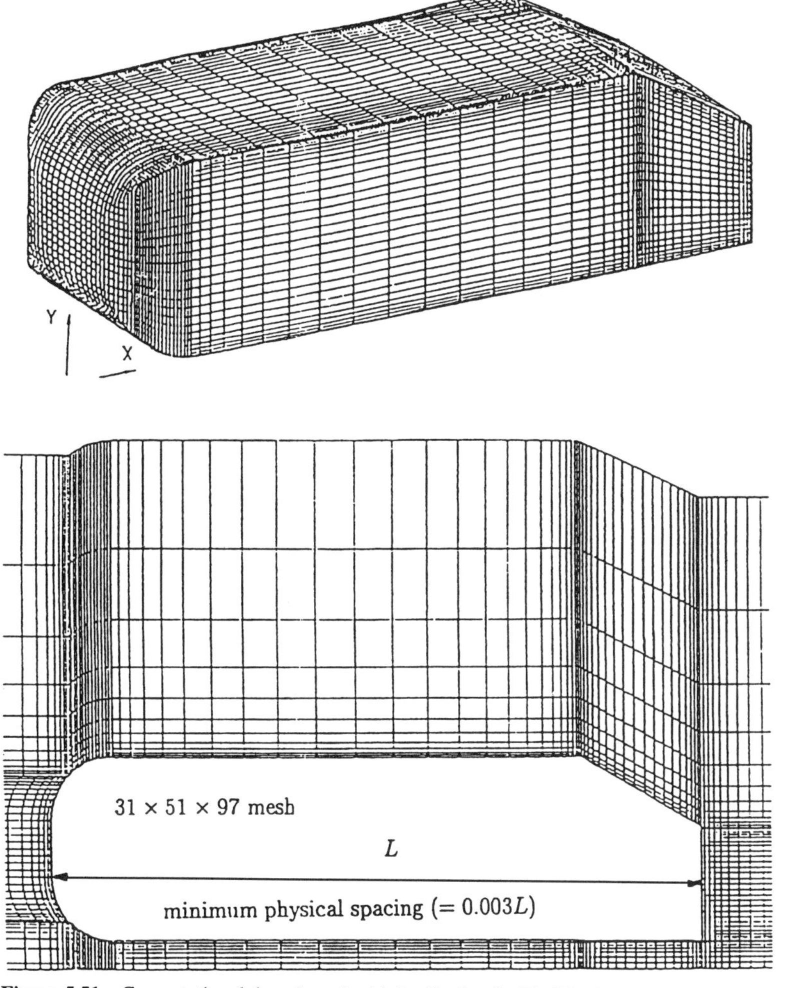

Figure 5.51 Computational domain and grid distribution for bluff body. (Adapted from Han (1987).)

5.6.5 Flow Around a Bluff Body

This work was done by Han (1987)[54]. Figure 5.51 shows the computational domain consisting of 31 × 51 × 97 nodes. The governing equations are

- the continuity equation

$$\frac{\partial U_i}{\partial X_i} = 0, \tag{5.84}$$

and

- the momentum equation,

$$\frac{\partial}{\partial X_i}(U_i U_j) = -\frac{1}{\rho}\frac{\partial P}{\partial X_j} + \frac{\partial}{\partial X_i}\left[(\nu + \nu_t)\left(\frac{\partial U_i}{\partial X_j} + \frac{\partial U_j}{\partial X_i}\right) - \frac{2}{3}\delta_{ij}k\right], \tag{5.85}$$

where U_i, P, and ρ are the mean velocity, the mean pressure, and the density, respectively.

In order to obtain closure of the above-indicated system of equations, the eddy viscosity, ν_t, is related to the turbulent kinetic energy k and its rate of dissipation, ϵ, by

$$\nu_t = C_\mu \frac{k^2}{\epsilon}.$$

Also, the Reynolds stress equation is given by

$$-\overline{u_i u_j} = \nu_t \left(\frac{\partial U_i}{\partial X_j} + \frac{\partial U_j}{\partial X_i} \right) - \frac{2}{3}\delta_{ij} k, \tag{5.86}$$

where C_μ is a constant, and k and ϵ are obtained from the standard k-ϵ turbulence model equations.

In order to facilitate the treatment of the boundary conditions for general curved boundaries, the independent space variables (X, Y, Z) in the above-indicated equations are transformed to a body-fitted coordinate system (ξ, η, ζ), leaving the velocity components (U, V, W) in the (X, Y, Z) coordinates.

Using the standard transformation formulas, the continuity equation becomes

$$U_\xi + V_\eta + W_\zeta = 0, \tag{5.87}$$

and the five transport equations for $\phi = U,\ V,\ W,\ k$, and ϵ can be expressed in the compact form as follows:

$$\begin{aligned}(U\phi)_\xi &+ (V\phi)_\eta + (W\phi)_\zeta \\ &= \left[\frac{\Gamma\phi}{J}(D_{11}^2 + D_{12}^2 + D_{13}^2)\phi_\xi\right]_\xi + \left[\frac{\Gamma\phi}{J}(D_{21}^2 + D_{22}^2 + D_{23}^2)\phi_\eta\right]_\eta \\ &\quad + \left[\frac{\Gamma\phi}{J}(D_{31}^2 + D_{32}^2 + D_{33}^2)\phi_\zeta\right]_\zeta + S + S^\phi J\end{aligned} \tag{5.88}$$

Boundary conditions. Three-component velocities are specified at the inlet and at the outer boundaries from the potential flow solution. The exit plane is located at roughly ten hydraulic diameters downstream of the body; the constant pressure boundary condition is imposed. At the symmetry plane, the symmetry boundary conditions are enforced. At the body surface and at the ground plane, no-slip boundary conditions are imposed. Near the wall, the standard log-law wall functions are used for the velocity. The turbulent kinetic energy and its dissipation rate are specified at the inlet, assuming local equilibrium of turbulence within the boundary layer.

Computational grids. A simple algebraically generated sheared grid system was used for the numerical computation. The grid resolution near the sharp gradients of the flow field is necessary for the accuracy of numerical solutions. Exponential and hyperbolic tangent stretching transformations were used while the desired minimum physical spacing ($=0.003L$) was achieved near the body. The grid distribution near the body in the symmetry plane is shown in Fig. 5.51 for a $31 \times 51 \times 97$ mesh.

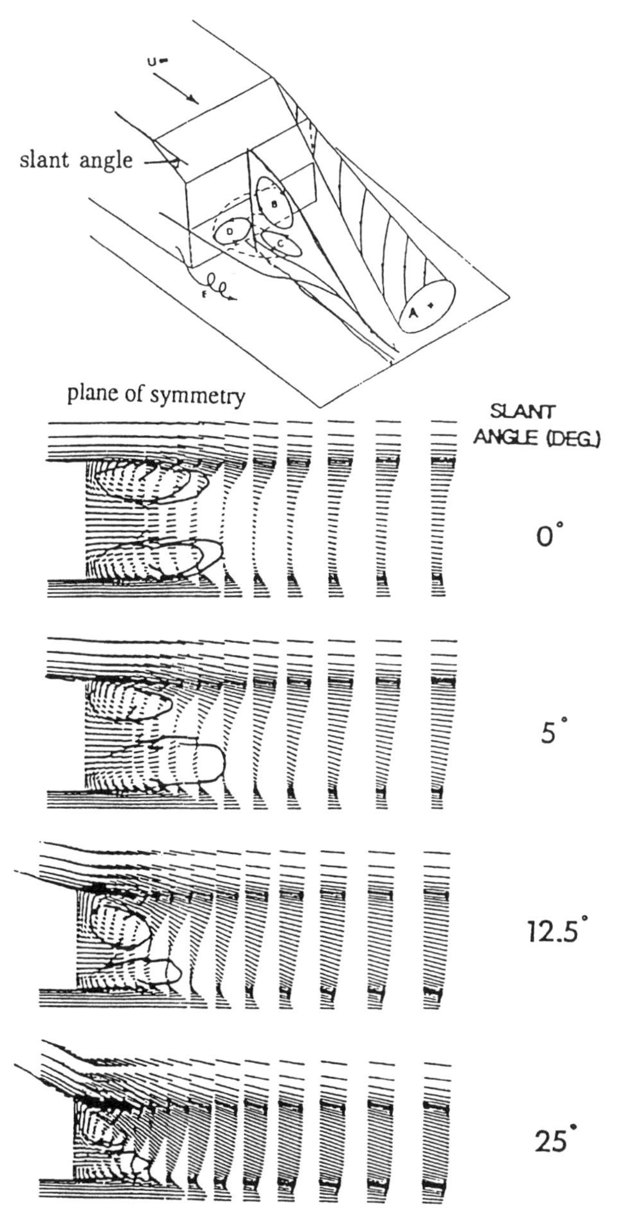

Figure 5.52 Wake pattern in the plane of symmetry. (Adapted from Han (1987).)

5.6.6 Results

Figure 5.52 shows the three-dimensional wake pattern for the body with slanted rear surface and the wake pattern in the plane of symmetry for a variety of slant angles for the rear surface. Figure 5.53 shows the wake patterns in the horizontal plane for the same

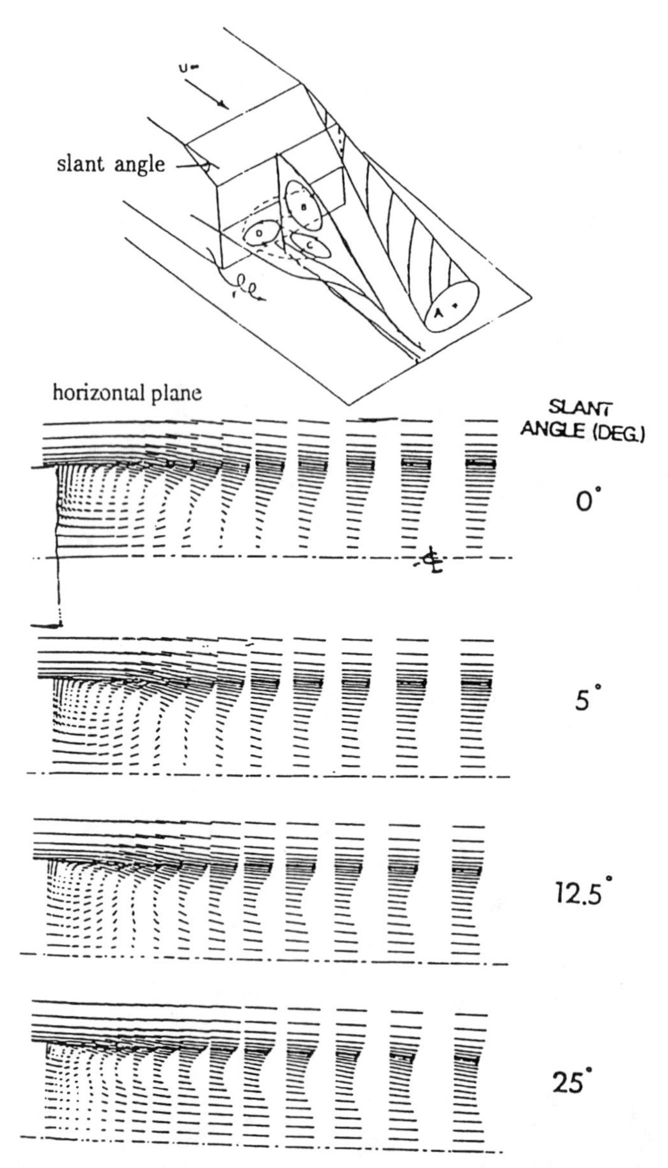

Figure 5.53 Wake pattern in the horizontal plane. (Adapted from Han (1987).)

range of slant angles. Figure 5.54 compares the computed and experimentally obtained wake profiles in the plane of symmetry for a body with a 5° slant angle. Figure 5.55 compares the experimentally and computationally obtained cross-flow velocity distribution in the transverse plane at $x/L = 1.115$ for a 12.5° slant angle. Figure 5.55 also compares the experimentally and computationally obtained wake in the plane of symmetry and compares the experimental and computational cross-flow velocity distribution in the transverse plane at $x/L = 1.479$ for a 25° slant angle body.

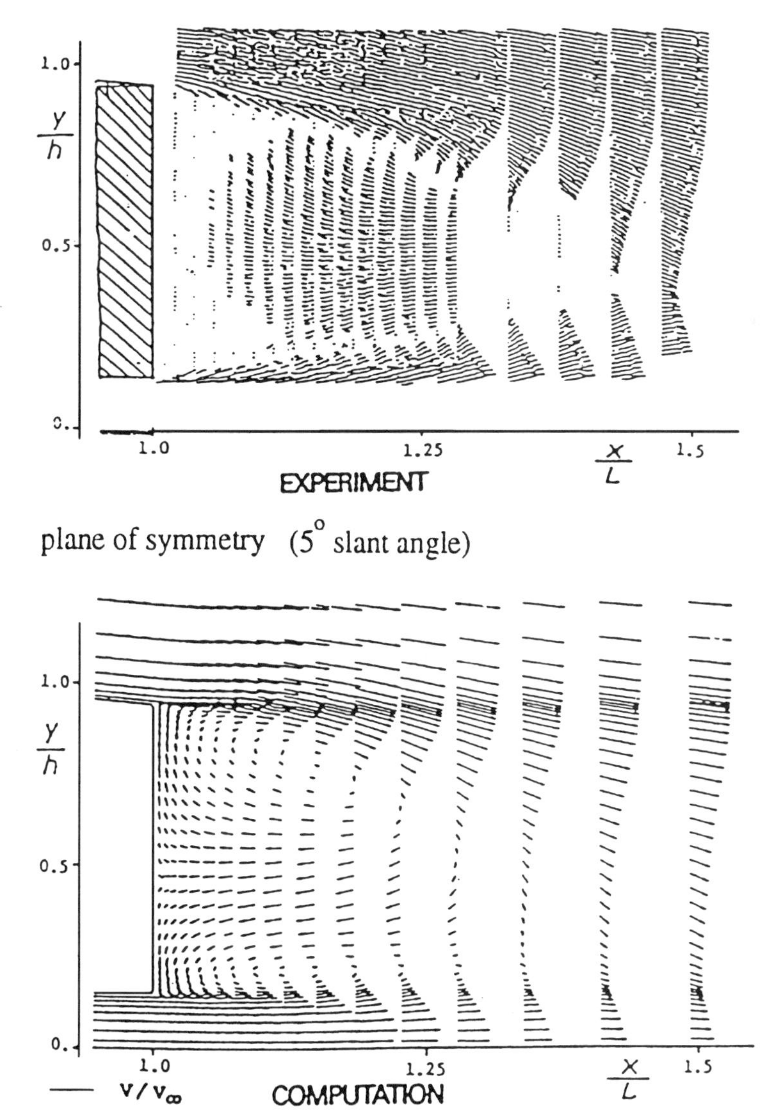

Figure 5.54 Comparison of computation and experiment in the plane of symmetry. (Adapted from Han (1987).)

5.7 ANISOTROPIC TURBULENCE MODELS

Predictions of turbulent wall-shear flows by an anisotropic turbulence model were studied by Jaw (1991)[67] are presented and compared in this section. Three problems are considered, namely, two-dimensional channel flow, driven square cavity flow, and flow past a symmetric backward facing step. These flows have been used by many researchers as benchmark comparisons because many experimental studies are available. Besides, the flow patterns of driven cavity flow and backward facing step flow are complex, although

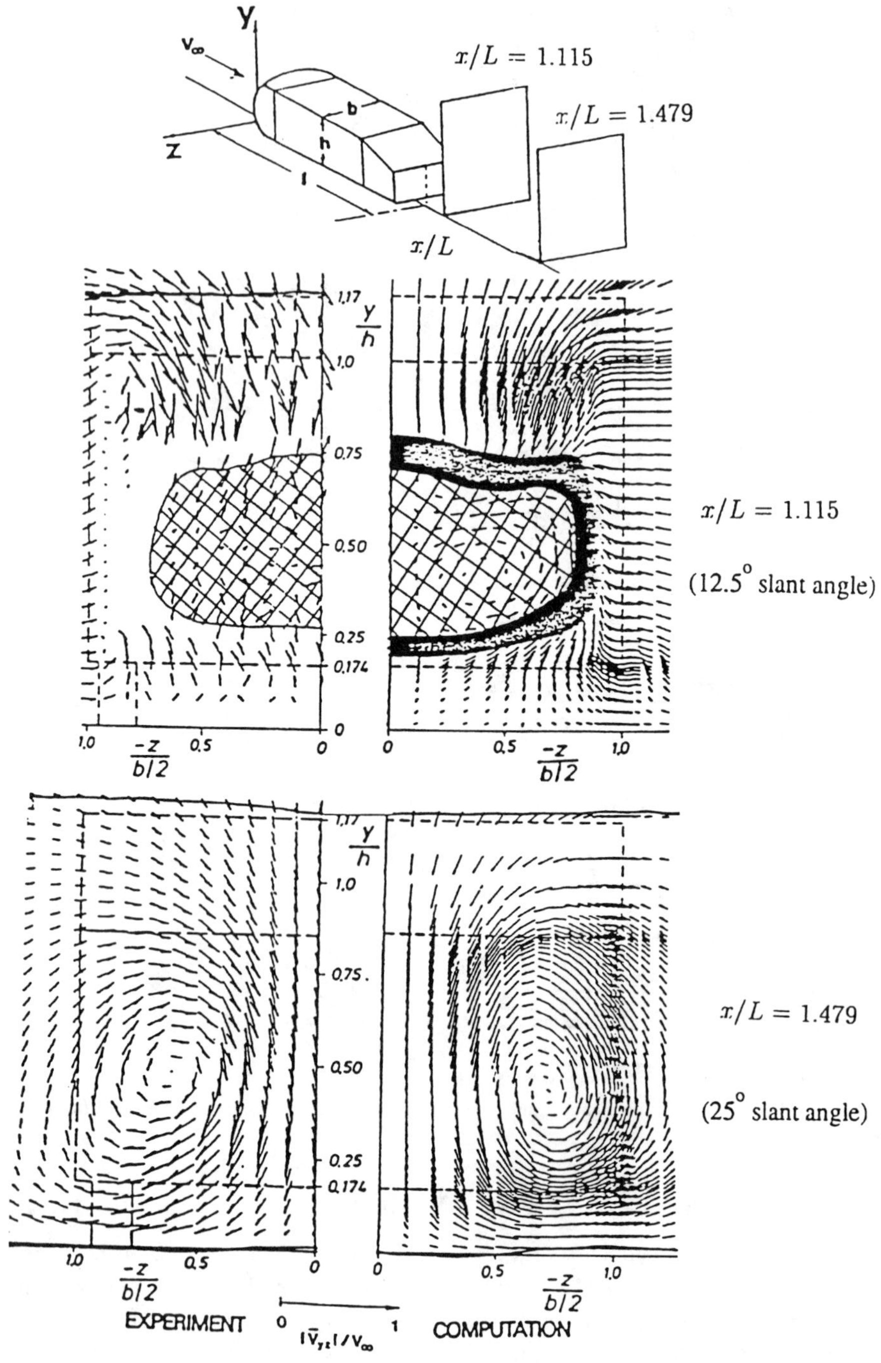

Figure 5.55 Comparison of computation and experiment in the horizontal plane in wake. (Adapted from Han (1987).)

their geometries are simple. Predictions of these flows may verify the predictability of the turbulence models for a flow including wall effects, or even separation, without introducing errors from complex boundary geometry.

Because the tested flows may involve recirculation or separation, the full continuity, momentum, and turbulent transport equations for the two-dimensional incompressible turbulent flow must be used in making predictions. These are as follows:

- The continuity equation is

$$\frac{\partial U}{\partial X} + \frac{\partial V}{\partial Y} = 0. \tag{5.89}$$

- The X-momentum equation is

$$U\frac{\partial U}{\partial X} + V\frac{\partial U}{\partial Y} = \frac{\partial}{\partial X}\left(\nu\frac{\partial U}{\partial X}\right) + \frac{\partial}{\partial Y}\left(\nu\frac{\partial U}{\partial Y}\right) - \frac{\partial}{\partial X}(\overline{u^2}) - \frac{\partial}{\partial Y}(\overline{uv}) - \frac{1}{\rho}\frac{\partial P}{\partial X}. \tag{5.90}$$

- The Y-momentum equation is

$$U\frac{\partial V}{\partial X} + V\frac{\partial V}{\partial Y} = \frac{\partial}{\partial X}\left(\nu\frac{\partial V}{\partial X}\right) + \frac{\partial}{\partial Y}\left(\nu\frac{\partial V}{\partial Y}\right) - \frac{\partial}{\partial X}(\overline{uv}) - \frac{\partial}{\partial Y}(\overline{v^2}) - \frac{1}{\rho}\frac{\partial P}{\partial Y}. \tag{5.91}$$

- The $\overline{uv}$ transport equation is

$$\begin{aligned} U\frac{\partial \overline{uv}}{\partial X} + V\frac{\partial \overline{uv}}{\partial Y} &= \frac{\partial}{\partial X}\left[\left(C_k\frac{k^2}{\epsilon} + \nu\right)\frac{\partial \overline{uv}}{\partial X}\right] + \frac{\partial}{\partial Y}\left[\left(C_k\frac{k^2}{\epsilon} + \nu\right)\frac{\partial \overline{uv}}{\partial Y}\right] \\ &\times (1 + C_1)\frac{\epsilon}{k}\overline{uv} - (1 - C_2)\left(\overline{u^2}\frac{\partial V}{\partial X} + \overline{v^2}\frac{\partial U}{\partial Y}\right). \end{aligned} \tag{5.92}$$

- The $\overline{u^2}$ transport equation is

$$\begin{aligned} U\frac{\partial \overline{u^2}}{\partial X} + V\frac{\partial \overline{u^2}}{\partial Y} &= \frac{\partial}{\partial X}\left[\left(C_k\frac{k^2}{\epsilon} + \nu\right)\frac{\partial \overline{u^2}}{\partial X}\right] + \frac{\partial}{\partial Y}\left[\left(C_k\frac{k^2}{\epsilon} + \nu\right)\frac{\partial \overline{u^2}}{\partial Y}\right] \\ &+ \frac{2}{3}C_1\epsilon - (1 + C_1)\frac{\epsilon}{k}\overline{u^2} - \frac{2}{3}C_2\left[\overline{uv}\left(\frac{\partial U}{\partial Y} + \frac{\partial V}{\partial X}\right)\right] \\ &- 2\left(1 - \frac{2}{3}C_2\right)\left(\overline{u^2}\frac{\partial U}{\partial X} + \overline{uv}\frac{\partial U}{\partial Y}\right). \end{aligned} \tag{5.93}$$

- The $\overline{v^2}$ transport equation is

$$\begin{aligned} U\frac{\partial \overline{v^2}}{\partial X} + V\frac{\partial \overline{v^2}}{\partial Y} &= \frac{\partial}{\partial X}\left[\left(C_k\frac{k^2}{\epsilon} + \nu\right)\frac{\partial \overline{v^2}}{\partial X}\right] + \frac{\partial}{\partial Y}\left[\left(C_k\frac{k^2}{\epsilon} + \nu\right)\frac{\partial \overline{v^2}}{\partial Y}\right] \\ &+ \frac{2}{3}C_1\epsilon - (1 + C_1)\frac{\epsilon}{k}\overline{v^2} - \frac{2}{3}C_2\left[\overline{uv}\left(\frac{\partial U}{\partial Y} + \frac{\partial V}{\partial X}\right)\right] \\ &- 2\left(1 - \frac{2}{3}C_2\right)\left(\overline{uv}\frac{\partial V}{\partial X} + \overline{v^2}\frac{\partial V}{\partial Y}\right). \end{aligned} \tag{5.94}$$

- The kinetic energy, k, equation is given by

$$U\frac{\partial k}{\partial X}+V\frac{\partial k}{\partial Y}=\frac{\partial}{\partial X}\left[\left(C_k\frac{k^2}{\epsilon}+\nu\right)\frac{\partial k}{\partial X}\right]+\frac{\partial}{\partial Y}\left[\left(C_k\frac{k^2}{\epsilon}+\nu\right)\frac{\partial k}{\partial Y}\right]$$
$$-\left[\overline{uv}\left(\frac{\partial U}{\partial Y}+\frac{\partial V}{\partial X}\right)\right]-\overline{u^2}\frac{\partial U}{\partial X}-\overline{v^2}\frac{\partial V}{\partial Y}-\epsilon. \tag{5.95}$$

- The rate of dissipation, ϵ, equation including the cross diffusion term of turbulent kinetic energy, k, is given by Jaw (1991)[67]

$$U\frac{\partial \epsilon}{\partial X}+V\frac{\partial \epsilon}{\partial Y}=\frac{\partial}{\partial X}\left[\left(C_\epsilon\frac{k^2}{\epsilon}+\nu\right)\frac{\partial \epsilon}{\partial X}\right]+\frac{\partial}{\partial Y}\left[\left(C_\epsilon\frac{k^2}{\epsilon}+\nu\right)\frac{\partial \epsilon}{\partial Y}\right]$$
$$-C_{\epsilon1}\frac{\epsilon}{k}\left[\overline{uv}\left(\frac{\partial U}{\partial Y}+\frac{\partial V}{\partial X}\right)+\overline{u^2}\frac{\partial U}{\partial X}+\overline{v^2}\frac{\partial V}{\partial Y}\right]-C_{\epsilon2}\frac{\epsilon^2}{k}$$
$$+C_{\epsilon3}\frac{\epsilon}{k}\left\{\frac{\partial}{\partial X}\left[\left(C_k\frac{k^2}{\epsilon}+\nu\right)\frac{\partial k}{\partial X}\right]\right\}$$
$$+C_{\epsilon3}\frac{\epsilon}{k}\left\{\frac{\partial}{\partial Y}\left[\left(C_k\frac{k^2}{\epsilon}+\nu\right)\frac{\partial k}{\partial Y}\right]\right\}. \tag{5.96}$$

For the eddy viscosity model, the Reynolds stresses for each component read

$$-\overline{uu}=2\nu_t\left(\frac{\partial U}{\partial X}\right)-\frac{2}{3}k \tag{5.97}$$

$$-\overline{vv}=2\nu_t\left(\frac{\partial V}{\partial Y}\right)-\frac{2}{3}k \tag{5.98}$$

$$-\overline{uv}=\nu_t\left(\frac{\partial V}{\partial X}+\frac{\partial U}{\partial Y}\right). \tag{5.99}$$

The above-indicated k-ϵ-E model with the cross-diffusion term of k in the ϵ is termed as the k-ϵ-EX model by Jaw (1991)[67]. The model coefficients adopted are listed in Table 5.2. The turbulence model coefficients adopted for wall-shear flows are different from those adopted for free-shear flows (chapter 3, Section 3.2.8). As mentioned in chapter 2, Section 2.6, determining model coefficients from some specific experiments clearly indicates that several model coefficients may be variables or functions of some parameters and are correlated with one another. Although at present the available

Table 5.2 Turbulence model coefficients for wall-shear flows

Models	C_μ	C_k	C_1	C_2	C_ϵ	$C_{\epsilon1}$	$C_{\epsilon2}$	$C_{\epsilon3}$
RSM		0.09	2.14	0.14	0.06	1.55	1.92	0.01
k-ϵ-EX	0.09	0.09			0.06	1.55	1.92	0.01
k-ϵ-E	0.09	0.09			0.07	1.44	1.92	0.00

experimental data are not detailed enough to determine precisely the functional form, calculating turbulent flows with a unique set of model coefficients seems to be difficult. Using different model coefficients for wall-shear and free-shear flows may improve their overall performance. In this section, the model coefficients of C_ϵ, $C_{\epsilon 1}$, $C_{\epsilon 2}$, and C_2 adopted for RSM and k-ϵ-EX model are determined in a systematic way for the wall-shear flows by Jaw [67]. For the examples illustrated in this section, the wall function approach is adopted. It should also be pointed out that the wall function dominated region is large enough to overlay the near-wall diffusion effective region. The effects of the cross-diffusion term in the RSM and the k-ϵ-EX models may be quite small. In this situation the k-ϵ-EX and k-ϵ-E models have the same eddy viscosity models but with different model coefficients.

5.7.1 Prediction of Two-Dimensional Channel Flow

In predicting the channel flow, the models are tested for their ability to model the turbulence production, dissipation, and redistribution of Reynolds stresses. The prediction of the mean-velocity profile for the U component was, in general, satisfactory. Adopting the wall function approach to provide the near-wall turbulent behavior is considered to be appropriate. Therefore, the prediction of the flow is designed to test critically the turbulence model away from the wall. Because the pressure is constant in cross-section, the distributions of the turbulent transport quantities, such as k, ϵ, and $\overline{u_i u_j}$, are essentially the result of turbulence production, dissipation, and redistribution.

Computations were performed in a $W \times 24\,W$ two-dimensional channel with the Reynolds number, $Re = U_m W/\nu = 65{,}600$, the same as that of Telbany and Reynolds's (1981)[175] experiment. Here W is the channel half width, and U_m is the mean-inlet velocity. Figure 5.56 compares the predicted mean-velocity profiles with the experimental data at three different locations. One is in the developing flow region, $X = 3.5$, and the other two are in the fully developed flow region, $X = 13.5$ and $X = 22$. It is found that the mean velocity profiles predicted by all three models almost collapse into one curve and agree well with the experimental data.

Figure 5.57 compares the predicted turbulent kinetic energy and Reynolds stress distribution with experimental data at the fully developed location, $X/W = 22$. All three turbulence models predict satisfactorily the Reynolds stress $\overline{uv}$ when compared with experimental measurements. For this simple flow, the prediction results of the k-ϵ-EX and k-ϵ-E models are close to each other. However, for k, especially $\overline{u^2}^{1/2}$ and $\overline{v^2}^{1/2}$ profiles, the RSM predicts better results than the k-ϵ-EX and k-ϵ-E models.

5.7.2 Prediction of Square Cavity Flow

In the case of a cavity flow, the turbulence models are tested for their ability to predict both the U and V velocity components when both velocities have significant magnitudes. The turbulent flow in a cavity, in general, recirculates and does not separate. The wall function approach is considered to be applicable, although the wall functions may introduce some errors at the cavity corner where the flow is not parallel to the surface and thus violates the wall function assumption (Chen, 1983 [12]). The square cavity flow predicted is

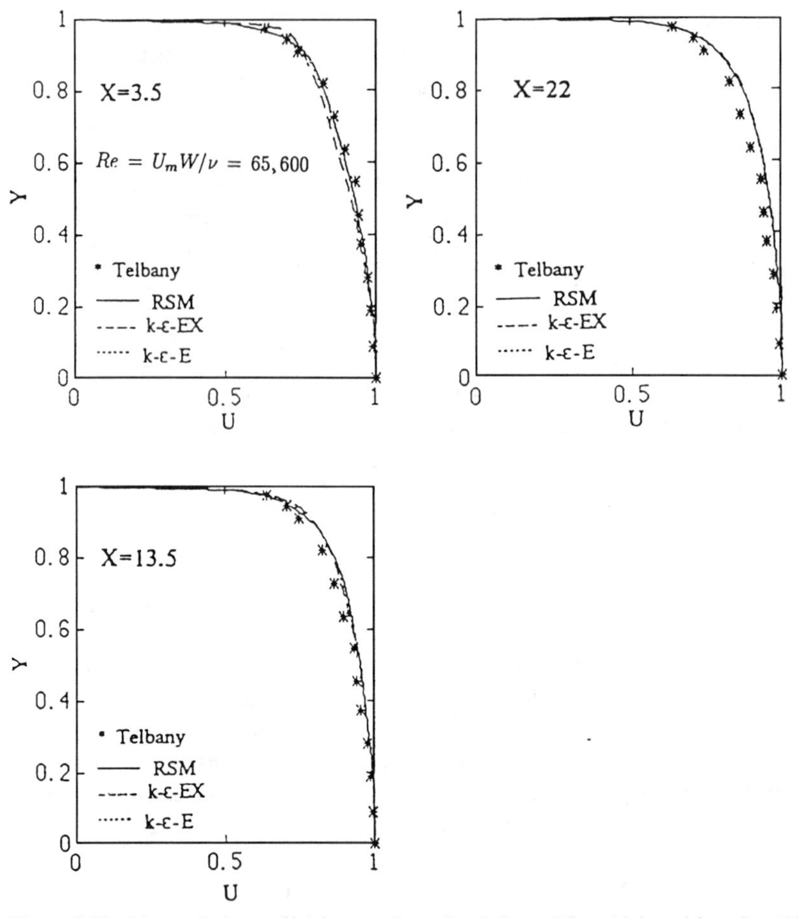

Figure 5.56 Mean velocity profiles for two-dimensional channel flow. (Adapted from Jaw (1991).)

a turbulent flow with a Reynolds number of 480,000. The numerical results of mean velocities on the horizontal and vertical centerlines and the Reynolds stress $\overline{v^2}$ on the horizontal centerline are compared with the experimental data of Grand (1975)[53], Girard and Curlet (1975)[52], and Mills (1965)[108].

Figure 5.58 presents the predicted profiles of the horizontal mean velocity, U/U_m, along the vertical centerline. The grid sizes of 50×50 were used by Jaw (1991)[67]. The symbol * in this figure represents the experimental data of Mills [108] for $Re = 100{,}000$. This figure shows that all three turbulence models predict almost identical U profiles. Mill's [108] measurement did not provide the turbulence values.

Figure 5.58 also presents the predicted results of V/U_m and $\overline{v^2}^{1/2}/U_m$ along the horizontal centerline. The symbols * and + in the figure are the experimental data of Grand (1975)[53] and Girard and Curlet (1975)[52], respectively. The predicted mean-velocity profiles, both U/U_m and V/U_m in Fig. 5.58, are closer to the experimental data in the near-wall region but become flatter than the experimental measurements in the interior

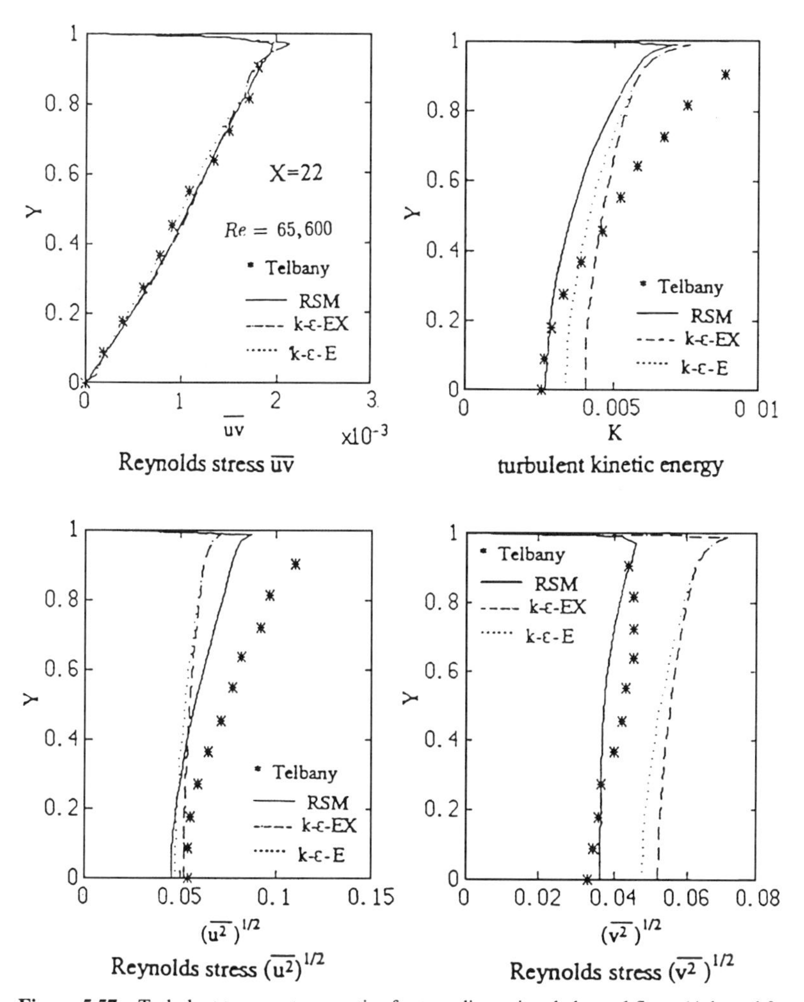

Figure 5.57 Turbulent transport properties for two-dimensional channel flow. (Adapted from Jaw (1991).)

region. This implies that all three turbulence models predict smaller rotating motions. A similar result was obtained by Bernatz (1991)[5], who also used the FA method but he computed with 60×60 grid points. The discrepancy between the calculated results and the experimental data is probably caused by the inaccurate modeling and the use of wall functions that are not appropriate at the cavity corners. The straight profile of velocity over the interior portions shows that the fluid is essentially rotating as a solid body. Hence, the rate of strain and the Reynolds stress near the center are nearly constant, as compared with near the walls, as shown in Fig. 5.58. Prediction of the velocity profile by the RSM also presents similar behavior except that the velocity magnitude is smaller than the experimental data. Both the k-ϵ-EX and the k-ϵ-E models predict rounded distributions and are considered to be worse than the RSM.

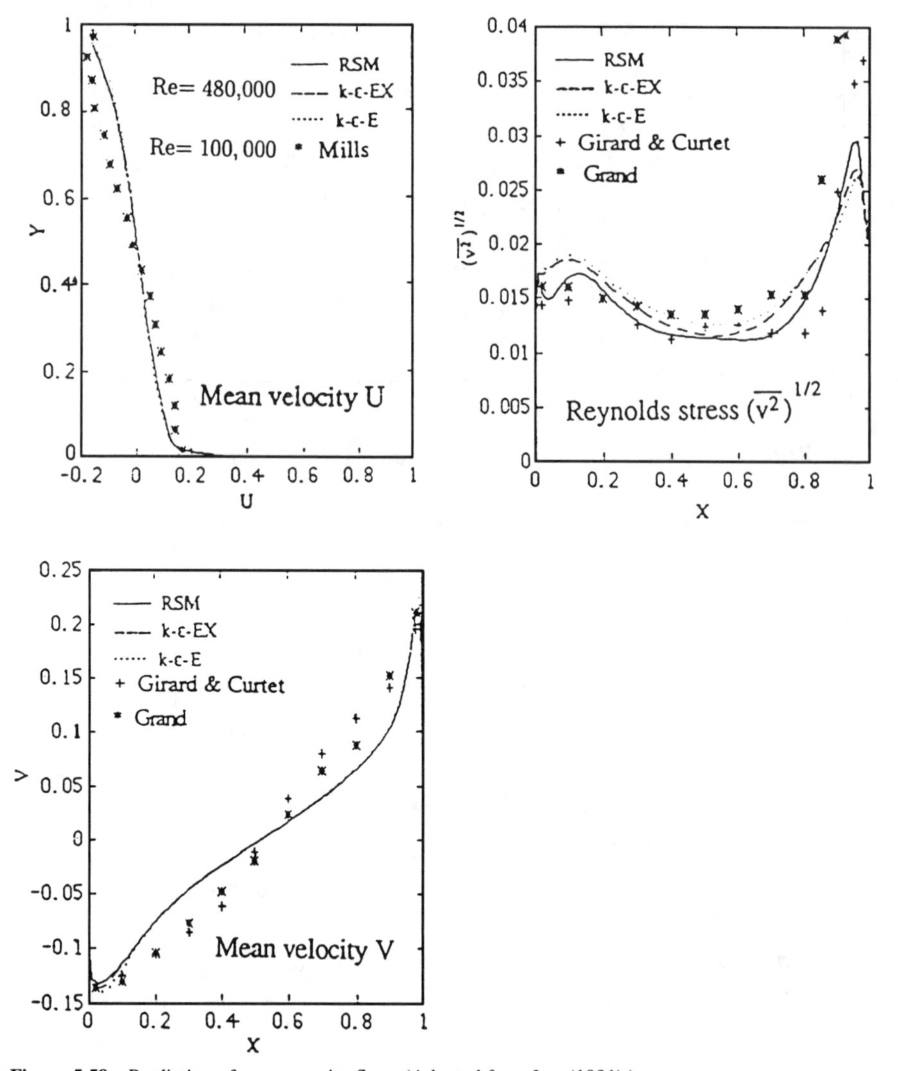

Figure 5.58 Prediction of square cavity flow. (Adapted from Jaw (1991).)

5.7.3 Prediction of Backward Facing Step Flow

A turbulent flow past a backward facing step has extensively been measured because it is considered by many researchers and engineers as a fundamental configuration of internal flow. This flow can provide an additional domain of test for turbulence models because the flow separates at the step and reattaches downstream. The flow then recirculates behind the step. Predictions of this flow thus can examine the capability of turbulence models in predicting not only distribution of turbulent transport quantities but also the mean-velocity distribution, particularly near the reattachment where

Table 5.3 Comparison of lengths of separation zone

Models	Length of separation zone (no. of step height)
Smyth (1979)[157] Experiment data	4.5
RSM	3.9
k-ϵ-EX	4.5
k-ϵ-E	3.0

the turbulence model is used in conjunction with the wall functions. Use of wall functions near the points of separation or reattachment is known to be inappropriate because the flow is no longer parallel and thus violates the wall functions' assumption. The use of the wall functions in predicting the backward facing step flow is largely due to engineering simplification. As an engineering approximation, the wall functions are considered appropriate for the near-wall flow both inside and outside the recirculation zone except at or in the close neighborhood of reattachment or separation points. The following is an attempt to examine the validity of the three turbulence models with the wall function approach in predicting flow separation as it passes a backward facing step.

The backward facing step flow considered is the same as Smyth's experiment (1979)[157], with a channel expansion ratio of 1.5 and a Reynolds number $Re = U_m W/\nu = 30{,}210$. Here W is the step height, and U_m is the mean-inlet velocity. In predicting the separation flows, a criterion that can immediately distinguish models' performance is to check the predicted length of the separation zone with the experiment. Predicted separation lengths compared with experimental data of Smyth (1979)[157] are listed in Table 5.3. It is found that the k-ϵ-EX model and the RSM predict much better results than the k-ϵ-E model. Comparing the velocity distributions will also reveal the same results. Figure 5.59 presents three velocity profiles within the separation zone, at $X = 0.4$, 0.8, and 1.2. Note that the reverse flow predicted by the k-ϵ-E model decays too quickly, hence, its separation length is too short. Velocity distributions of the RSM and the k-ϵ-EX model are much closer to the experimental data, but the RSM decays a little faster than the k-ϵ-EX model. The experimental separation length, 4.5 times step height, corresponds to the longitudinal distance $X = 1.5$ in the computational domain. Figure 5.60 is the velocity profile outside the separation zone at $X = 4.0$. All of the predicted profiles in this location are considered satisfactory.

Figure 5.61 presents the normal stress, $\overline{v^2}^{1/2}$, distributions at four locations. It is quite clear that the RSM predicts much better results than the k-ϵ-EX or the k-ϵ-E models.

5.7.4 Summary

From the predictions of the turbulent wall-shear flows presented above, it can be concluded that the RSM and the k-ϵ-EX model provide better prediction capability than the k-ϵ-E model. For the two-dimensional channel flow and the square cavity flow, the

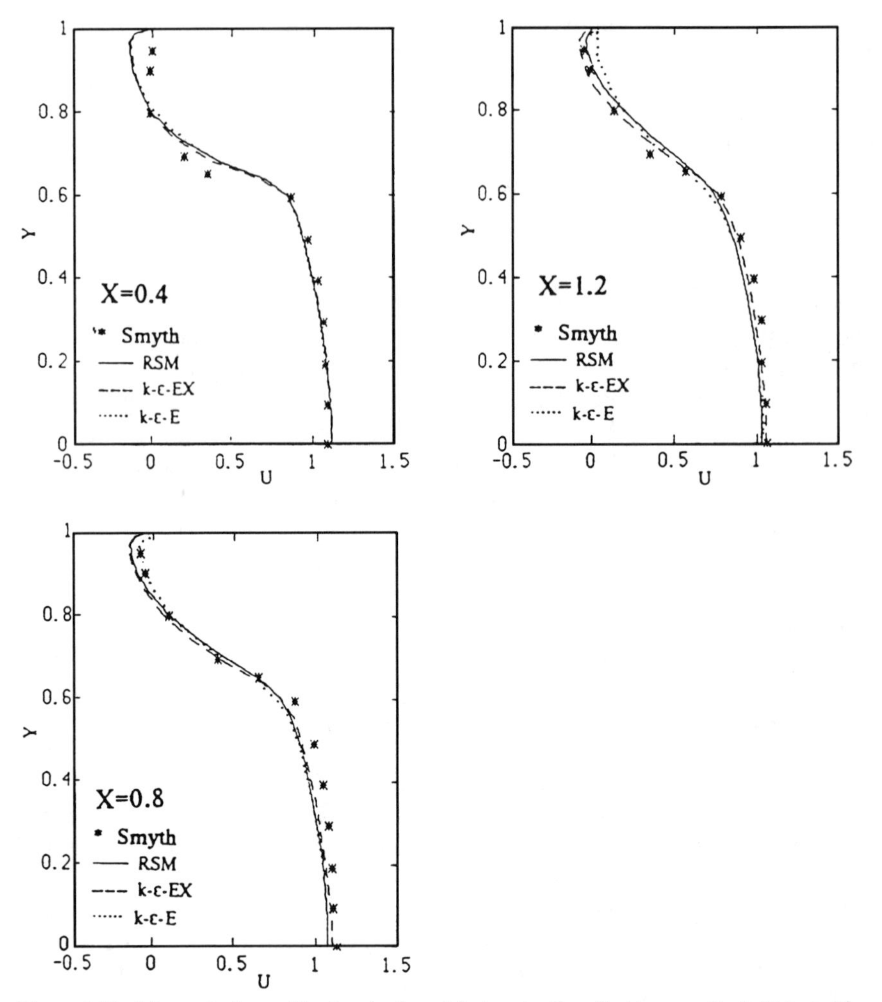

Figure 5.59 Mean velocity profiles for a backward facing step flow (Inside separation). (Adapted from Jaw (1991).)

k-ϵ-EX and k-ϵ-E models predict similar results in both mean velocities and turbulent quantities. But for the backward facing step flow, the k-ϵ-EX model predicts much better results than the k-ϵ-E model. The k-ϵ-EX model is able to predict satisfactorily the separation length and the reverse flow in the separation region, whereas the k-ϵ-E model underpredicts these quantities too much. The RSM in general predicts similar mean-flow quantities as those of the k-ϵ-EX model because the same model coefficients for k and ϵ equations are adopted. However, the RSM prediction of Reynolds stress distributions is better, especially the $\overline{v^2}^{1/2}$ component, but at the expense of solving an additional three differential equations. It should be remarked that these improvements

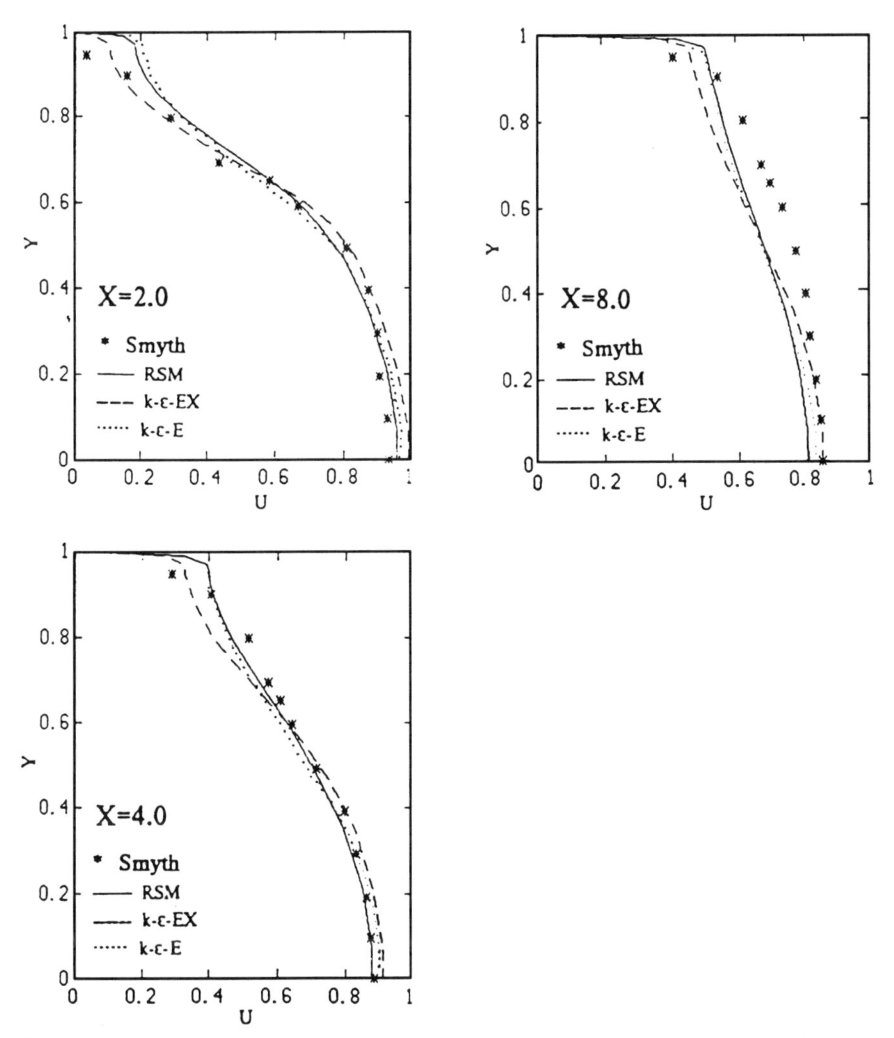

Figure 5.60 Mean-velocity profiles for a backward facing step flow (Downstream from separation). (Adapted from Jaw (1991).)

are limited because the mean-flow quantities predicted by the k-ϵ-EX model are even better than those predicted by the RSM.

As mentioned earlier, applying the wall function approach in the region close to the reattachment, or the separation, points is inappropriate. However, the number of nodes for which the wall functions are not appropriate is far less than the total number of wall nodes. Thus, for engineering applications, the wall function approach is still popular among users because of its simplicity. Besides, by carefully determining the turbulence model coefficients, the use of the wall functions approach can still predict satisfactorily some complex flows, such as flow with separation.

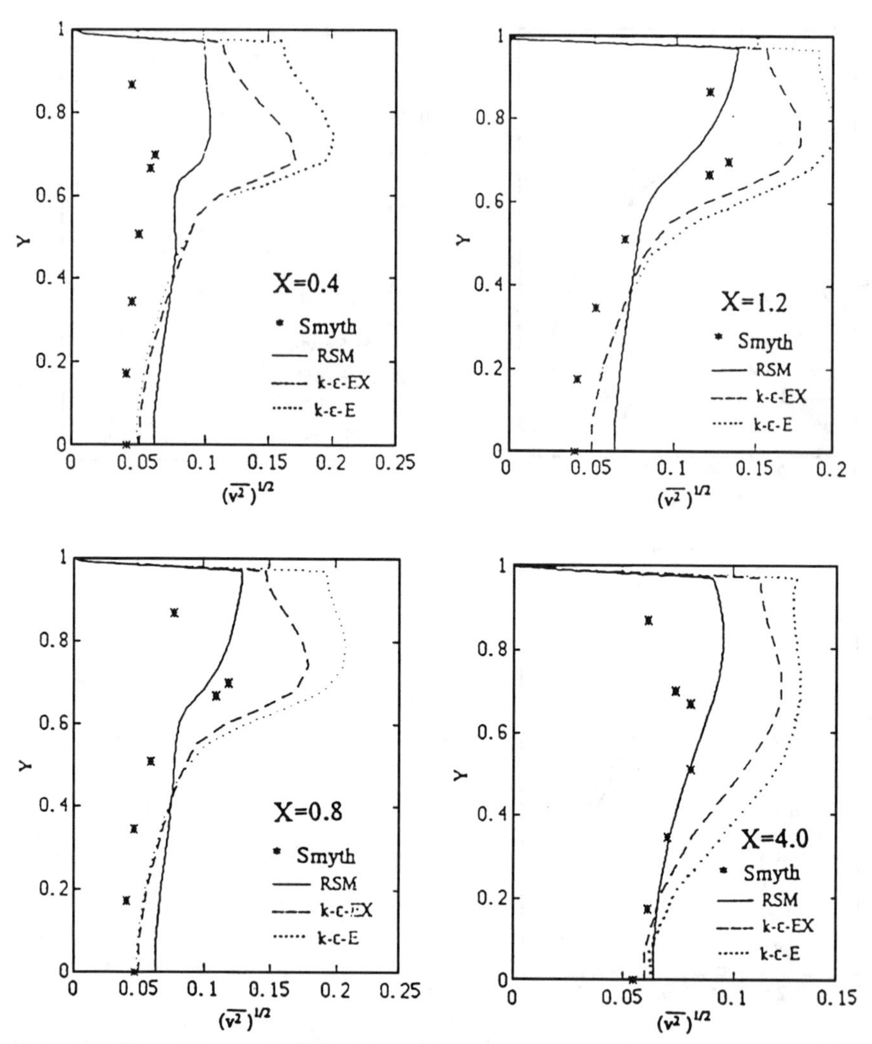

Figure 5.61 Reynolds stress distribution for a backward facing step flow. (Adapted from Jaw (1991).)

5.8 CONCLUSION

5.8.1 The Differential Model (RSM)

This is the most rigorous second-order, one-point turbulence model. Although not tested for many flows, when used, it seems to yield the best predictions as compared with any other simplified second-order or lower order (i.e., zero-equation) model. The prediction capability is greatly improved over the other models, even though the modeling of the ϵ equation is weak.

It is hoped that the inclusion of the two scale concept (k, ϵ)–(ν, ϵ) will further improve the second-order closure model. The disadvantage of this model is only temporary in that it requires a fair amount of computational effort.

5.8.2 The Algebraic Stress Model (ASM or k-ϵ-A)

This is the most popular model. It has the advantage of simplicity over the RSM closure model. Its performance is generally good, and sometimes one cannot determine the real advantage of using more complex models (RSM) than the k-ϵ-A model.

The use of Rodi's hypothesis to generate algebraic stress model seems to improve the k-ϵ-A model. This model is recommended for most engineering applications.

5.8.3 The Eddy Viscosity Model (k-ϵ-E)

The k-ϵ-E model is also one of the more popular turbulence models in use by engineers. It is less accurate than the k-ϵ-A model, but the use of the k-ϵ-E model will simplify the numerical effort even further from the use of the k-ϵ-A model. For a complex engineering problem, the k-ϵ-E model may be sufficient to produce engineering data such as mean-flow quantities and flow patterns. The k-ϵ-EX model may improve the prediction capability.

The turbulence models simpler than the k-ϵ-E model may suffer from decreased predictability. However, if the repeated study of similar geometry is required, one equation model may be used.

CHAPTER

SIX

TURBULENT BUOYANT FLOWS

6.1 INTRODUCTION

All flows on earth are subject to the gravitational force. Even incompressible fluids are affected by gravity because they are subject to differences in hydrostatic force. Hydraulic power generation is a good example of utilizing gravitational potential energy.

Flow patterns are further complicated when the density of a fluid is variable, either because of a temperature difference or a concentration difference. The force produced by gravity and density differences is called *buoyant force*. In this chapter we study the turbulence modeling of turbulent buoyant flows and its applications.

6.2 EQUATION OF STATE

Although the variable density appears naturally in continuity and momentum equations, it is difficult or cumbersome to devise an equation for conservation of energy with density as the dependent variable, which would require the state equation to relate the density to temperature. For example, an ideal gas relation is $\rho = p/RT$ (where p is the pressure, R is the gas constant, and T is the absolute temperature). For a liquid, the equation of state is complicated.

We choose to keep the energy equation with T as the dependent variable and to replace the density ρ by the temperature T in the continuity and momentum equations.

Because the density is considered to be variable, it becomes an unknown variable. In order to close the problem, an additional independent equation other than the continuity, momentum, and energy equations must be supplied. This is the equation of state, and it is an empirical equation. Depending on the fluid and the range of parameters (pressure and temperature), the equation of state differs.

For gases at low pressure (less than ten atmospheres) and low temperature (less than 1000 °F), the equation of state is approximately

$$\rho = \frac{p}{RT}, \tag{6.1}$$

where R is the gas constant, p is the absolute pressure, and T is the absolute temperature.

For a liquid, the relation among p, ρ, and T is not a simple one. In general,

$$\rho = f(p, T) \tag{6.2}$$

or

$$d\rho = \left(\frac{\partial \rho}{\partial p}\right)_T dp + \left(\frac{\partial \rho}{\partial T}\right)_p dT. \tag{6.3}$$

For small variations of pressure and temperature from a reference state, ρ_s,P_s, and T_s, one may approximately write

$$\frac{\rho - \rho_s}{\rho_s} = \alpha(p - p_s) - \beta(T - T_s), \tag{6.4}$$

where

$$\alpha = \frac{1}{\rho}\left(\frac{\partial \rho}{\partial P}\right)_T\bigg|_s > 0$$

$$\beta = -\frac{1}{\rho}\left(\frac{\partial \rho}{\partial T}\right)_P\bigg|_s > 0.$$

α and β are called coefficients of expansion and are evaluated at the reference state, T_s, and p_s, of the problem. α and β are thus functions of reference state variables and often have a positive value.

For problems when the change in pressure is small compared with change in temperature, one often finds $\alpha \Delta p < -\beta \Delta T$. Thus,

$$\frac{\rho - \rho_s}{\rho_s} \approx -\beta(T - T_s).$$

For ideal gas, one has

$$\beta = -\frac{1}{\rho}\left(\frac{\partial \rho}{\partial T}\right)_p\bigg|_s = \frac{1}{T_s},$$

where T is the absolute temperature. For liquids, it is not easy to have a simple expression for β. However, one may use data, such as those shown in Fig. 6.1. For example, at one atmosphere $\rho(\text{g/cm}^3)$ may be written as a function of T (°C).

- Water: $\rho = 1 - 2.76787 \times 10^{-5} \times (T - 4) - 5.74221 \times 10^{-6} \times (T - 4)^2 + 1.60668 \times 10^{-6} \times (T - 4)^3$.
- Sodium as reported by Moriya (1983): $\rho = 0.95 - 2.3 \times 10^{-4} \times T - 1.46 \times 10^{-8} \times T^2 + 5.64 \times 10^{-12} \times T^3$.

In general, thus, one writes $\rho = a_o - a_1 T - a_2 T^2 - a_3 T^3$.

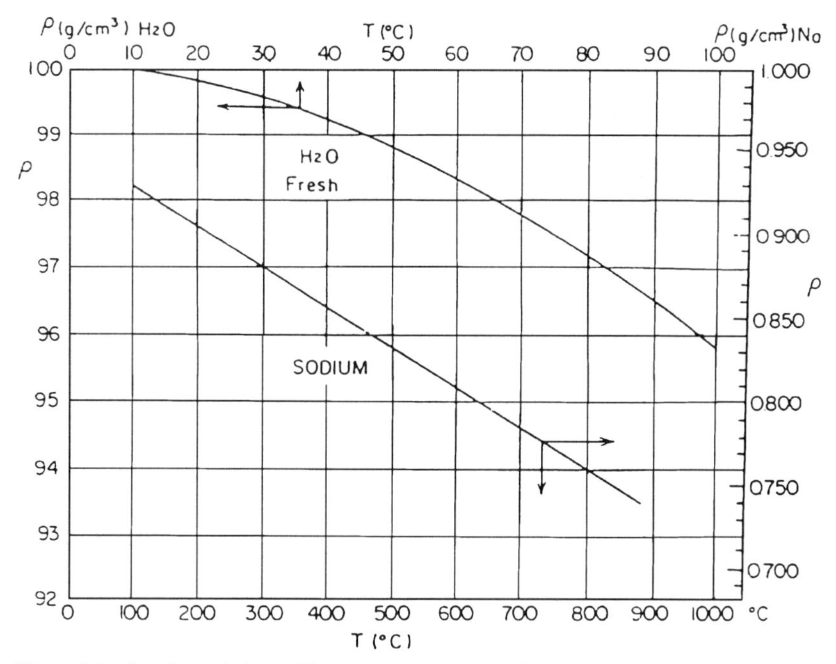

Figure 6.1 Density variations with respect to temperature change.

For small ΔT and Δp, $\rho \approx a_o - a_1 T$ for liquids. Thus, for $\beta > 0$,

$$\beta = -\frac{1}{\rho}\left(\frac{\partial \rho}{\partial T}\right)_p\bigg|_s \approx \frac{a_1}{\rho_s} \approx \frac{a_1}{a_o - a_1 T_s}.$$

6.3 BOUSSINESQ APPROXIMATION

For problems in which the density gradient is not large, the governing equations can be simplified by invoking the Boussinesq approximation (Boussinesq, 1877 [6]). In the Boussinesq approximation, the density variation is considered only significant in the gravitational term. In other words, the density variations in the conservation of mass, in the time rate change of momentum, and in the work done because of density change, are considered negligible.

The governing equations for instantaneous flow under the Boussinesq approximation are the

- continuity equation,

$$\frac{\partial U_i^*}{\partial X_i} = 0, \tag{6.5}$$

- momentum equations,

$$\rho_s \frac{DU_i^*}{Dt} = -\frac{\partial p^*}{\partial X_i} + \rho^* g_i + \mu \frac{\partial^2 U_i^*}{\partial X_j \partial X_j}, \tag{6.6}$$

and

- energy equation,

$$\rho_s C_p \frac{DT^*}{Dt} = K \frac{\partial^2 T^*}{\partial X_j \partial X_j} + \phi, \tag{6.7}$$

where ρ_s, C_p, K, and μ are considered to be constants, whereas ρ^*, U_i^* , and T^* are instantaneous variables.

Let p^* and ρ^* at the static state ($U_i^* = 0$) be P_s, ρ_s. Then the momentum equation at static state is

$$0 = -\frac{\partial P_s}{\partial x_i} + \rho_s g_i. \tag{6.8}$$

Subtracting the two momentum equations (Eqs. 6.6–6.8), we have

$$\rho_s \frac{DU_i^*}{Dt} = -\frac{\partial (p^* - P_s)}{\partial X_i} + (\rho^* - \rho_s) g_i + \mu \frac{\partial^2 U_i^*}{\partial X_j \partial X_j}, \tag{6.9}$$

where the term containing $(p^* - P_s)$ represents the pressure term without the hydrostatic force, and the term $(\rho^* - \rho_s)g_i$ represents the buoyant force.

6.4 AVERAGED TURBULENCE EQUATIONS

With the Boussinesq approximation we let $P^* = p^* - P_s$,

$$\frac{\rho^* - \rho_s}{\rho_s} = -\beta(T^* - T_s) = -\beta \Delta T^*$$

and T_s = constant, $\rho_s = \rho$ = constant, where the suffix s denotes the reference static state. Then the equations of motion can be written as the

- continuity equation,

$$\frac{\partial U_i^*}{\partial X_i} = 0, \tag{6.10}$$

- momentum equation,

$$\frac{DU_i^*}{Dt} = -\frac{1}{\rho}\frac{\partial P^*}{\partial X_i} - \beta \Delta T^* g_i + \nu \frac{\partial^2 U_i^*}{\partial X_j \partial X_j}, \tag{6.11}$$

and

- energy equation,

$$\frac{DT^*}{Dt} = \alpha \frac{\partial^2 T^*}{\partial X_j \partial X_j} + \tau_{ij}^* \frac{\partial U_i^*}{\partial X_j}. \tag{6.12}$$

Taking a short time or ensemble average of the governing equation and letting $U_i^* = U_i + u_i$, $P^* = P + p$, $T^* = T + \theta$, and $\tau_{ij}^* = \tau_{ij} + \tau_{ij}'$, the turbulence equations can be written as the

- continuity equation,

$$\frac{\partial U_i}{\partial X_i} = 0, \tag{6.13}$$

- momentum equation,

$$\frac{DU_i}{Dt} = -\frac{1}{\rho}\frac{\partial P}{\partial X_i} - \beta \Delta T g_i + \nu \frac{\partial^2 U_i}{\partial X_j \partial X_j} - \frac{\partial \overline{u_i u_j}}{\partial X_i}, \tag{6.14}$$

and

- energy equation,

$$\frac{DT}{Dt} = \alpha \frac{\partial^2 T}{\partial X_j \partial X_j} - \frac{\partial \overline{u_i \theta}}{\partial X_i} + \frac{\overline{\tau'_{ij}\frac{\partial u_i}{\partial X_j}} + \phi}{C_p \rho}, \tag{6.15}$$

where the terms $\overline{u_i u_j}$, $\overline{u_i \theta}$, and $\overline{\tau'_{ij}\frac{\partial u_i}{\partial X_j}}$ need modeling.

In order to model $\overline{u_i u_j}$ and $\overline{u_i \theta}$, we have the equations for the fluctuating components as the following:

- For the fluctuating component of the velocity, we have

$$\frac{\partial u_i}{\partial t} + U_l \frac{\partial u_i}{\partial X_l} + u_l \frac{\partial U_i}{\partial X_l} + u_l \frac{\partial u_i}{\partial X_l} = -\frac{1}{\rho}\frac{\partial p}{\partial X_i} - \beta \theta g_i + \nu \frac{\partial^2 u_i}{\partial X_j \partial X_j} + \frac{\partial \overline{u_i u_j}}{\partial X_j}. \tag{6.16}$$

- For the fluctuating component of the temperature, we have

$$\frac{\partial \theta}{\partial t} + U_l \frac{\partial \theta}{\partial X_l} + u_l \frac{\partial T}{\partial X_l} + u_l \frac{\partial \theta}{\partial X_l} = \alpha \frac{\partial^2 \theta}{\partial X_j \partial X_j} + \frac{\partial \overline{u_i \theta}}{\partial X_i} + \frac{1}{\rho C_p}(\Phi' - \Phi), \tag{6.17}$$

where

$$\Phi' = \tau_{ij}\frac{\partial u_i}{\partial X_j} + \tau'_{ij}\frac{\partial U_i}{\partial X_l} + \tau'_{ij}\frac{\partial u_i}{\partial X_j}.$$

6.5 TURBULENT TRANSPORT EQUATIONS

Under the Boussinesq approximation (Boussinesq, 1877 [6]), the exact transport equations for $\overline{u_i u_j}$, k, ϵ, and $\overline{u_i \theta}$ are derived below. The Reynolds-stress equation with the buoyant force is

$$\frac{D\overline{u_i u_j}}{Dt} = \frac{\partial}{\partial X_l}\left(-\overline{u_i u_j u_l} - \overline{\frac{p}{\rho}(\delta_{jl} u_i + \delta_{il} u_j)} + \nu \frac{\partial \overline{u_i u_j}}{\partial X_l}\right) - \left(\overline{u_i u_l}\frac{\partial U_j}{\partial X_l} + \overline{u_j u_l}\frac{\partial U_i}{\partial X_l}\right) - 2\nu \overline{\frac{\partial u_i}{\partial X_l}\frac{\partial u_j}{\partial X_l}} + \overline{\frac{p}{\rho}\left(\frac{\partial u_i}{\partial X_j} + \frac{\partial u_j}{\partial X_i}\right)} - \beta(g_i \overline{u_j \theta} + g_j \overline{u_i \theta}). \tag{6.18}$$

The k exact equation with

$$P_k = -\overline{u_i u_l}\frac{\partial U_i}{\partial x_l} \tag{6.19}$$

and

$$\frac{1}{2}\overline{u_i u_i} = k \tag{6.20}$$

is

$$\frac{Dk}{Dt} = \frac{\partial}{\partial X_l}\left[-\overline{\left(\frac{u_i u_i}{2} + \frac{p}{\rho}\right)u_l} + \nu\frac{\partial k}{\partial X_l}\right] + P_k - \beta g_i \overline{u_i \theta} - \epsilon. \tag{6.21}$$

For the ϵ equation, where

$$\epsilon = \nu\overline{\frac{\partial u_i}{\partial X_l}\frac{\partial u_i}{\partial X_l}}, \tag{6.22}$$

we have

$$\begin{aligned}\frac{D\epsilon}{Dt} &= \frac{\partial}{\partial X_l}\left(-\overline{\epsilon' u_l} - \frac{2\nu}{\rho}\overline{\frac{\partial u_l}{\partial X_j}\frac{\partial p}{\partial X_j}} + \nu\frac{\partial \epsilon}{\partial X_l}\right)\\ &\quad - 2\nu\overline{u_l\frac{\partial u_i}{\partial X_j}}\frac{\partial^2 U_i}{\partial X_l \partial X_j} - 2\nu\frac{\partial U_i}{\partial X_j}\left(\overline{\frac{\partial u_l}{\partial X_i}\frac{\partial u_l}{\partial X_j}} + \overline{\frac{\partial u_i}{\partial X_l}\frac{\partial u_j}{\partial X_l}}\right)\\ &\quad - 2\nu\overline{\frac{\partial u_i}{\partial X_j}\frac{\partial u_i}{\partial X_l}\frac{\partial u_j}{\partial X_l}} - 2\overline{\left(\nu\frac{\partial^2 u_i}{\partial X_l \partial X_l}\right)^2} - 2\beta g_i \nu\overline{\frac{\partial u_i}{\partial X_j}\frac{\partial \theta}{\partial X_j}}.\end{aligned} \tag{6.23}$$

The $\overline{u_i\theta}$ exact equation is

$$\begin{aligned}\frac{D\overline{u_i\theta}}{Dt} &= \frac{\partial}{\partial X_l}\left(-\overline{u_i u_l \theta} - \delta_{il}\frac{\overline{P\theta}}{\rho} + \alpha\overline{u_i\frac{\partial \theta}{\partial X_l}} + \nu\overline{\theta\frac{\partial u_i}{\partial X_l}}\right)\\ &\quad - \left(\overline{u_i u_l}\frac{\partial \theta}{\partial X_l} + \overline{u_l\theta}\frac{\partial U_i}{\partial X_l}\right) - (\alpha + \nu)\overline{\frac{\partial u_i}{\partial X_l}\frac{\partial \theta}{\partial X_l}}\\ &\quad + \overline{\frac{P\partial\theta}{\rho\partial X_i}} + \frac{1}{\rho C_p}\overline{\Phi' u_i} - \beta g_i \overline{\theta^2}.\end{aligned} \tag{6.24}$$

New quantities that describe the effect of the buoyant force on the Reynolds stress, the turbulent kinetic energy, the dissipation, and the Reynolds heat flux appear in each exact equation. In the limit, when the buoyant effect is negligible, the exact equations will retain their appearance as in chapter 2. Some new terms that appear in the Reynolds-stress equation, the kinetic energy equation, and the rate of dissipation equation will need additional modeling. The new quantity, $\overline{\theta^2}$, that is introduced in the Reynolds heat flux equation will need additional derivation. An exact equation for this quantity can be obtained by multiplying the fluctuating temperature equation with the fluctuating temperature and taking the average. Then the equation for $\overline{\theta^2}$ can be derived from Eq. 6.17 as

$$\frac{D\overline{\theta^2}}{Dt} = \frac{\partial}{\partial X_l}\left(-\overline{u_l\theta^2} + \alpha\frac{\partial\overline{\theta^2}}{\partial X_l}\right) - 2\overline{u_l\theta}\frac{\partial T}{\partial x_l} - 2\alpha\overline{\frac{\partial\theta}{\partial X_l}\frac{\partial\theta}{\partial X_l}} + \frac{1}{\rho C_p}\overline{\theta\phi'}. \tag{6.25}$$

Another new quantity is introduced in the $\overline{\theta^2}$ equation, which is ϵ_θ, where

$$\epsilon_\theta = 2\alpha \overline{\frac{\partial \theta}{\partial X_l} \frac{\partial \theta}{\partial X_l}}.$$

An equation for this quantity can be obtained by taking the average of the gradient of the temperature equation, Eq. 6.17 multiplied by the gradient of the fluctuation temperature, such that

$$\frac{D\epsilon_\theta}{Dt} = \frac{\partial}{\partial X_l}\left(\overline{-\epsilon'_\theta u_l + \alpha \frac{\partial \epsilon_\theta}{\partial X_l}}\right) - 2\alpha \frac{\partial U_l}{\partial X_i} \overline{\frac{\partial \theta}{\partial X_i} \frac{\partial \theta}{\partial X_l}} - 2\alpha \frac{\partial T}{\partial X_l} \overline{\frac{\partial \theta}{\partial X_i} \frac{\partial u_l}{\partial X_i}}$$
$$- 2\alpha \overline{\frac{\partial u_l}{\partial X_i} \frac{\partial \theta}{\partial X_i} \frac{\partial \theta}{\partial X_l}} - 2\left(\overline{\alpha \frac{\partial^2 \theta}{\partial X_l \partial X_i}}\right)^2 + 2\alpha \overline{\frac{\partial \phi'}{\partial X_i} \frac{\partial \theta}{\partial X_i}}. \quad (6.26)$$

Both $\overline{\theta^2}$ and ϵ_θ are new equations needed in buoyant flow modeling because $\overline{\theta^2}$ appears in the $\overline{u_i\theta}$ equation. One may see that the appearance of ϵ_θ is only for the $\overline{\theta^2}$ equation. Therefore, one expects that ϵ_θ will play a less important role than ϵ (Eq. 6.23) in determining mean-flow quantities such as U_l, P, and T.

The $\overline{u_i\theta}$ terms now appear in both the $\overline{u_i u_j}$ and k equations. Therefore, the turbulent momentum transfer and the thermal energy transfer are now coupled. The last term in the ϵ_θ equation is considered small and is dropped in most applications. Gravitational terms now also appear in the $\overline{u_i u_j}$, ϵ, and $\overline{u_i\theta}$ equations. These terms require new modeling. The $\overline{\theta^2}$ and ϵ_θ equations are also completely new and must be modeled.

6.6 TURBULENCE MODELING OF TURBULENT BUOYANT FLOWS

6.6.1 Turbulence Closure Postulations

The postulations are the same as in Section 2.1.3, except that we expand the closure to include buoyancy related terms.

1. The average Navier–Stokes (N-S) and energy equations are applicable to turbulent flows.
2. Diffusion of turbulent transport quantity ϕ is proportional to its gradient (i.e., grad ϕ).
3. Small eddies are isotropic.
4. All turbulent transport quantities are functions of $\overline{u_i u_j}$, $\overline{u_i\theta}$, k, ϵ, $\overline{\theta^2}$, ϵ_θ, U_i, P, T, ρ, ν, and α.
5. Modeled turbulent phenomena must be consistent in invariance, symmetry, permutations, and observations.
6. In turbulent flow, turbulent scales are functions of k, ϵ, and ν (turbulent scale hypothesis).
   - For large eddies (k, ϵ) scale $\rightarrow u = \sqrt{k}$, $l = k^{3/2}/\epsilon$, and $t = k/\epsilon$.
   - For small eddies (ν, ϵ) scale $\rightarrow u = (\nu\epsilon)^{1/4}$, $l = (\nu^3/\epsilon)^{1/4}$, and $t = (\nu/\epsilon)^{1/2}$.
7. Model coefficients and constants are determined from experiments.

It should be remarked here that Postulation 3 can be removed to include some anisotropic effects. All nonbuoyancy-related terms are modeled exactly the same as before in chapter 2. Only new terms and equations are considered here.

6.6.2 Modeling of $\overline{u_iu_j}$ and k Equations

The exact equation for $\overline{u_iu_j}$ is

$$\frac{D\overline{u_iu_j}}{Dt} = \frac{\partial}{\partial X_l}\left(-\overline{u_iu_ju_l} - \overline{\frac{p}{\rho}\left(\delta_{jl}u_i + \delta_{il}u_j\right)} + \nu\frac{\partial\overline{u_iu_j}}{\partial X_l}\right) - \left(\overline{u_iu_l}\frac{\partial U_j}{\partial X_l} + \overline{u_ju_l}\frac{\partial U_i}{\partial X_l}\right)$$
$$- 2\nu\overline{\frac{\partial u_i}{\partial X_l}\frac{\partial u_j}{\partial X_l}} + \overline{\frac{p}{\rho}\left(\frac{\partial u_i}{\partial X_j} + \frac{\partial u_j}{\partial X_i}\right)} - \beta(g_i\overline{u_j\theta} + g_j\overline{u_i\theta}) \tag{6.27}$$

The $\overline{u_iu_j}$ equation is modeled as before, and no new modeling is required except for the pressure-strain (PS) term. To model the PS term let us first consider p/ρ. The divergence of the momentum equation m_i gives

$$\nabla^2\frac{p}{\rho} = -\frac{\partial^2(U_lu_m - \overline{u_lu_m})}{\partial X_l \partial X_m} - 2\frac{\partial U_l}{\partial X_m}\frac{\partial u_m}{\partial X_l} - \beta\frac{\partial\theta}{\partial X_l}g_l = -(F). \tag{6.28}$$

Note that in this equation, there is an additional term that reflects the effect of the buoyancy force, as compared with the modeling of the PS as in chapter 2. The solution of the above-mentioned Poisson equation is

$$\frac{p}{\rho} = \frac{1}{4\pi}\int_{\text{vol}}(F)\frac{\text{dvol}}{\boldsymbol{r}} - \frac{1}{4\pi}\int_s\left[\frac{p}{\rho}\frac{\partial}{\partial n}\frac{1}{r} - \frac{1}{r}\frac{\partial}{\partial n}\left(\frac{p}{\rho}\right)\right]ds, \tag{6.29}$$

where the last term is small when the surface is far from the integration point. Thus, for flows away from a solid wall, the last term is negligible. The PS term thus becomes

$$\overline{\frac{p}{\rho}\left(\frac{\partial u_i}{\partial X_j} + \frac{\partial u_j}{\partial X_i}\right)} = \frac{1}{4\pi}\int_{\text{vol}}(\Phi_{ij,1} + \Phi_{ij,2} + \Phi_{ij,3})\frac{\text{dvol}}{\boldsymbol{r}}, \tag{6.30}$$

where

$$\Phi_{ij,1} = \int_{\text{vol}}\overline{\left(\frac{\partial^2 u_lu_m}{\partial X_l\partial X_m}\right)^*\left(\frac{\partial u_i}{\partial X_j} + \frac{\partial u_j}{\partial X_i}\right)}\frac{\text{dvol}}{4\pi\boldsymbol{r}^*}, \tag{6.31}$$

$$\Phi_{ij,2} = \int_{\text{vol}}2\frac{\partial U_l}{\partial X_m}\overline{\frac{\partial u_m}{\partial X_l}\left(\frac{\partial u_i}{\partial X_j} + \frac{\partial u_j}{\partial X_i}\right)}\frac{\text{dvol}}{4\pi\boldsymbol{r}^*}, \tag{6.32}$$

and

$$\Phi_{ij,3} = \int_{\text{vol}}\beta g_l\overline{\frac{\partial\theta^*}{\partial X_l}\left(\frac{\partial u_i}{\partial X_j} + \frac{\partial u_j}{\partial X_i}\right)}\frac{\text{dvol}}{4\pi\boldsymbol{r}^*}. \tag{6.33}$$

Modeling of $\phi_{ij,1}$ and $\phi_{ij,2}$ is the same as before (see Section 2.2.2)

$$\Phi_{ij,1} + \Phi_{ij,2} = -C_1\frac{\epsilon}{k}\left(\overline{u_iu_j} - \frac{2}{3}\delta_{ij}k\right) - C_2\left(P_{ij} - \frac{2}{3}\delta_{ij}P_k\right). \tag{6.34}$$

Modeling of $\phi_{ij.3}$ can be done similarly by shrinking the integral to a small volume and noting that we can approximate $\phi_{ii.3} = 0$ with $i = j$ because of the incompressibility,

$$\frac{\partial u_i}{\partial x_i} = 0. \tag{6.35}$$

Thus

$$\begin{aligned}
\Phi_{ij.3} &= \frac{1}{4\pi}\int_{\text{vol}} \beta g_l \overline{\frac{\partial \theta^*}{\partial X_l}\left(\frac{\partial u_i}{\partial X_j}+\frac{\partial u_j}{\partial X_i}\right)}\frac{\text{dvol}}{\boldsymbol{r}^*} \\
&= \frac{1}{4\pi}\beta g_l \overline{\frac{\partial \theta}{\partial X_l}\left(\frac{\partial u_i}{\partial X_j}+\frac{\partial u_j}{\partial X_i}\right)} l^2 \\
&= C_3\left(\beta g_j \overline{u_i\theta} + \beta g_i \overline{u_j\theta} - \frac{2}{3}\delta_{ij}\beta g_i \overline{u_i\theta}\right) \\
&= -C_3\left(P_{ij.b} - \frac{2}{3}\delta_{ij} P_b\right),
\end{aligned} \tag{6.36}$$

with

$$P_{ij.b} = -\beta(g_i\overline{u_j\theta} + g_j\overline{u_i\theta}) \tag{6.37}$$

$$P_b = -\beta g_i \overline{u_i\theta} \tag{6.38}$$

Then the modeled $\overline{u_i u_j}$ equation becomes

$$\begin{aligned}
\frac{D\overline{u_iu_j}}{Dt} &= \frac{\partial}{\partial X_l}\left[\left(C_k\frac{k^2}{\epsilon}+\nu\right)\frac{\partial \overline{u_iu_j}}{\partial X_l}\right] + P_{ij} + P_{ij.b} - \frac{2}{3}\delta_{ij}\epsilon \\
&\quad - C_1\frac{\epsilon}{k}\left(\overline{u_iu_j} - \frac{2}{3}\delta_{ij}k\right) - C_2\left(P_{ij} - \frac{2}{3}\delta_{ij}P_k\right) \\
&\quad - C_3\left(P_{ij.b} - \frac{2}{3}\delta_{ij}P_b\right).
\end{aligned} \tag{6.39}$$

The modeling of the k equation can be obtained by setting $i = j$ and summing, or

$$\frac{Dk}{Dt} = \frac{\partial}{\partial X_l}\left[\left(C_k\frac{k^2}{\epsilon}+\nu\right)\frac{\partial k}{\partial X_l}\right] + P_k + P_b - \epsilon. \tag{6.40}$$

The turbulent constants are $C_k = 0.09$, $C_1 = 1.8 \approx 2.8$, $C_2 = 0.4 \approx 0.6$, and $C_3 = 0.3 \approx 0.5$.

6.6.3 Modeling of the ϵ Equation

The exact ϵ equation is

$$\begin{aligned}
\frac{D\epsilon}{Dt} &= \frac{\partial}{\partial X_l}\left(-\overline{\epsilon' u_l} - \frac{2\nu}{\rho}\overline{\frac{\partial u_l}{\partial X_j}\frac{\partial p}{\partial X_j}} + \nu\frac{\partial \epsilon}{\partial X_l}\right) - 2\nu \overline{u_l\frac{\partial u_i}{\partial X_j}}\frac{\partial^2 U_i}{\partial X_l \partial X_j} \\
&\quad - 2\nu\frac{\partial U_i}{\partial X_j}\left(\overline{\frac{\partial u_l}{\partial X_i}\frac{\partial u_l}{\partial X_j}} + \overline{\frac{\partial u_i}{\partial X_l}\frac{\partial u_j}{\partial X_l}}\right) - 2\nu\overline{\frac{\partial u_i}{\partial X_j}\frac{\partial u_i}{\partial X_l}\frac{\partial u_j}{\partial X_l}} \\
&\quad - 2\overline{\left(\nu\frac{\partial^2 u_i}{\partial X_l \partial X_l}\right)^2} - 2\beta g_i \nu \overline{\frac{\partial u_i}{\partial X_j}\frac{\partial \theta}{\partial X_j}}.
\end{aligned} \tag{6.41}$$

The new gravitational term is approximately zero because of the isotropic dissipation (small eddies) postulation. The modeling of the production terms (with mean-quantities U_i) is again set equal to zero because of incompressibility,

$$\frac{\partial U_i}{\partial X_i}, \tag{6.42}$$

and isotropic dissipation. The destruction term again (see Section 2.2.4) is modeled on the basis of Lumley's assumption (1970)[98]:

$$-2\nu\overline{\frac{\partial u_i}{\partial X_j}\frac{\partial u_i}{\partial X_l}\frac{\partial u_j}{\partial X_l}} - 2\left(\overline{\frac{\partial^2 u_i}{\partial X_l \partial X_j}}\right)^2 = \text{constant} \times \left(\frac{\epsilon}{t}\right) \times \left(\frac{\text{Prod.k}}{\epsilon} - 1\right). \tag{6.43}$$

However, the production of kinetic energy, Prod.k, now consists of shear force (P_k) and buoyant force (P_b). Thus, by noting that

$$\frac{Dk}{Dt} = D_k + P_k + P_b - \epsilon, \tag{6.44}$$

we have the destruction of ϵ or Eq. 6.43 as

$$\text{constant} \times [1/t] \times (P_k + P_b - \epsilon). \tag{6.45}$$

The model of the ϵ equation becomes, with the (k, ϵ) scale for $1/t$,

$$\frac{D\epsilon}{Dt} = \frac{\partial}{\partial X_l}\left[\left(C_\epsilon \frac{k^2}{\epsilon} + \nu\right)\frac{\epsilon}{\partial X_l}\right] + C_{\epsilon 1}\frac{\epsilon}{k}P_k + C_{\epsilon 3}\frac{\epsilon}{k}P_b - C_{\epsilon 2}\frac{\epsilon^2}{k} \tag{6.46}$$

The constants are $C_\epsilon = 0.07$, $C_{\epsilon 1} = 1.44$, $C_{\epsilon 2} = 1.92$, and $C_{\epsilon 3} = 1.44 \sim 1.92$. The $C_{\epsilon 3}$ value is reasoned by Chen (1983) on the basis of an equilibrium argument. Its value needs to be tested. The ϵ equation can be improved with the two-scale turbulent hypothesis (see Section 3.4.2).

6.6.4 Modeling of the $\overline{u_i\theta}$ Equation (for $\alpha = \nu$)

The exact $\overline{u_i\theta}$ equation is

$$\begin{aligned}\frac{D\overline{u_i\theta}}{Dt} &= \frac{\partial}{\partial X_l}\left(-\overline{u_i u_l \theta} - \delta_{il}\frac{\overline{p\theta}}{\rho} + \alpha\overline{u_i\frac{\partial\theta}{\partial X_l}} + \nu\overline{\theta\frac{\partial u_i}{\partial X_l}}\right) - \left(\overline{u_i u_l}\frac{\partial\theta}{\partial X_l} + \overline{u_l\theta}\frac{\partial U_i}{\partial X_l}\right) \\ &\quad - (\alpha+\nu)\overline{\frac{\partial u_i}{\partial X_l}\frac{\partial\theta}{\partial X_l}} + \overline{\frac{p\partial\theta}{\rho\partial X_i}} + \frac{1}{\rho C_p}\overline{\Phi' u_i} - \beta g_i\overline{\theta^2}.\end{aligned} \tag{6.47}$$

With isotropic dissipation, we let

$$\overline{\frac{\partial u_i}{\partial X_l}\frac{\partial\theta}{\partial X_l}} = 0 \tag{6.48}$$

with small frictional heat flux, $\overline{\phi'\theta} = 0$.

$$P_T = -\left(\overline{u_i u_l}\frac{\partial T}{\partial X_l} + \overline{u_l\theta}\frac{\partial U_i}{\partial X_l}\right). \tag{6.49}$$

P_T is the production of $\overline{u_i\theta}$. The diffusion term is modeled as before (see Section 2.2.5). The term

$$\overline{\frac{p}{\rho}\frac{\partial\theta}{\partial X_i}}$$

now includes a buoyant term, or

$$\overline{\frac{p\partial\theta}{\rho\partial X_i}} = \frac{1}{4\pi}\int_{\text{vol}}\left(\overline{\frac{\partial^2 u_l u_m^*}{\partial X_l \partial X_m}\frac{\partial\theta}{\partial X_i}} + 2\frac{\partial U_l^*}{\partial X_m}\overline{\frac{\partial u_m}{\partial X_l}\frac{\partial\theta}{\partial X_i}} + \beta g_l \overline{\frac{\partial\theta^*}{\partial X_l}\frac{\partial\theta}{\partial X_i}}\right)\frac{\text{dvol}}{r^*}. \tag{6.50}$$

Shrinking to a small volume, we model each term as before

$$\begin{aligned}
\overline{\frac{p\partial\theta}{\rho\partial X_i}} &= \text{constant} \times \left(\overline{\frac{\partial^2 u_l u_m}{\partial X_l \partial X_m}\frac{\partial\theta}{\partial X_i}} + 2\frac{\partial U_l}{\partial X_m}\overline{\frac{\partial u_m}{\partial X_l}\frac{\partial\theta}{\partial X_i}} + \beta g_l \overline{\frac{\partial\theta}{\partial X_l}\frac{\partial\theta}{\partial X_i}}\right) l^2 \\
&= -C_{T1}\left(\frac{1}{t}\right)\overline{u_i\theta} + C_{T2}\frac{\partial U_i}{\partial X_m}\overline{u_m\theta} - C_{T3}\beta g_i \overline{\theta^2} \\
&= -C_{T1}\frac{\epsilon}{k}\overline{u_i\theta} + C_{T2}\frac{\partial U_i}{\partial X_m}\overline{u_m\theta} - C_{T3}\beta g_i \overline{\theta^2}.
\end{aligned} \tag{6.51}$$

The modeled $\overline{u_i\theta}$ equation becomes

$$\begin{aligned}
\frac{D\overline{u_i\theta}}{Dt} &= \frac{\partial}{\partial X_l}\left[\left(C_T\frac{k^2}{\epsilon} + \nu\right)\frac{\partial\overline{u_i\theta}}{\partial X_l}\right] + P_T - (1 + C_{T3})\beta g_i \overline{\theta^2} - C_{T1}\frac{\epsilon}{k}\overline{u_i\theta} \\
&\quad + C_{T2}\frac{\partial U_i}{\partial X_m}\overline{u_m\theta}.
\end{aligned} \tag{6.52}$$

The model constants are $C_T = 0.07$, $C_{T1} = 3.2$, and $C_{T2} = C_{T3} = 0.5$. Constant C_{T3} needs to be verified. Some workers set $C_{T3} = 0$.

6.6.5 Modeling of the $\overline{\theta^2}$ Equation

Because $\overline{\theta^2}$ appears in the $\overline{u_i\theta}$ equation, it must be modeled. The $\overline{\theta^2}$ equation is like the k equation in that it controls the turbulent thermal energy. $C_p(\overline{\theta^2})^{1/2}$ is the turbulent thermal energy.

The exact $\overline{\theta^2}$ equation is

$$\frac{D\overline{\theta^2}}{Dt} = \frac{\partial}{\partial X_l}\left(-\overline{u_l\theta^2} + \alpha\frac{\partial\overline{\theta^2}}{\partial X_l}\right) - 2\overline{u_l\theta}\frac{\partial T}{\partial x_l} - 2\alpha\overline{\frac{\partial\theta}{\partial X_l}\frac{\partial\theta}{\partial X_l}} + \frac{1}{\rho C_p}\overline{\theta\phi'}. \tag{6.53}$$

The diffusion term is modeled as

$$-\overline{u_l\theta^2} = C_\theta\frac{k^2}{\epsilon}\frac{\partial\overline{\theta^2}}{\partial X_l}. \tag{6.54}$$

For the dissipation term, Launder (1980) made a further approximation by letting

$$\alpha\overline{\frac{\partial\theta}{\partial X_l}\frac{\partial\theta}{\partial X_l}} = \epsilon_\theta = \text{constant} \times \left(\frac{\theta^2}{u^2}\right) \times \epsilon = C_{\theta 1}\frac{\overline{\theta^2}}{k}\epsilon \tag{6.55}$$

and used $C_{\theta 1} \approx 0.62$. However, if ϵ_θ is solved from the ϵ_θ equation, then there is no need to model ϵ_θ here. Also

$$\frac{1}{\rho C_p}\overline{\theta \phi'} \approx 0 \tag{6.56}$$

because of the low order of magnitude. The modeled $\overline{\theta^2}$ equation thus can be summarized as the following three equations, the choice of which depends on the level of the modeling.

- $$\frac{D\overline{\theta^2}}{Dt} = \frac{\partial}{\partial X_l}\left[\left(C_\theta \frac{k^2}{\epsilon} + \nu\right)\frac{\partial \overline{\theta^2}}{\partial X_l}\right] - 2\overline{u_l\theta}\frac{\partial T}{\partial X_l} - 2\epsilon_\theta \tag{6.57}$$

- $$\frac{D\overline{\theta^2}}{Dt} = \frac{\partial}{\partial X_l}\left[\left(C_\theta \frac{k^2}{\epsilon} + \nu\right)\frac{\partial \overline{\theta^2}}{\partial X_l}\right] - 2\overline{u_l\theta}\frac{\partial T}{\partial X_l} - 2C_{\theta 1}\frac{\overline{\theta^2}}{k}\epsilon \tag{6.58}$$

By omitting the differential terms in Eq. (6.58) we have the algebraic model as

$$\overline{\theta^2} = -\frac{1}{C_{\theta 1}}\frac{k}{\epsilon}\overline{u_l\theta}\frac{\partial T}{\partial X_l}. \tag{6.59}$$

The constants are $C_\theta = 0.13$ and $C_{\theta 1} = 0.62$.

6.6.6 Modeling of the ϵ_θ Equation

If in the $\overline{\theta^2}$ equation, ϵ_θ is modeled, then there is no need to consider the ϵ_θ equation. The reasons are (1) the modeling of the ϵ_θ complicates modeling and computation, and (2) the mean flow is probably very insensitive to ϵ_θ variations. Because ϵ_θ affects $\overline{\theta^2}$ and $\overline{\theta^2}$ affects $\overline{u_i\theta}$ and $\overline{u_i\theta}$ affects T and U_i, the effect of the ϵ_θ equation on the mean flow becomes secondary when one is interested in T and U_i. The exact ϵ_θ equation is

$$\begin{aligned}\frac{D\epsilon_\theta}{Dt} &= \frac{\partial}{\partial X_l}\left(-\overline{\epsilon_\theta' u_l} + \alpha\frac{\partial \epsilon_\theta}{\partial X_l}\right) - 2\alpha\frac{\partial U_l}{\partial X_i}\overline{\frac{\partial \theta}{\partial X_i}\frac{\partial \theta}{\partial X_l}} - 2\alpha\frac{\partial T}{\partial X_l}\overline{\frac{\partial \theta}{\partial X_i}\frac{\partial u_l}{\partial X_i}}\\ &\quad - 2\alpha\overline{\frac{\partial u_l}{\partial X_i}\frac{\partial \theta}{\partial X_i}\frac{\partial \theta}{\partial X_l}} - 2\overline{\left(\alpha\frac{\partial^2 \theta}{\partial X_l \partial X_i}\right)^2} + 2\alpha\overline{\frac{\partial \phi'}{\partial X_i}\frac{\partial \theta}{\partial X_i}}.\end{aligned} \tag{6.60}$$

The diffusion term is

$$-\overline{\epsilon_\theta' u_l} = C_e\frac{k^2}{\epsilon}\frac{\partial \epsilon_\theta}{\partial X_l} \tag{6.61}$$

and

$$\alpha\frac{\partial U_l}{\partial X_i}\overline{\frac{\partial \theta}{\partial X_i}\frac{\partial \theta}{\partial X_l}} = 0 \tag{6.62}$$

because of isotropic dissipation and incompressibility. Finally,

$$\alpha \overline{\frac{\partial \theta}{\partial X_i} \frac{\partial u_l}{\partial X_i}} = 0 \tag{6.63}$$

and

$$\alpha \overline{\frac{\partial \phi'}{\partial X_i} \frac{\partial \theta}{\partial X_i}} = 0 \tag{6.64}$$

because of isotropic dissipation. Then, borrowing Lumley's assumption for the ϵ equation (see Section 2.2.4),

$$\begin{aligned} -2\alpha \overline{\frac{\partial u_l}{\partial X_i} \frac{\partial \theta}{\partial X_i} \frac{\partial \theta}{\partial X_l}} - 2\left(\alpha \overline{\frac{\partial^2 \theta}{\partial X_l \partial X_i}}\right)^2 &= \text{constant} \times \left(\frac{\epsilon_\theta}{t}\right) \times \left(\frac{\text{Prod}.\overline{\theta^2}}{\epsilon_\theta} - 1\right) \\ &= -C_{e1}\frac{\epsilon}{k}\overline{u_l \theta}\frac{\partial T}{\partial X_l} - C_{e2}\frac{\epsilon}{k}\epsilon_\theta, \end{aligned} \tag{6.65}$$

and the modeled ϵ_θ equation becomes

$$\frac{D\epsilon_\theta}{Dt} = \frac{\partial}{\partial X_l}\left[\left(C_e \frac{k^2}{\epsilon} + \nu\right)\frac{\partial \epsilon_\theta}{\partial X_l}\right] - C_{e1}\frac{\epsilon}{k}\overline{u_l \theta}\frac{\partial T}{\partial X_l} - C_{e2}\frac{\epsilon}{k}\epsilon_\theta. \tag{6.66}$$

The constants are $C_e = ?(C_\epsilon)$, $C_{e1} = ?(2.5)$, and $C_{e2} = 2.5$ (given by Lumley). The model constants for the ϵ_θ equation are rather uncertain since this equation is seldom used.

6.7 SUMMARY OF THE TURBULENCE MODEL

6.7.1 Differential Model

The differential model uses fully modeled $\overline{u_i u_j}$, k, ϵ, $\overline{u_i \theta}$, $\overline{\theta^2}$, and ϵ_θ equations. The complete modeled equations are as follows:

- The Reynolds-stress equation,

$$\begin{aligned} \frac{D\overline{u_i u_j}}{Dt} &= \frac{\partial}{\partial X_l}\left[\left(C_k \frac{k^2}{\epsilon} + \nu\right)\frac{\partial \overline{u_i u_j}}{\partial X_l}\right] + P_{ij} + P_{ij,b} - \frac{2}{3}\delta_{ij}\epsilon \\ &\quad - C_1 \frac{\epsilon}{k}\left(\overline{u_i u_j} - \frac{2}{3}\delta_{ij} k\right) - C_2\left(P_{ij} - \frac{2}{3}\delta_{ij} P_k\right) - C_3\left(P_{ij,b} - \frac{2}{3}\delta_{ij} P_b\right), \end{aligned} \tag{6.67}$$

where

$$P_{ij} = -\left(\overline{u_i u_l}\frac{\partial U_j}{\partial X_l} + \overline{u_j u_l}\frac{\partial U_i}{\partial X_l}\right) \tag{6.68}$$

$$P_k = -\overline{u_i u_l}\frac{\partial U_i}{\partial X_l} \tag{6.69}$$

$$P_{ij,b} = -\beta(g_i \overline{u_j \theta} + g_j \overline{u_i \theta}) \tag{6.70}$$

$$P_b = -\beta g_i \overline{u_i \theta}. \tag{6.71}$$

- The kinetic energy equation is

$$\frac{Dk}{Dt} = \frac{\partial}{\partial X_l}\left[\left(C_k\frac{k^2}{\epsilon} + \nu\right)\frac{\partial k}{\partial X_l}\right] + P_k + P_b - \epsilon. \tag{6.72}$$

- The rate of dissipation, ϵ, equation is

$$\frac{D\epsilon}{Dt} = \frac{\partial}{\partial X_l}\left[\left(C_\epsilon\frac{k^2}{\epsilon} + \nu\right)\frac{\epsilon}{\partial X_l}\right] + C_{\epsilon 1}\frac{\epsilon}{k}P_k + C_{\epsilon 3}\frac{\epsilon}{k}P_b - C_{\epsilon 2}\frac{\epsilon^2}{k}. \tag{6.73}$$

- The Reynolds heat flux equation is

$$\frac{D\overline{u_i\theta}}{Dt} = \frac{\partial}{\partial X_l}\left[\left(C_T\frac{k^2}{\epsilon} + \nu\right)\frac{\partial \overline{u_i\theta}}{\partial X_l}\right] + P_T - (1 + C_{T3})\beta g_i\overline{\theta^2}$$
$$- C_{T1}\frac{\epsilon}{k}\overline{u_i\theta} + C_{T2}\frac{\partial U_i}{\partial X_m}\overline{u_m\theta}, \tag{6.74}$$

where

$$P_T = -\overline{u_iu_l}\frac{\partial T}{\partial X_l} - \overline{u_l\theta}\frac{\partial U_i}{\partial X_l}. \tag{6.75}$$

- The $\overline{\theta^2}$ equation is

$$\frac{D\overline{\theta^2}}{Dt} = \frac{\partial}{\partial X_l}\left[\left(C_\theta\frac{k^2}{\epsilon} + \nu\right)\frac{\partial\overline{\theta^2}}{\partial X_l}\right] - 2\overline{u_l\theta}\frac{\partial T}{\partial X_l} - 2\epsilon_\theta. \tag{6.76}$$

- The ϵ_θ equation is

$$\frac{D\epsilon_\theta}{Dt} = \frac{\partial}{\partial X_l}\left[\left(C_e\frac{k^2}{\epsilon} + \nu\right)\frac{\partial\epsilon_\theta}{\partial X_l}\right] - C_{e1}\frac{\epsilon}{k}\overline{u_l\theta}\frac{\partial T}{\partial X_l} - C_{e2}\frac{\epsilon}{k}\epsilon_\theta. \tag{6.77}$$

The various scalar coefficients in these equations are $C_k = 0.09$, $C_\epsilon = 0.07$, $C_\theta = 0.13$, $C_T = 0.07$, $C_1 = 1.82 \sim 2.8$, $C_{\epsilon 1} = 1.44$, $C_e = ?0.1$, $C_{T1} = 3.2$, $C_2 = 0.4 \sim 0.6$, $C_{\epsilon 2} = 1.92$, $C_{e1} = ?2.5$, $C_{T2} = 0.5$, $C_3 = 0.3 \sim 0.5$, $C_{\epsilon 3} = 1.44 \sim 1.92$, $C_{e2} = 2.5$, and $C_{T3} \approx 0.5$.

6.7.2 k-ϵ-$\overline{\theta^2}$-(A or E) Model

With Rodi's hypothesis (1972)[134] for algebraic stress and heat flux approximation and modeling $\epsilon_\theta = C_{\theta 1}\overline{\theta^2}/k\epsilon$, we have the k-ϵ-$\overline{\theta^2}$ model. The complete equations are

- the $\overline{u_iu_j}$ equation,

$$-\overline{u_iu_j} = \frac{k}{C_1\epsilon}\left\{\frac{(C_2 - 1)P_{ij} + (C_3 - 1)P_{ij,b} - \frac{2}{3}\delta_{ij}\left[C_2P_k + C_3P_b + (C_1 - 1)\epsilon\right]}{1 + \left(\frac{P_k + P_b}{\epsilon} - 1\right)/C_1}\right\}, \tag{6.78}$$

- the k equation,

$$\frac{Dk}{Dt} = \frac{\partial}{\partial X_l}\left[\left(C_k\frac{k^2}{\epsilon} + \nu\right)\frac{\partial k}{\partial X_l}\right] + P_k + P_b - \epsilon, \tag{6.79}$$

- the ϵ equation,

$$\frac{D\epsilon}{Dt} = \frac{\partial}{\partial X_l}\left[\left(C_\epsilon \frac{k^2}{\epsilon} + \nu\right)\frac{\epsilon}{\partial X_l}\right] + C_{\epsilon 1}\frac{\epsilon}{k}P_k + C_{\epsilon 3}\frac{\epsilon}{k}P_b - C_{\epsilon 2}\frac{\epsilon^2}{k}, \tag{6.80}$$

- the $\overline{u_i\theta}$ equation,

$$-\overline{u_i\theta} = \frac{k}{C_{T1}\epsilon}\left[\frac{-P_T + (1 + C_{T3})\,\beta g_i \overline{\theta^2} - C_{T2}\overline{u_m\theta}\frac{\partial U_i}{\partial X_m}}{1 + \left(\frac{P_k + P_b}{\epsilon} - 1\right)/C_{T1}}\right], \tag{6.81}$$

and

- the $\overline{\theta^2}$ equation,

$$\frac{D\overline{\theta^2}}{Dt} = \frac{\partial}{\partial X_l}\left[\left(C_\theta \frac{k^2}{\epsilon} + \nu\right)\frac{\partial\overline{\theta^2}}{\partial X_l}\right] - 2\overline{u_l\theta}\frac{\partial T}{\partial X_l} - 2C_{\theta 1}\overline{\theta^2}/k\epsilon. \tag{6.82}$$

The thirteen constants are $C_\epsilon = 0.07$, $C_\theta = 0.13$, $C_{T1} = 3.2$, $C_1 = 1.82 \sim 2.8$, $C_{\epsilon 1} = 1.44$, $C_{\theta 1} = 0.62$, $C_{T2} = 0.5$, $C_2 = 0.4 \sim 0.6$, $C_{\epsilon 2} = 1.92$, $C_{T3} = 0.5$, $C_3 = 0.3 \sim 0.5$, and $C_{\epsilon 3} = 1.44 \sim 1.92$.

Data on the determination of these constants can be found from Launder (1975)[75], who determined that $C_{T3} = C_{T2}$ and that $C_{\theta 1} = 0.62$. Other constants are based on data from Gibson and Schwartz (1963)[47].

6.7.3 *k*-ϵ-(*A* or *E*) Model

In the k-ϵ-A and k-ϵ-E models, the $\overline{\theta^2}$ equation (6.82) is made algebraic by omitting the convection and diffusion terms. The complete equations are as follows:

- The $\overline{u_iu_j}$ equation is

$$-\overline{u_iu_j} = \frac{k}{C_1\epsilon}\left\{\frac{(C_2 - 1)\,P_{ij} + (C_3 - 1)\,P_{ij,b} - \frac{2}{3}\delta_{ij}\left[C_2P_k + C_3P_b + (C_1 - 1)\,\epsilon\right]}{1 + \left(\frac{P_k + P_b}{\epsilon} - 1\right)/C_1}\right\}. \tag{6.83}$$

- The k equation is

$$\frac{Dk}{Dt} = \frac{\partial}{\partial X_l}\left[\left(C_k \frac{k^2}{\epsilon} + \nu\right)\frac{\partial k}{\partial X_l}\right] + P_k + P_b - \epsilon. \tag{6.84}$$

- The ϵ equation is

$$\frac{D\epsilon}{Dt} = \frac{\partial}{\partial X_l}\left[\left(C_\epsilon \frac{k^2}{\epsilon} + \nu\right)\frac{\epsilon}{\partial X_l}\right] + C_{\epsilon 1}\frac{\epsilon}{k}P_k + C_{\epsilon 3}\frac{\epsilon}{k}P_b - C_{\epsilon 2}\frac{\epsilon^2}{k}. \tag{6.85}$$

- The $\overline{u_i\theta}$ equation is

$$-\overline{u_i\theta} = \frac{k}{C_{T1}\epsilon}\left[\frac{-P_T + (1 + C_{T3})\,\beta g_i \overline{\theta^2} - C_{T2}\overline{u_m\theta}\frac{\partial U_i}{\partial X_m}}{1 + \left(\frac{P_k + P_b}{\epsilon} - 1\right)/C_{T1}}\right]. \tag{6.86}$$

• The $\overline{\theta^2}$ equation is

$$\overline{\theta^2} = -\frac{k}{C_{\theta 1}\epsilon}\overline{u_l\theta}\frac{\partial T}{\partial X_l}. \tag{6.87}$$

The twelve constants are $C_k = 0.09$, $C_\epsilon = 0.07$, $C_{T1} = 3.2$, $C_1 = 1.82 \sim 2.8$, $C_{\epsilon 1} = 1.44$, $C_{T2} = 0.5$, $C_2 = 0.4 \sim 0.6$, $C_{\epsilon 2} = 1.92$, $C_{T3} = 0.5$, $C_3 = 0.3 \sim 0.5$, $C_{\epsilon 3} = 1.44 \sim 1.92$, and $C_{\theta 1} = 0.62$.

6.8 TURBULENT FLOW PREDICTIONS: FIVE (BUOYANT FLOWS)

6.8.1 Density Stratified Shear Flow (k-ϵ-A)

Let us consider two-dimensional steady flow of air in a wind tunnel with linear velocity and temperature profiles, as shown in Fig. 6.2, where $0 < U_c < 2.5$ m/s, $300K < T_c < 330K$, and $0 < Ri < 0.8$. Ri is the Richardson number, which is the measure of the rate at which the turbulence energy is removed by working against the gravitational field related to the rate at which it is created by mean shear and can be written as

$$Ri = \frac{g}{T}\frac{\frac{\partial T}{\partial Y}}{\left(\frac{\partial U}{\partial Y}\right)^2} = g\beta\frac{\frac{\partial T}{\partial Y}}{\left(\frac{\partial U}{\partial Y}\right)^2} \tag{6.88}$$

or

$$Ri = \frac{g\beta\Delta TL}{\Delta U^2} = \frac{\text{Buoyant Force}}{\text{Inertia Force}}. \tag{6.89}$$

Let us examine the $\overline{u_iu_j}$ equation without convection and diffusion terms. Then

$$-\overline{u_iu_j} = \frac{k}{C_1\epsilon}\left[(C_2-1)P_{ij} + (C_3-1)P_{ij,b} - \frac{2}{3}\delta_{ij}(C_2P_k + C_3P_b)\right] - \frac{2}{3}\delta_{ij}\frac{(C_1-1)}{C_1}k, \tag{6.90}$$

where

$$P_{ij} = -\left(\overline{u_iu_l}\frac{\partial U_j}{\partial X_l} + \overline{u_ju_l}\frac{\partial U_i}{\partial X_l}\right), \tag{6.91}$$

$$P_k = -\overline{u_iu_l}\frac{\partial U_i}{\partial X_l}, \tag{6.92}$$

$$P_{ij,b} = -\beta(g_i\overline{u_j\theta} + g_j\overline{u_i\theta}) \tag{6.93}$$

$$P_b = -\beta g_i\overline{u_i\theta}. \tag{6.94}$$

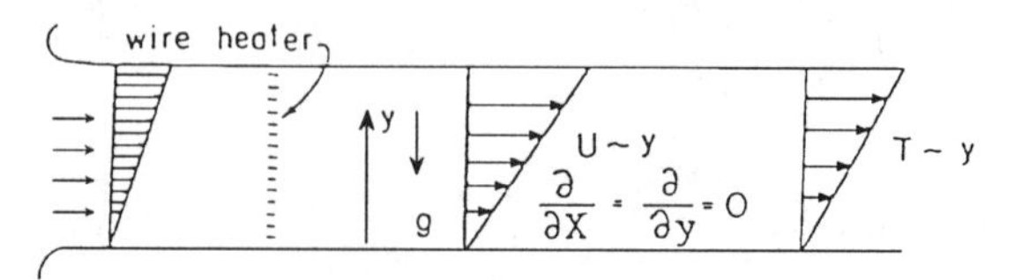

Figure 6.2 Density stratified shear flow.

Let us also examine the k equation. In this case $U \sim Y, T \sim Y$, and

$$\frac{\partial \phi}{\partial X} = \frac{\partial \phi}{\partial Y} = 0, \tag{6.95}$$

where $\phi = \overline{u_i u_j}, k, \epsilon$, or $\overline{u_i \theta}$. $P_k = -\overline{uv}\frac{\partial U}{\partial Y}$ and $P_b = \beta g \overline{v\theta}$. Substituting in the k equation yields

$$\epsilon = -\overline{uv}\frac{\partial U}{\partial Y} - \beta g \overline{v\theta}. \tag{6.96}$$

Therefore, we see that the buoyant force can amplify ($\beta g \overline{v\theta} < 0$) or damp ($\beta g \overline{v\theta} > 0$) the turbulence, depending on the direction of gravitational force.

The governing equations are

- the $\overline{u^2}$ equation,

$$\frac{\overline{u^2}}{k} = \frac{-\overline{uv}\frac{\partial U}{\partial Y}}{\epsilon}\left(\frac{6-4C_2}{3C_1}\right) - \frac{2C_3 \beta g \overline{v\theta}}{3C_1 \epsilon} + \frac{2(C_1-1)}{3C_1}, \tag{6.97}$$

- the $\overline{uv}$ equation,

$$\overline{uv} = \frac{\overline{v^2}\frac{\partial U}{\partial Y}}{\epsilon}\left(\frac{C_2-1}{C_1}\right) - \frac{g\beta \overline{u\theta}}{\epsilon}\left(\frac{C_3-1}{C_1}\right), \tag{6.98}$$

- the $\overline{v^2}$ equation,

$$\frac{\overline{v^2}}{k} = \frac{-\overline{uv}\frac{\partial U}{\partial Y}}{\epsilon}\left(\frac{2C_2}{3C_1}\right) + \left(\frac{3-C_3}{3C_1\epsilon}\right)\beta g \overline{v\theta} + \frac{2(C_1-1)}{3C_1}, \tag{6.99}$$

and

- the $\overline{w^2}$ equation,

$$\frac{\overline{w^2}}{k} = \frac{-\overline{uv}\frac{\partial U}{\partial y}}{\epsilon}\left(\frac{2C_2}{3C_1}\right) + \frac{2C_3}{3C_1\epsilon}\beta g \overline{v\theta} + \frac{2(C_1-1)}{3C_1}. \tag{6.100}$$

Examining $\overline{u_i\theta}$ and $\overline{\theta^2}$ without convection and diffusion terms, we have

- the $\overline{u_i\theta}$ equation,

$$-\overline{u_i\theta} = \frac{k}{C_{T1}\epsilon}\left[-P_T + (1+C_{T3})\,\beta g_i \overline{\theta^2} - C_{T2}\overline{u_l\theta}\frac{\partial U_i}{\partial Xl}\right] \tag{6.101}$$

$$P_T = -\left(\overline{u_i u_l}\frac{\partial T}{\partial X_l} + \overline{u_l\theta}\frac{\partial U_i}{\partial X_l}\right), \tag{6.102}$$

and

- the $\overline{\theta^2}$ equation,

$$\overline{\theta^2} = -\frac{k}{C_{\theta 1}\epsilon}\overline{u_l\theta}\frac{\partial T}{\partial X_l}, \tag{6.103}$$

or

$$-\overline{u\theta} = \frac{k}{C_{T1}\epsilon}\left[\overline{uv}\frac{\partial T}{\partial Y} + (1 - C_{T2})\,\overline{v\theta}\frac{\partial U}{\partial Y}\right] \tag{6.104}$$

$$-\overline{v\theta} = \frac{k}{C_{T1}\epsilon}\left[\overline{v^2}\frac{\partial T}{\partial Y} + (1 + C_{T3})\,\beta g\overline{\theta^2}\right] \tag{6.105}$$

$$\overline{\theta^2} = -\frac{k}{C_{\theta 1}\epsilon}\overline{v\theta}\frac{\partial T}{\partial Y}. \tag{6.106}$$

With $C_1 = 2$, $C_2 = C_3 = 0.6$, $C_{T1} = 3.2$, $C_{T2} = C_{T3} = 0.5$, and $C_{\theta 1} = 0.625$,

$$\frac{\overline{u^2}}{k} = 0.94 + 0.41\frac{Ri}{1 - Ri} \tag{6.107}$$

$$\frac{\overline{w^2}}{k} = 0.53 \tag{6.108}$$

$$\frac{\overline{v^2}}{k} = 0.53 - 0.41\frac{Ri}{1 - Ri}. \tag{6.109}$$

Figure 6.3 compares the variation of the normal stresses with the Richardson's number utilizing Eqs. 6.107–6.109 and Webster's (1964)[182] experimental data. From Eqs. 6.107–6.109, a stable gravitational field leads to a relative gain in the level of streamwise fluctuations at the expense of the vertical fluctuations. The relative magnitude of the lateral fluctuations is unaffected. These features are generally in accord with Webster's measurements.

The predicted correlation coefficients for the vertical and horizontal heat fluxes are shown in Fig. 6.4(a) and (b), the comparison being made with the experimental data by Webster (1964)[182]. Comparison between the experimental data and the model

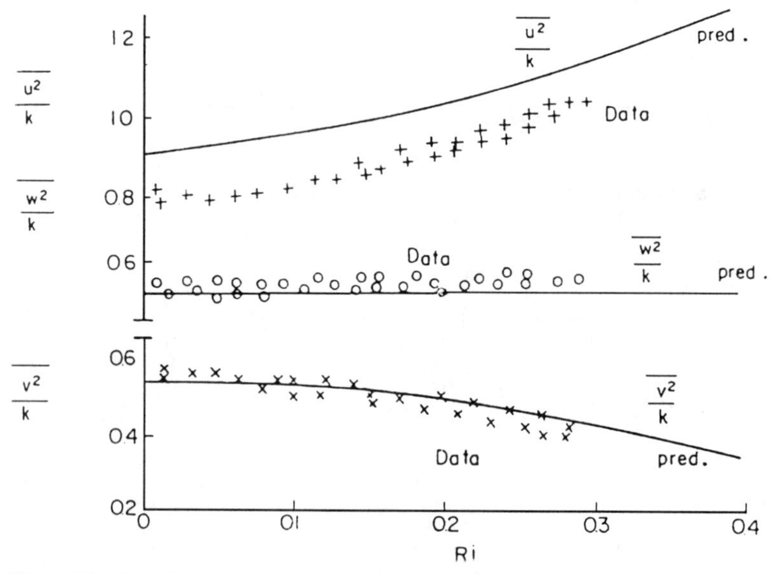

Figure 6.3 Turbulent intensity distribution. (Adapted from Webster (1964).)

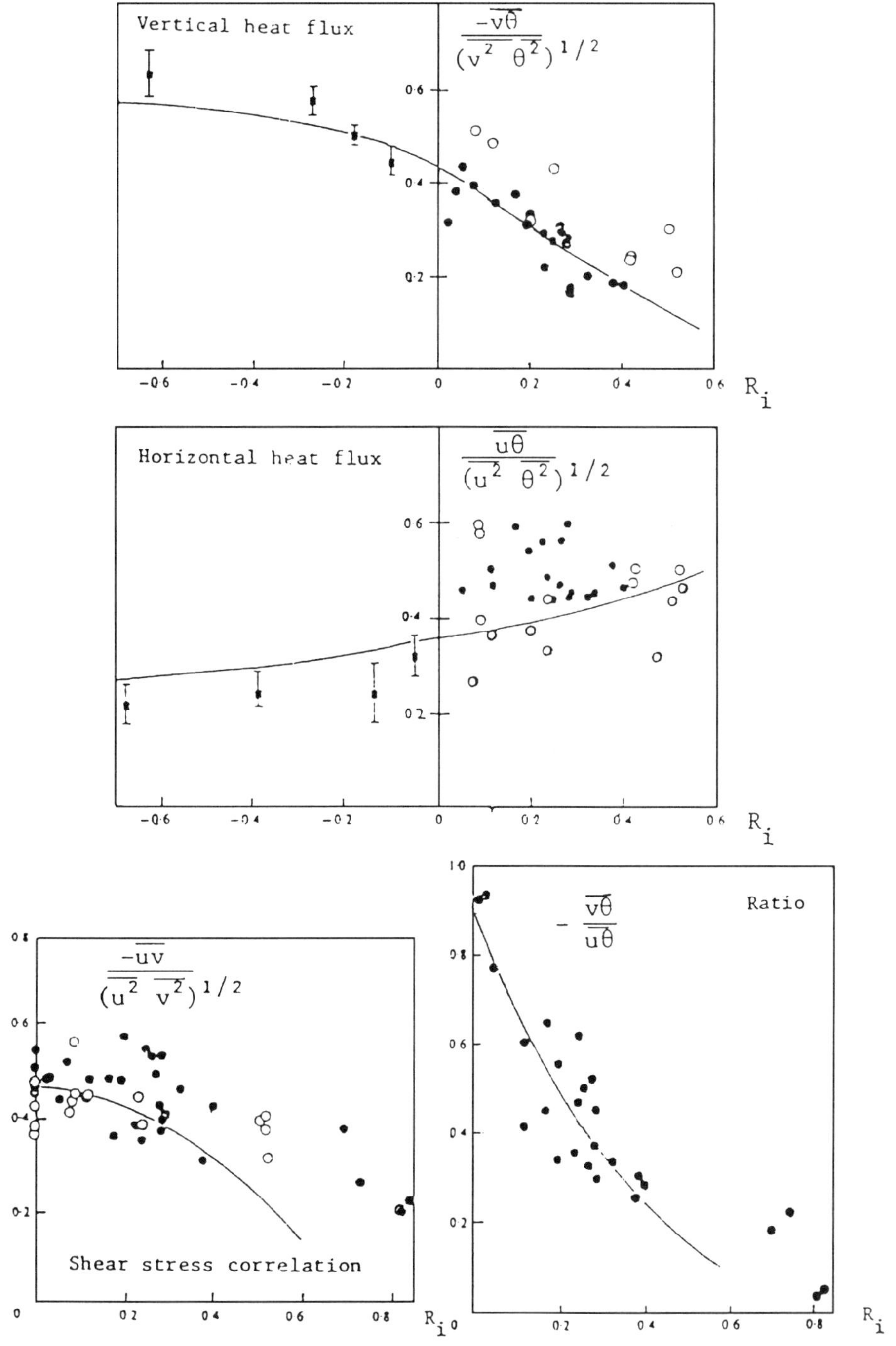

Figure 6.4 Reynolds-stress and heat flux distribution. Data ○ and • are Webster's data at two stations. (Adapted from Webster (1964).)

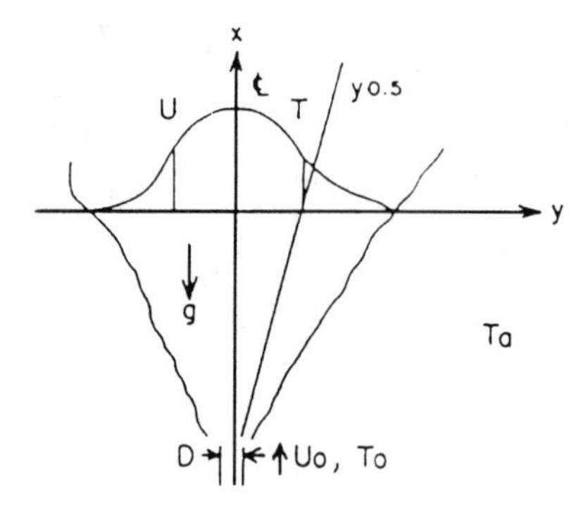

Figure 6.5 Vertical buoyant jet.

predictions (solid lines) shows a fair agreement. For stable flow, ($Ri > 0$) the calculated coefficient for the vertical heat flux falls steeply as Ri increases while that for $\overline{u\theta}$ rises gradually. The present model is strictly inapplicable to flows near walls because the model neglects the influence of a wall on the pressure terms. However, it is known that these correlation coefficients are relatively unaffected by the presence of a wall. Figure 6.4 shows that the shear stress coefficient is virtually constant for Ri up to 0.25. In contrast, the ratio $-\overline{v\theta}/\overline{u\theta}$, shown in Fig 6.4, falls from 0.9 under natural conditions to 0.4 over the same span of Richardson numbers. Webster's experimental data certainly support the behavior predicted by the model.

6.8.2 (k-ϵ-$\overline{\theta^2}$) Model

The turbulence models test for problems with no wall interaction should be tested. One simple, yet important, application is the vertical jet. Chen and Rodi (1975)[19] and Chen and Chen (1979)[15] studied the vertical-jet discharging into a neutral or stratified environment as shown in Fig. 6.5.

With the free-shear flow, $\partial/\partial X = \partial/\partial Z \ll \partial/\partial Y$ and steady-state assumption, we may obtain boundary-layer like equations. The mean governing equations with the Boussinesq approximation and the k-ϵ-$\overline{\theta^2}$-A model are as follows:

- The continuity equation is

$$\frac{\partial U}{\partial X} + \frac{\partial V}{\partial Y} = 0. \tag{6.110}$$

- The momentum equation is

$$U\frac{\partial U}{\partial X} + V\frac{\partial U}{\partial Y} = -\frac{\partial}{\partial Y}(-\overline{uv}) + \beta g (T - T_a). \tag{6.111}$$

- The energy equation is

$$U\frac{\partial T}{\partial X} + V\frac{\partial T}{\partial Y} = -\frac{\partial}{\partial Y}\left(-\overline{v\theta}\right). \tag{6.112}$$

- The kinetic energy equation is

$$U\frac{\partial K}{\partial X} + V\frac{\partial K}{\partial Y} = -\frac{\partial}{\partial Y}\left(C_k\frac{k^2}{\epsilon}\frac{\partial k}{\partial Y}\right) - \overline{uv}\frac{\partial U}{\partial Y} + \beta g\overline{v\theta} - \epsilon. \tag{6.113}$$

- The ϵ equation is

$$U\frac{\partial \epsilon}{\partial X} + V\frac{\partial \epsilon}{\partial Y} = -\frac{\partial}{\partial Y}\left(C_\epsilon \frac{k^2}{\epsilon}\frac{\partial \epsilon}{\partial Y}\right) - C_{\epsilon 1}\frac{\epsilon}{k}\left(-\overline{uv}\frac{\partial U}{\partial Y} + \beta g \overline{v\theta}\right) - C_{\epsilon 2}\frac{\epsilon^2}{k}. \tag{6.114}$$

- The $\overline{\theta^2}$ equation is

$$U\frac{\partial \overline{\theta^2}}{\partial X} + V\frac{\partial \overline{\theta^2}}{\partial Y} = \frac{\partial}{\partial Y}\left(C_\theta \frac{k^2}{\epsilon}\frac{\partial \overline{\theta^2}}{\partial Y}\right) - 2\overline{v\theta}\frac{\partial T}{\partial Y} - 2C_{\theta 1}\frac{\epsilon}{k}\overline{\theta^2}. \tag{6.115}$$

- The $\overline{uv}$ equation is

$$-\overline{uv} = \left(\frac{1-C_2}{C_1}\right)\frac{\overline{v^2}}{k}\left(1 + \frac{\beta g k \frac{\partial T}{\partial Y}}{C_{T1}\epsilon \frac{\partial U}{\partial Y}}\right)\frac{k^2}{\epsilon}\frac{\partial U}{\partial Y} = \nu_t \frac{\partial U}{\partial Y}. \tag{6.116}$$

- The $\overline{v^2}$ equation is

$$\overline{v^2} = \frac{2}{3}\frac{C_1 - 1 + C_2}{C_1}k = 0.53k. \tag{6.117}$$

- The $\overline{v\theta}$ equation is

$$-\overline{v\theta} = \frac{1}{C_{T1}}\frac{\overline{v^2}}{k}\frac{k^2}{\epsilon}\frac{\partial T}{\partial Y} = \alpha_t \frac{\partial T}{\partial Y}. \tag{6.118}$$

The constants are $C_k = 0.09$, $C_1 = 2.2$, $C_2 = 0.55$, $C_3 = 0.3 \sim 0.5$, $C_{\epsilon 1} = 1.43$, $C_{\epsilon 2} = 1.92$, $C_\theta = 0.13$, $C_{\theta 1} = 0.625$, and $C_{T1} = 3.2$. $Pr_t = \nu_t/\alpha_t \doteq 0.7 \sim 0.9$ is the turbulent Prandtl number. If the model is to be fairly general, it should be able to predict correctly jets without buoyancy influence and buoyant jets with a limiting case as pure plumes with the same empirical constants.

The numerical calculations of the rates of velocity and temperature spread for both cases are compared with the experimental results. For pure plumes, the calculations were compared with the experimental results of Rouse, Yih, and Humphreys (1952)[145]; whereas for nonbuoyant jets, the calculations are compared with the experimental results in the review of Rodi (1972)[134]. The predicted half-width spreading, the $dy_{0.5u}/dx$, and the $dy_{0.5T}/dx$ for two-dimensional buoyant jets in uniform ambient density are given in Table 6.1. The agreement between experiment and prediction is good. Transverse variations for uniform ambient temperature are shown in Fig. 6.6. The mean velocity and temperature profiles for buoyant jets are almost identical. The profiles for the kinetic energy and the shear stress are similar in shape to those for nonbuoyant jets,

Table 6.1 Predicted and experimental half-width spreading for two-dimensional buoyant jets

Comparison	Jet: $dy_{0.5u}/dx$	Plume: $dy_{0.5u}/dx$	Plume: $dy_{0.5T}/dx$
Experimental	0.11	0.13	0.12
Predicted	0.11	0.12	0.116

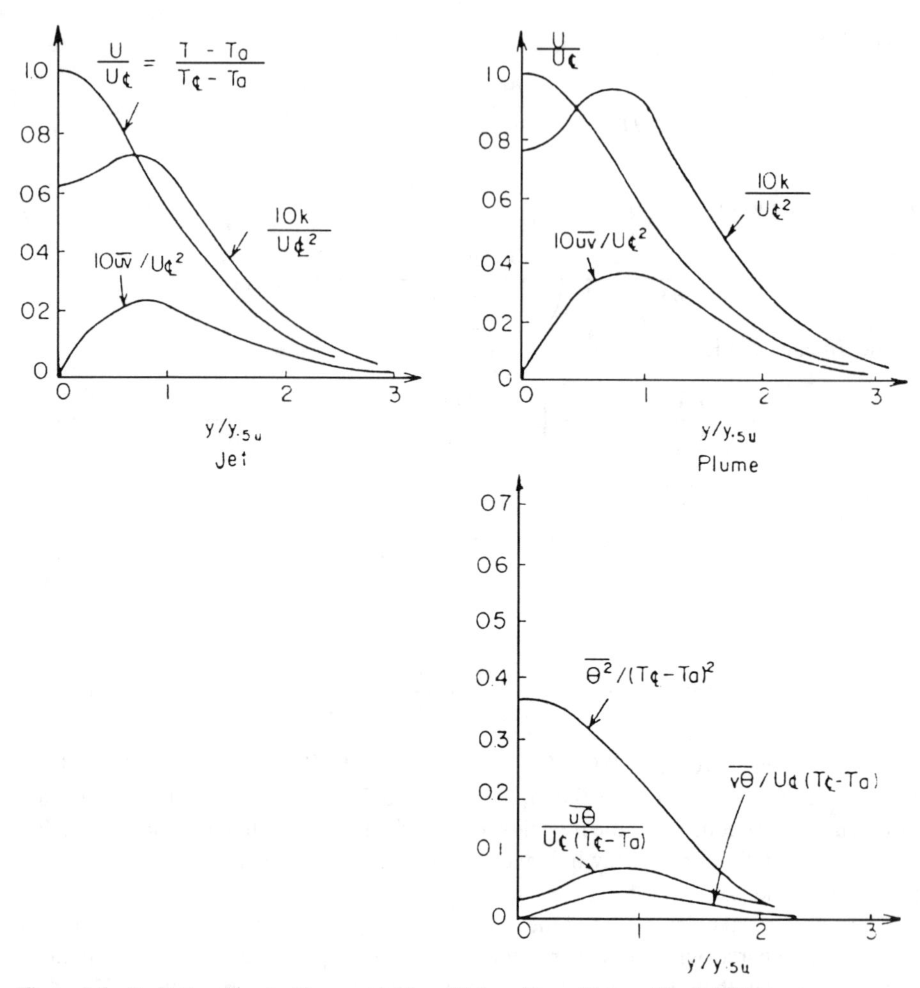

Figure 6.6 Prediction of vertical buoyant jet flow. (Adapted from Chen and Rodi (1975).)

but the magnitudes of the kinetic energy and the shear stress are larger because of the buoyant action. With the increase of turbulent intensity, one expects an increase in the rate of spread for the buoyant jets over that of nonbuoyant jets. A substantial increase in $\overline{\theta^2}$ is concentrated near the center of the buoyant jet. The longitudinal variations for uniform ambient conditions are shown in Fig. 6.7, where

$$F = \frac{U_o^2}{g\frac{\rho}{2}\frac{\rho_a-\rho_o}{\rho_o}} = \text{Froude number.}$$

The calculation is compared with experimental results for different Froude numbers. The calculations show a good agreement with the experimental data. An example of round buoyant jet in a linearly stratified ambient was also investigated by Chen and Rodi (1975)[19], in which the flow situation corresponds to that of Sneck and Brown's

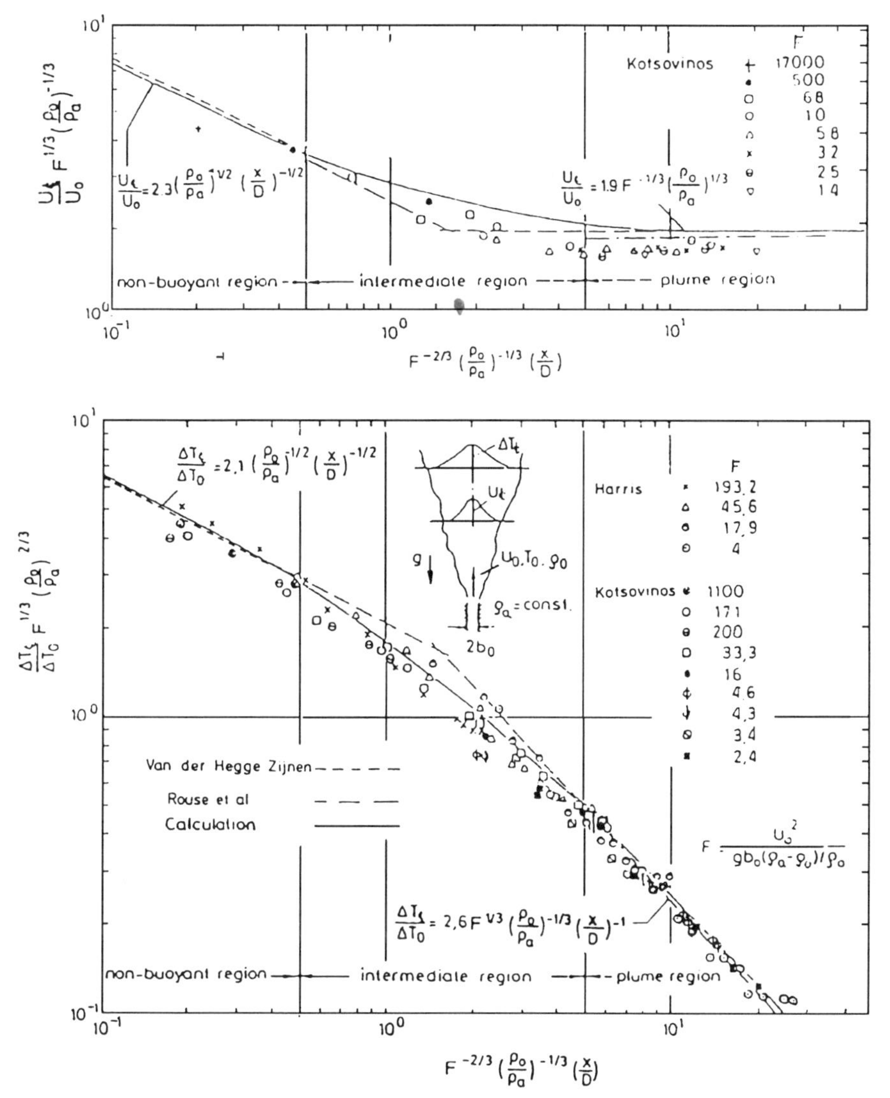

Figure 6.7 Central velocity and temperature decay. (Adapted from Chen and Rodi (1975).)

(1974)[158] experiments. Figure 6.8 compares the predicted and measured results. The predicted results at the axis agree well with the experiments, even in the region where the jet is deflected horizontally.

6.8.3 Horizontal Buoyant Surface Jet

Using the k-ϵ model and the shear flow approximation, Hossain (1979)[62] obtained the governing equations to be the same as those given in Section 6.8.2, except that $\overline{\theta^2}$ is

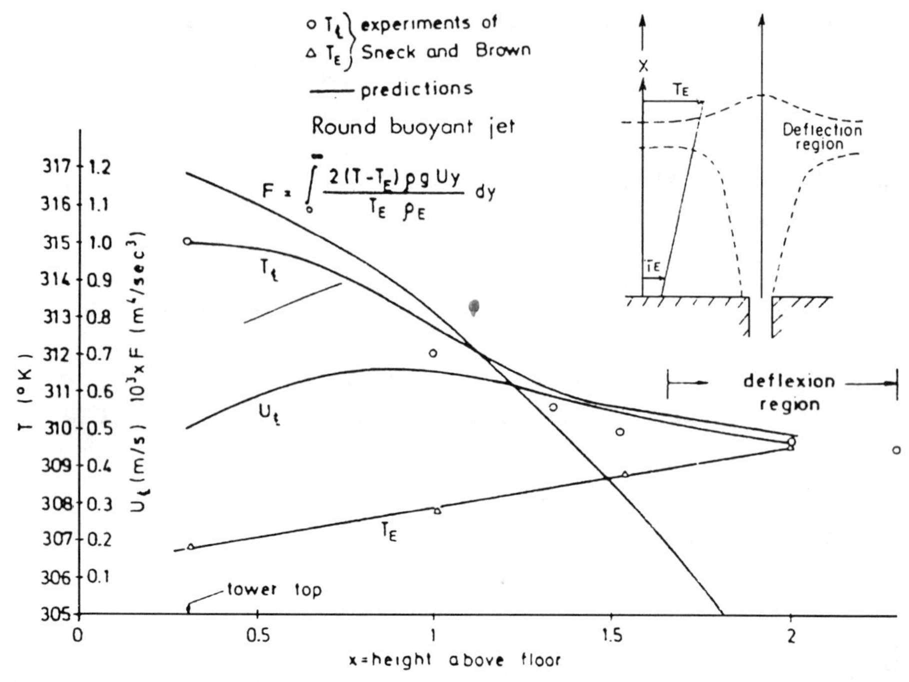

Figure 6.8 Buoyant round jet in stable, stratified surroundings. (Adapted from Chen and Rodi (1975).)

closed by

$$\overline{\theta^2} = -\frac{k}{C_{\theta 1}} \frac{\overline{v\theta}}{\epsilon} \frac{\partial T}{\partial y}. \tag{6.119}$$

A two-dimensional jet predicted by Hossain (1979)[62] was compared with experimental data as shown in Fig. 6.9. Generally fair agreement is obtained.

6.8.4 Three-Dimensional Surface Jet

McGuirk and Rodi (1978)[105] studied the three-dimensional heated jets. The jet is discharged from a slot of $h_o \times 2b_o$ cross-sectional area to an ambient condition. The Froude number considered in the calculation was 2.56. The results are shown in Fig. 6.10. The predictions appear to agree well with the experimental results.

6.8.5 Horizontal Submerged Buoyant Jets

The (k-ϵ-A) and (k-ϵ-E) models were tested in a buoyant flow of coaxial slot discharge. This work was done by McGuirk and Spalding (1975)[106]. The results are shown in Fig. 6.11.

The coaxial jet is issuing from a slot. McGuirk and Spalding (1975)[106] also calculated the flow and scalar fields resulting from the coaxial discharge of heavier and lighter water from a round pipe into a straight rectangular open channel flow [see

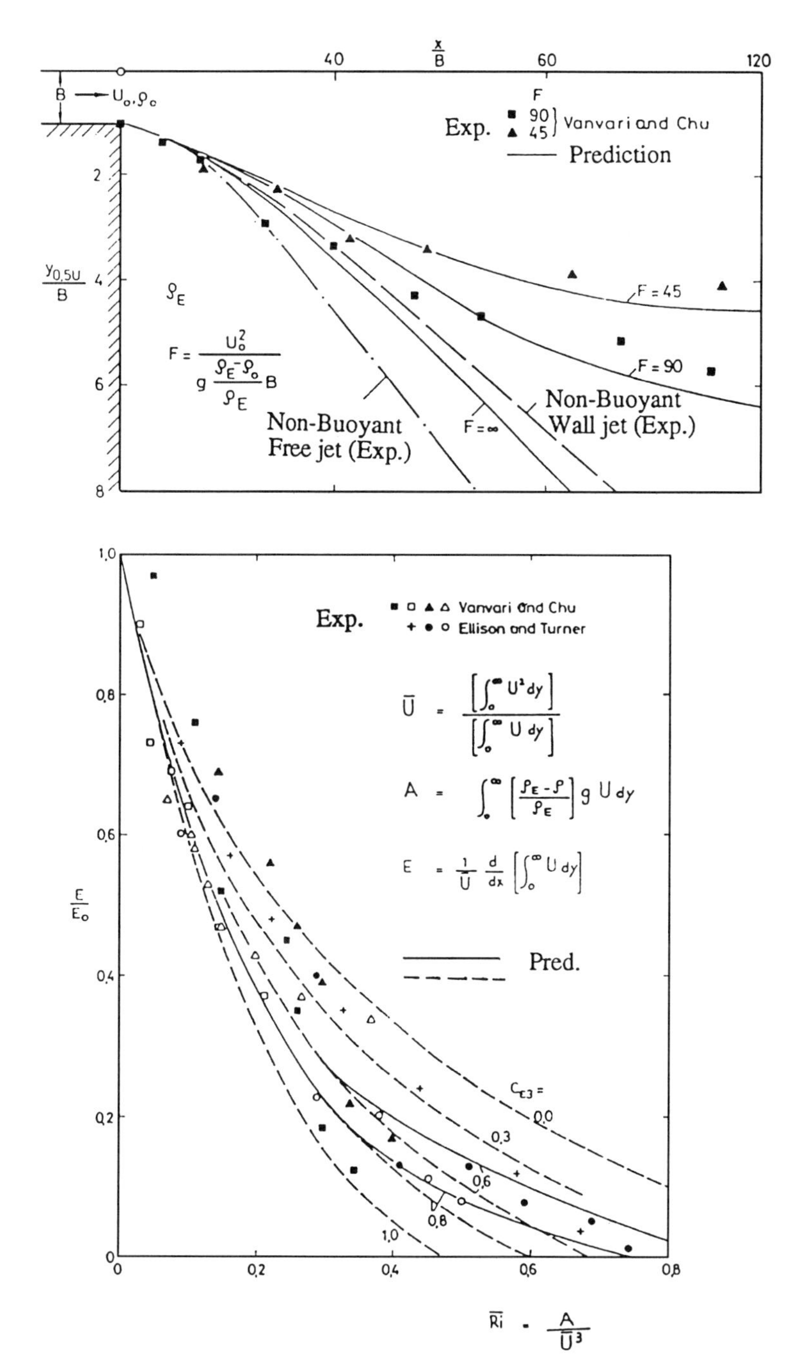

Figure 6.9 Prediction of horizontal buoyant surface jet. (a) Flow configuration and development of half width and (b) entrainment rate. (Adapted from Hossain (1979).)

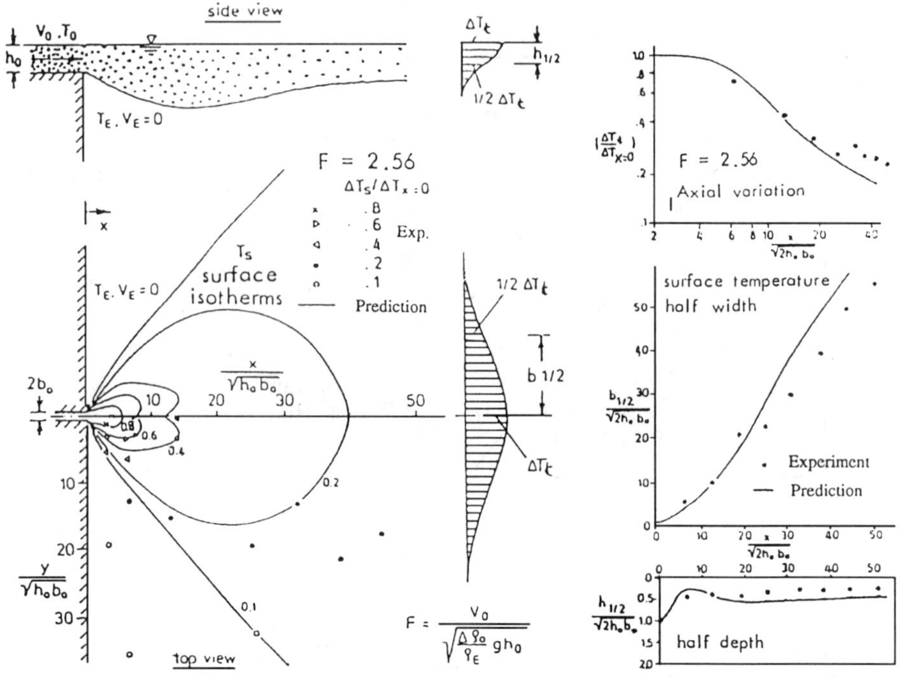

Figure 6.10 Prediction of three-dimensional surface jet. (a) Flow configuration and surface isotherms for $F = 2.56$. (b) Axial variation of maximum excess surface temperature half width and half depth for $F = 2.56$. (Adapted from McGuirk and Rodi (1978).)

Fig. 6.11]. In the experimental situations simulated, the densimetric Froude number F was never below 15 so that the influence of buoyancy forces on turbulence was weak. Accordingly, McGuirk and Spalding neglected this influence and performed the calculation with the standard k-ϵ model not involving any buoyancy terms; the influence of buoyancy in the calculation was therefore solely due to the buoyancy term in the vertical momentum equation.

Figure 6.11 compares calculated and measured jet trajectories for situations in which the discharged water was heavier than the channel water. In Fig. 6.12 the calculated and measured temperature contours are compared with various cross-sections downstream of the jet exit for a situation in which heated water was discharged. These figures show that the flows considered are simulated well by the mathematical model.

6.8.6 Jets in Cross-Flows

The k-ϵ-E model was used in the prediction of jets in cross-flows by Jones and McGuirk (1980)[71]. In their study, they considered a round jet of diameter, D, discharging into a tunnel with $H/D = 12$. The flow in the tunnel is axial, with a mean-velocity U_o, the density of the mean flow in the tunnel is ρ_o, and the jet fluid density is ρ_j. The study was carried out for four kinetic energy ratios, the kinetic energy ratio is defined as the ratio of the jet kinetic energy to the mean tunnel kinetic energy. Comparison between

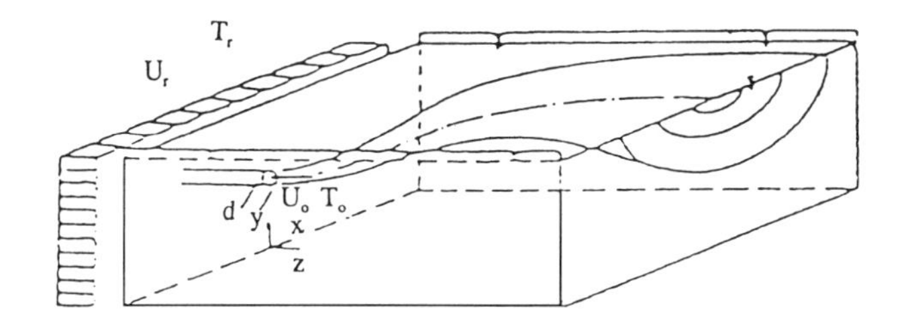

$$F = U_d / \sqrt{gd(\rho_r - \rho_d)/\rho_r}$$

$$R = U_d / U_r$$

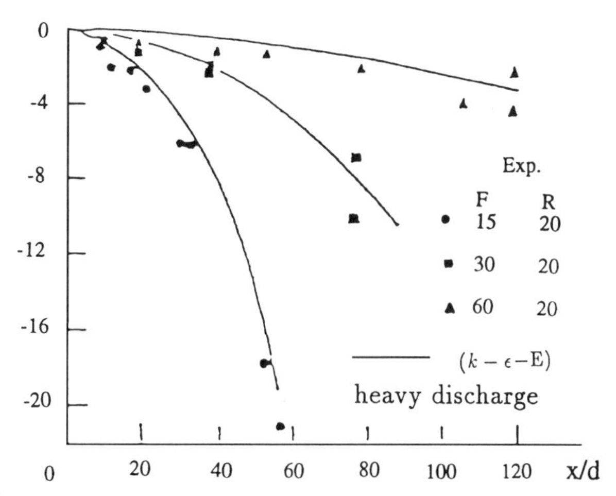

Figure 6.11 Buoyant flow in a coaxial pipe discharge: Heavy discharge. (Adapted from McGuirk and Spalding (1975).)

measured results and the predicted results is given in Fig. 6.13, although there is some discrepancy in predicting the jet trajectory and temperature profile. The difference may be due to the numerical solver treating the problem as parabolic while the flow near the jet exit is a strong elliptic flow.

6.9 TWO-SCALE TURBULENCE CONCEPT

Predictions of buoyant free-shear flows are obtained by the k-ϵ model on the basis of the two-scale turbulence concept. This concept was given by Chen and Singh (1990)[22]. The modeled equations are as follows:

- The continuity equation is

$$\frac{\partial U_i}{\partial X_i} = 0. \tag{6.120}$$

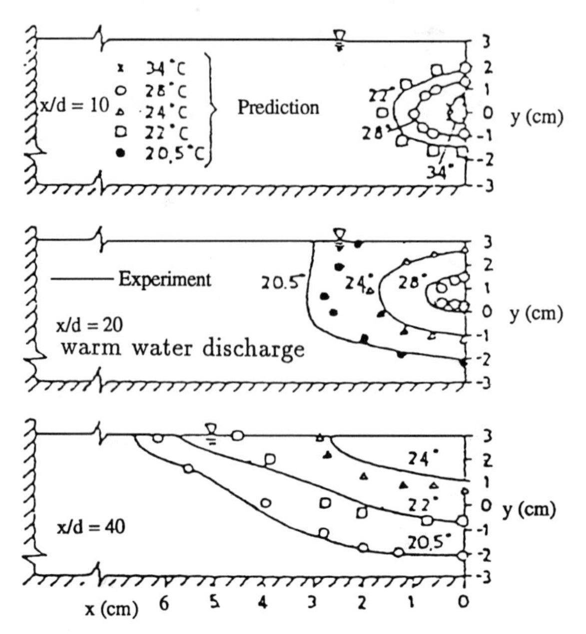

Figure 6.12 Buoyant round jet in a coaxial channel flow. (Adapted from McGuirk and Spalding (1975).)

- The momentum equations is

$$\frac{DU_i}{Dt} = -\frac{1}{\rho}\frac{\partial p}{\partial X_i} - \beta g_i \Delta T + \nu \frac{\partial^2 U_i}{\partial X_j \partial X_j} - \frac{\partial \overline{u_i u_j}}{\partial X_j}. \tag{6.121}$$

- The energy equation is

$$\frac{DT}{Dt} = \alpha \frac{\partial^2 T}{\partial X_j \partial X_j} - \frac{\partial \overline{u_i \theta}}{\partial X_i}. \tag{6.122}$$

- The Reynolds stress equation is

$$\begin{aligned} \frac{D\overline{u_i u_j}}{Dt} = {} & \frac{\partial}{\partial X_l}\left[\left(C_k \frac{k^2}{\epsilon} + \nu\right)\frac{\partial \overline{u_i u_j}}{\partial X_l}\right] + P_{ij} + P_{ij.b} - \frac{2}{3}\delta_{ij}\epsilon \\ & - C_1 \frac{\epsilon}{k}\left(\overline{u_i u_j} - \frac{2}{3}\delta_{ij} k\right) - C_2\left(P_{ij} - \frac{2}{3}\delta_{ij} P_k\right) \\ & - C_3\left(P_{ij.b} - \frac{2}{3}\delta_{ij} P_b\right), \end{aligned} \tag{6.123}$$

where

$$P_{ij} = -\left(\overline{u_i u_l}\frac{\partial U_j}{\partial X_l} + \overline{u_j u_l}\frac{\partial U_i}{\partial X_l}\right) \tag{6.124}$$

$$P_k = -\overline{u_i u_l}\frac{\partial U_i}{\partial X_l} \tag{6.125}$$

$$P_{ij.b} = -\beta(g_i \overline{u_j \theta} + g_j \overline{u_i \theta}) \tag{6.126}$$

$$P_b = -\beta g_i \overline{u_i \theta}. \tag{6.127}$$

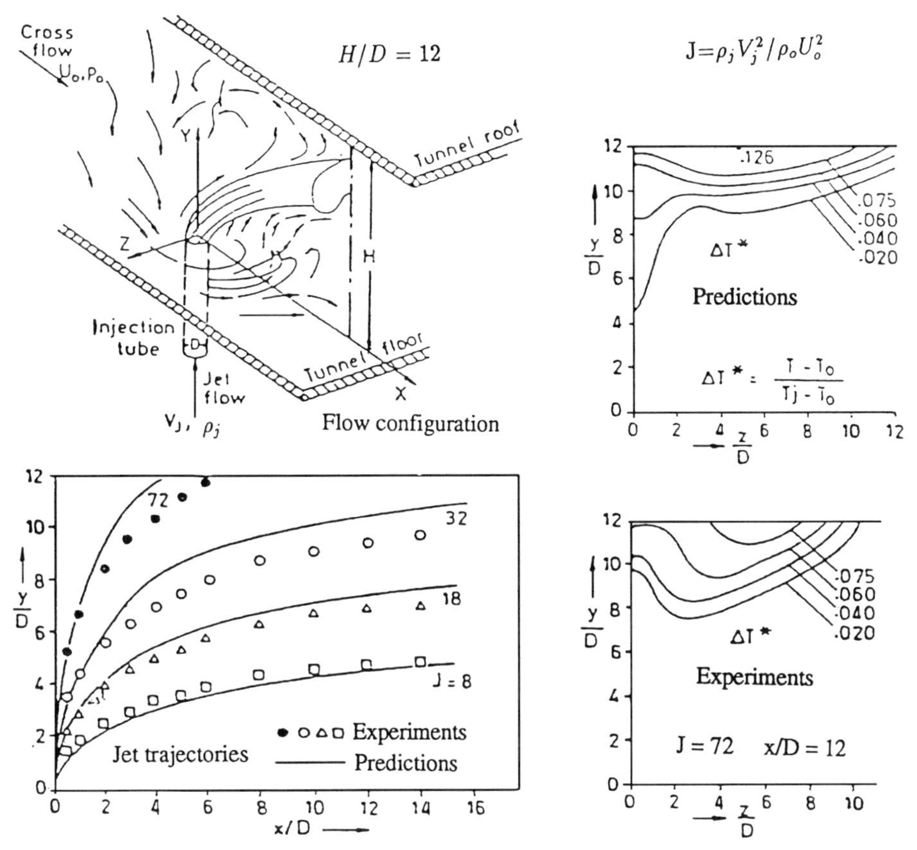

Figure 6.13 Prediction of a jet in cross-flow. (Adapted from Jones and McGuirk (1980).)

- The kinetic energy equation is

$$\frac{Dk}{Dt} = \frac{\partial}{\partial X_l}\left(\left(C_k\frac{k^2}{\epsilon} + \nu\right)\frac{\partial k}{\partial X_l}\right) + P_k + P_b - \epsilon. \tag{6.128}$$

- The rate of dissipation, ϵ, equation is

$$\frac{D\epsilon}{Dt} = \frac{\partial}{\partial X_l}\left[\left(C_\epsilon\frac{k^2}{\epsilon} + \nu\right)\frac{\partial \epsilon}{\partial X_l}\right] + C_{\epsilon 1}\left(\frac{\epsilon}{\nu}\right)^{1/2}(P_k + P_b) - C_{\epsilon 2}\left(\frac{\epsilon}{\nu}\right)^{1/2}\epsilon. \tag{6.129}$$

- The Reynolds heat flux equation is

$$\frac{D\overline{u_i\theta}}{Dt} = \frac{\partial}{\partial X_l}\left[\left(C_T\frac{k^2}{\epsilon} + \nu\right)\frac{\partial\overline{u_i\theta}}{\partial X_l}\right] + P_T - (1 + C_{T3})\beta g_i\overline{\theta^2}$$
$$- C_{T1}\frac{\epsilon}{k}\overline{u_i\theta} + C_{T2}\frac{\partial U_i}{\partial X_m}\overline{u_m\theta}. \tag{6.130}$$

- The $\overline{\theta^2}$ equation is

$$\overline{\theta^2} = -\frac{k}{C_{\theta 1}\epsilon}\overline{u_l\theta}\frac{\partial T}{\partial X_l}. \tag{6.131}$$

There are eleven turbulent moduli for twelve turbulent transport quantities ($\overline{u_iu_j}$, $\overline{u_i\theta}$, k, ϵ, $\overline{\theta^2}$). These moduli were determined by Hanjalic and Launder (1972)[55] and Chen and Singh (1990)[22] as

$C_k = 0.33$, $C_\epsilon = 0.74$, $C_T = 0.13$, $C_{T1} = 3.5$,
$C_1 = 2.8$, $C_{\theta 1} = 0.625$, $C_3 = C_2 = 0.47$, $C_{T3} = C_{T2} = 0.5$,
$C_{\epsilon 1} = 17.5/(Re)^{1/2}$, and $C_{\epsilon 2} = 18.9/(Re)^{1/2}$,

where Re is the Reynolds number that is based on mean velocity and mean characteristic length of the given problem.

The two-scale turbulence model is further approximated by reducing the $\overline{u_iu_j}$ and $\overline{u_i\theta}$ equations into algebraic equations for thin-shear flows such as jets, wakes, and mixing layer flows. The algebraic approximation is based on the fact that the convection and diffusion are approximately equal and small. The resulting model is called the two-scale k-ϵ-A model. The two-dimensional ($j=0$) and axisymmetric ($j=1$) free-shear flows are governed by

- the continuity equation,

$$\frac{\partial U}{\partial X} + \frac{\partial V}{\partial Y} = 0, \tag{6.132}$$

- the momentum equations,

$$U\frac{\partial U}{\partial X} + V\frac{\partial U}{\partial Y} = -\frac{1}{Y^j}\frac{\partial}{\partial Y}(Y^j - \overline{uv}) + g\frac{T - T_a}{T_a}, \tag{6.133}$$

- the energy equation,

$$U\frac{\partial T}{\partial X} + V\frac{\partial T}{\partial Y} = -\frac{1}{Y^j}\frac{\partial}{\partial Y}(Y^j - \overline{v\theta}), \tag{6.134}$$

- the kinetic energy equation,

$$U\frac{\partial K}{\partial X} + V\frac{\partial K}{\partial Y} = -\frac{1}{Y^j}\frac{\partial}{\partial Y}\left(Y^jC_k\frac{k^2}{\epsilon}\frac{\partial k}{\partial Y}\right) - \overline{uv}\frac{\partial U}{\partial Y} + g\frac{\overline{v\theta}}{T_a} - \epsilon, \tag{6.135}$$

- the rate of dissipation, ϵ, equation

$$U\frac{\partial \epsilon}{\partial X} + V\frac{\partial \epsilon}{\partial Y} = -\frac{1}{Y^j}\frac{\partial}{\partial Y}\left(Y^jC_\epsilon\frac{k^2}{\epsilon}\frac{\partial \epsilon}{\partial Y}\right) + C_{\epsilon 1}\left(\frac{\epsilon}{\nu}\right)^{1/2}\left(-\overline{uv}\frac{\partial U}{\partial Y}\right)$$
$$+ C_{\epsilon 1}\left(\frac{\epsilon}{\nu}\right)^{1/2} g\frac{\overline{u\theta}}{T_a} - C_{\epsilon 2}\left(\frac{\epsilon}{\nu}\right)^{1/2}\epsilon, \tag{6.136}$$

- the $\overline{uv}$ equation,

$$-\overline{uv} = \left(\frac{1-C_2}{C_1}\right)\frac{\overline{v^2}}{k}\left(1+\frac{gk\frac{\partial T}{\partial Y}}{C_{T1}\epsilon T_a\frac{\partial U}{\partial Y}}\right)\frac{k^2}{\epsilon}\frac{\partial U}{\partial Y} = \nu_t\frac{\partial U}{\partial Y}, \tag{6.137}$$

- the $\overline{v^2}$ equation,

$$\overline{v^2} = \frac{2}{3}\frac{C_1 - 1 + C_2 + C_3\beta g\frac{\overline{u\theta}}{\epsilon}}{C_1}k = 0.53k, \tag{6.138}$$

- the $\overline{v\theta}$ equation,

$$-\overline{v\theta} = \frac{1}{C_{T1}}\frac{\overline{v^2}}{k}\frac{k^2}{\epsilon}\frac{\partial T}{\partial Y} = \alpha_t\frac{\partial T}{\partial Y}, \tag{6.139}$$

- the $\overline{u\theta}$ equation,

$$\overline{u\theta} = \frac{k}{C_{T1}\epsilon}\left[-\overline{uv}\frac{\partial T}{\partial Y} - \overline{v\theta}\left(1-C_{T2}\right)\frac{\partial U}{\partial Y} - g\overline{\theta^2}\frac{1-C_{T3}}{T_a}\right], \tag{6.140}$$

and

- the $\overline{\theta^2}$ equation,

$$\overline{\theta^2} = -\frac{k}{C_{\theta 1}\epsilon}\overline{v\theta}\frac{\partial T}{\partial Y}. \tag{6.141}$$

where the subscript a refers to ambient condition.

6.9.1 Prediction of Spread Parameters

The two-scale turbulence model is applied to solve several free-shear flows including (1) plane jet, (2) round jet, (3) plane wake, (4) plane mixing layer, (5) plane buoyant jet, and (6) round buoyant jet. In the prediction of these flows, the turbulent moduli given in the previous section are kept unchanged so that the prediction capabilities of the two-scale turbulence model can really be tested. In order to make a comparison with one-scale turbulence models, the standard k-ϵ model is also considered. The turbulence constants for the one-scale model are also kept constant without modification in all six flows considered.

Table 6.2 shows the spreading rate. For buoyant and nonbuoyant jets, the spread rate S is defined as the slope of $Y_{0.5}$ in the flow direction, where $Y_{0.5}$ is the location in the normal direction of a point where the velocity or temperature is one half its centerline value U_c or T_c. For wakes, S is the spread rate times the free-stream velocity U_E divided by the velocity defect, w_o, or $(U_E - U_c)$. In the case of the mixing layer, the spread rate is obtained in terms of the width of the mixing layer. The width is defined as the distance between the edges of the mixing layer where the velocity is 10 percent and 90 percent of the free-stream velocity. From Table 6.2, it is seen that the values of S predicted by the one-scale model for a round jet and a plane wake, without alternation of the turbulent moduli, are significantly different from the experimental data for both nonbuoyant and buoyant flows; whereas the two-scale turbulence model predicts satisfactory results for all cases calculated. This demonstrates that the two-scale turbulence model indeed provides better prediction than the one-scale turbulence model.

Table 6.2 Spread parameters

Flow type	Spread parameter	One-scale	Experimental data	Two-scale
Round jet	$\frac{dY_{1/2U}}{dX}$	0.1189	0.08	0.081
Plane jet	$\frac{dY_{1/2U}}{dX}$	0.1125	0.11	0.109
Plane wake	$\frac{U_E}{W_o}\frac{dY_{1/2U}}{dX}$	0.068	0.098	0.0975
Mixing layer	$\frac{d(Y_{0.5}-Y_{0.9})}{dX}$	0.159	0.16	0.15
Buoyant plane jet	$\frac{dY_{1/2U}}{dX}$	0.11	0.12	0.11
	$\frac{dY_{1/2T}}{dX}$	0.116	0.13	0.135
Buoyant round jet	$\frac{dY_{1/2U}}{dX}$	0.11	0.112	0.1
	$\frac{dY_{1/2T}}{dX}$	0.12	0.1	0.115

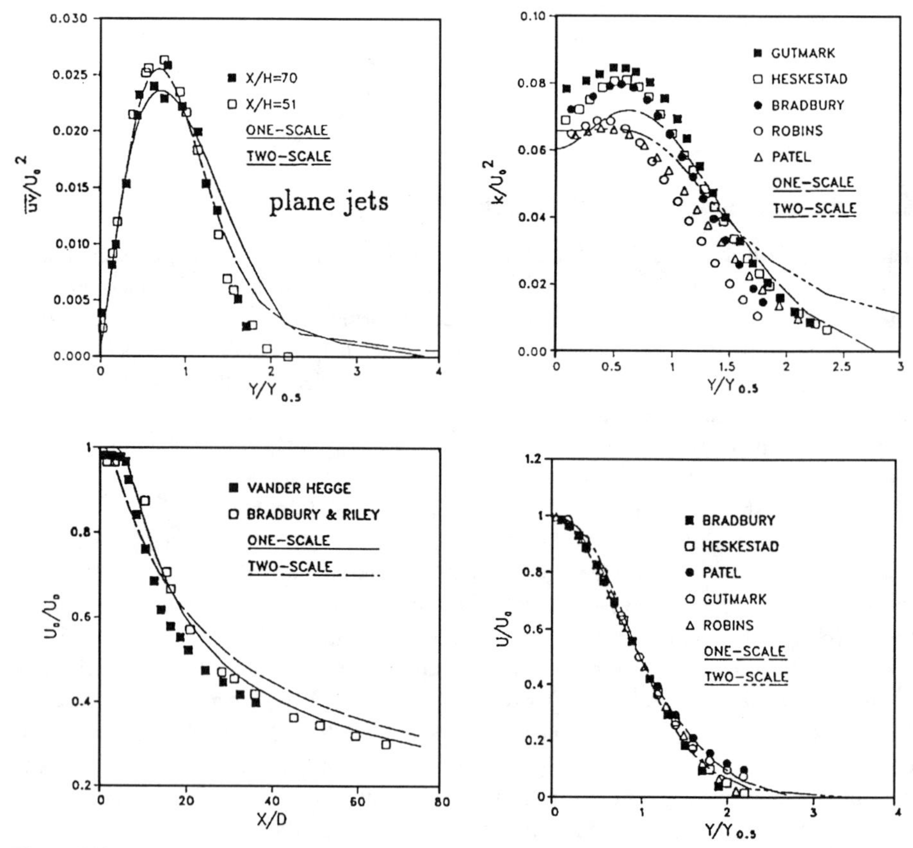

Figure 6.14 Prediction of plane jets by two scale model. (Adapted from Chen and Singh (1990).)

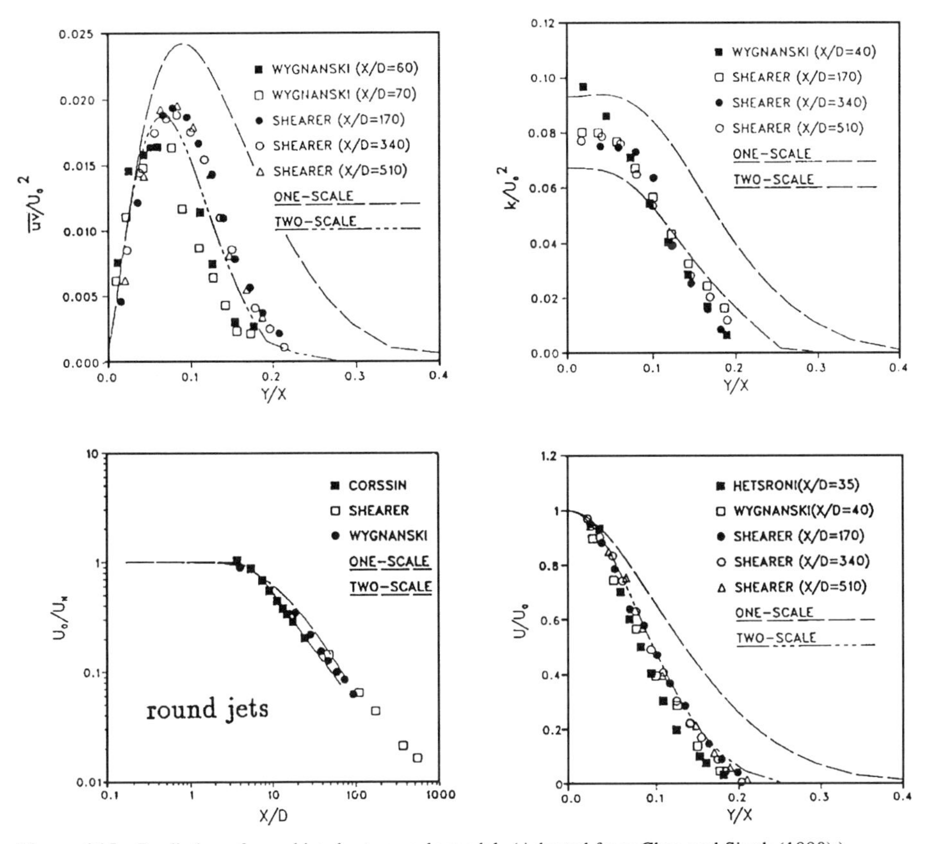

Figure 6.15 Prediction of round jets by two scale model. (Adapted from Chen and Singh (1990).)

6.9.2 Prediction of Mean and Turbulent Properties

Figures 6.14–6.17 are the predictions of detailed profiles. Figure 6.14 shows the predicted profile (in solid or dashed lines) for velocity, turbulent kinetic energy, Reynolds stress, and axial decay for a plane jet and also shows their comparison with experimental data (in square or circle symbols). Figure 6.15 shows the predictions of the two models for a round jet. Figure 6.16 shows the profiles for a plane buoyant jet. Figure 6.17 shows the profiles for a buoyant round jet. Because the predicted spread rate for the plane jet is about the same for both turbulence models, the transverse profiles are normalized with half-width $y_{0.5}$. The comparison of the profiles between the two models is about the same. The two-scale model seems to predict slower axial decay and diffuse turbulent kinetic energy and predicts good Reynolds shear profiles, particularly the peak value. For round jets, because the prediction of the spread rate is different between the models, the transverse profiles are plotted in the similarity variable Y/X. It is obvious that without the modification of turbulent constants in the ϵ equation for the one-scale k-ϵ model, the performance of the one scale k-ϵ model deviates appreciably from the measured data.

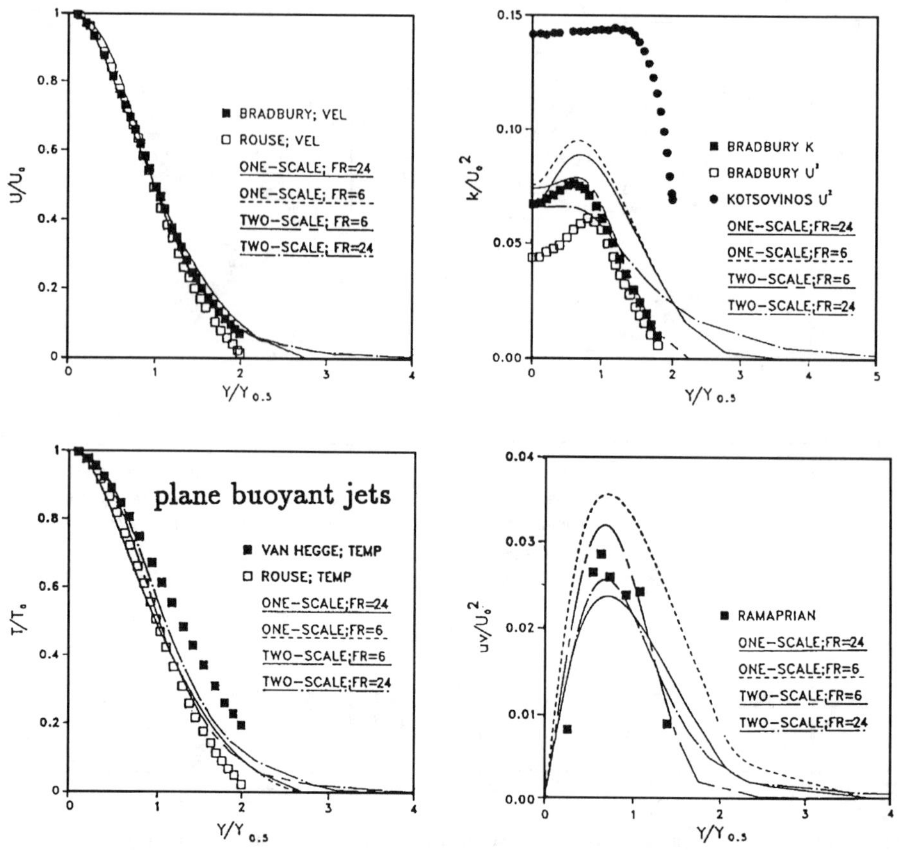

Figure 6.16 Prediction of plane buoyant jets by two scale model. (Adapted from Chen and Singh (1990).)

On the contrary, the two-scale model predicts quite satisfactorily results including the spread rate, the axial decay, and the shear stress. The predicted turbulent kinetic energy is fair with a slightly smaller centerline value.

For buoyant flow the Froude number Fr is defined as $\rho_o U_o^2/g(\rho_o - \rho_a)D$, where the subscribt o denotes the exit value, and a the ambient value. D is the width or diameter. The comparison given in Figs. 6.16 and 6.17 show that, in general, the two-scale k-ϵ model predicts satisfactory results both in mean and turbulent shear or kinetic energy. It should be mentioned that the predicted temperature spread rate shown in Table 6.2 for the buoyant jets by the one-scale k-ϵ model tends to be small for plane buoyant jets ($S_T = 0.116$) and large for round jets ($S_T = 0.12$), which contradicts the trend given by the experimental data ($S_T = 0.13$ (plane) and $S_T = 0.1$ (round). The prediction of spread rate by the two-scale model seems to give better agreement with experimental data and the trend from buoyant plane jet to buoyant round jet. The two-scale turbulence model also predicts good Reynolds shear stress profile when compared with experimental data. In general, both turbulence models tend to predict a wider spread of turbulent kinetic energy or Reynolds stress.

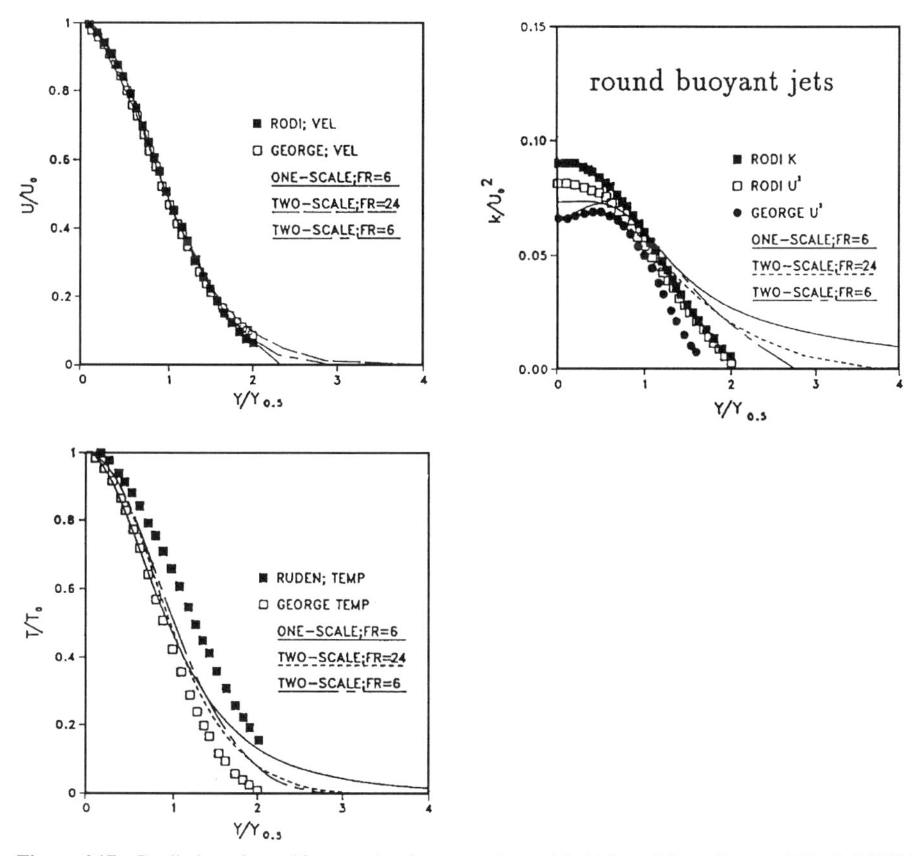

Figure 6.17 Prediction of round buoyant jets by two scale model. (Adapted from Chen and Singh (1990).)

6.9.3 Conclusion

In general, the second-order turbulence model predicts the general trend of buoyant flows. There are still some discrepancies in prediction and measurement. When the use of the k-ϵ-E model is not satisfactory, the k-ϵ-A model should be used. Also, the k-ϵ-$\overline{\theta^2}$ model should be used more.

CHAPTER

SEVEN

CLOSURE

After writing this book, we began to feel that although the second-order turbulence closure model still requires further study, it definitely has achieved some predictability. The confidence level of the predictability of the second-order model is much higher than that for the mixing-length or lower order models. The modeling is still ahead of computer capability in many problems and is also ahead of numerical methods.

It is safe to say that in the future there will be a minicomputer loaded with a turbulence model and a numerical method, neither of which a practical engineer will need to know—just like the present-day pocket calculator that is loaded for logarithmic functions and hyperbolic tangent evaluation. The minicomputer will provide answers to the laminar and turbulent flow problems. The engineer will thus be free from analyzing and studying turbulence models and numerical technique and will be able to devote his or her time in tackling other engineering problems and other duties.

We are optimistic that in the twenty-first century we may say to Leonardo da Vinci, "Our robot can predict it for you" (Fig. 7.1).

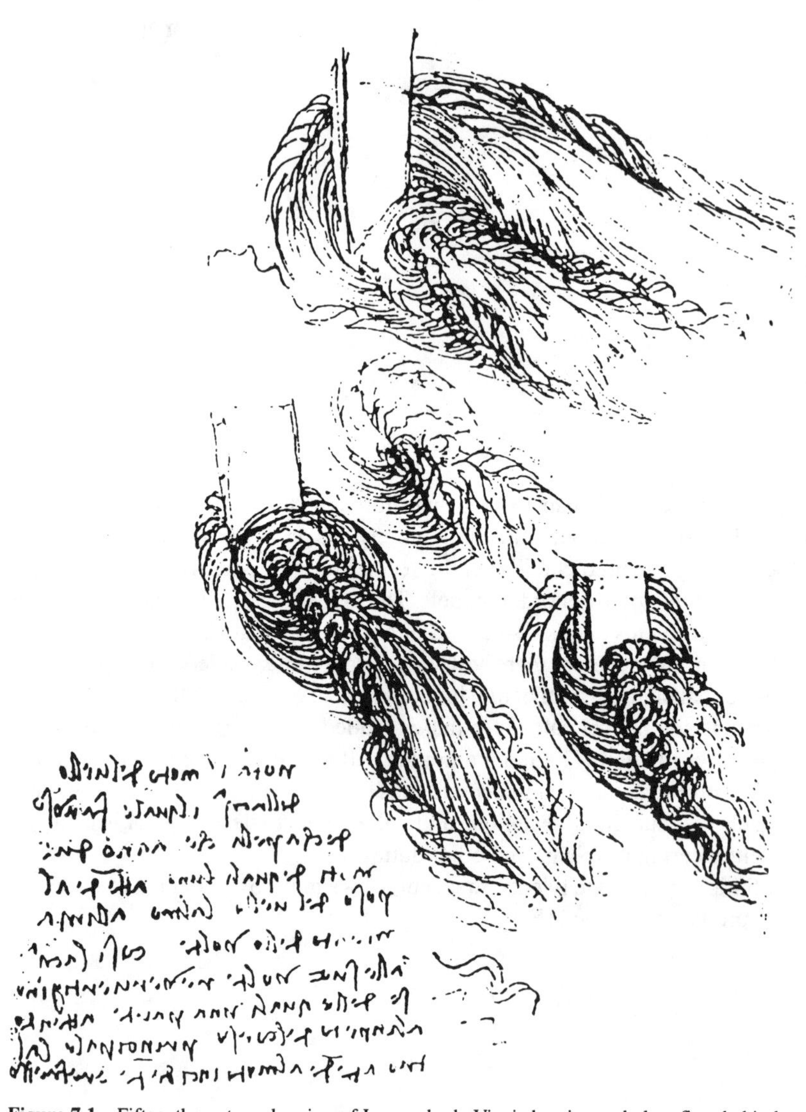

Figure 7.1 Fifteenth-century drawing of Leonardo da Vinci showing turbulent flow behind a pillar. (From Richter (1970).)

BIBLIOGRAPHY

1. K. S. Abdol-Hamid and R. G. Wilmoth. Multiscale turbulence effects in underexpanded supersonic jets. *American Institute of Aeronautics and Astronautics Journal* 27(3): 315–22, 1989.
2. D. E. Abott and S. J. Kline. Experimental investigation of subsonic flow over single and double backward facing steps. *J. of Basic Engineering* 84(September): 317–25, 1992.
3. R. S. Amano and P. Goel. A numerical study of a separating and reattaching flow by using Reynolds-stress turbulence closure. *Numerical Heat Transfer* 7: 343–57, 1984.
4. J. M. Barton, R. Rubinstein, and K. R. Kirtly. Nonlinear Reynolds stress model for turbulent shear flows. In *29th Aerospace Sciences Meeting*, Vol. AIAA 91-0609, Reno, NV, 1991.
5. R. A. Bernatz. *Finite analytic numerical solution of two-dimensional sea-breeze flow*. Ph.D. diss., University of Iowa, 1991.
6. J. Boussinesq. Essai sur la theorie des eaux courantes (Essay on the theory of water flow). *Memoires Academie de Science (Paris)* 23(1): 1–680, 1877.
7. P. Bradshaw, D. H. Ferriss, and R. F. Johnson. Turbulence in the noise producing region of circular jet. *Journal of Fluid Mechanics* 19: 591–624, 1964.
8. A. Brankovic and S. A. Syed. Validation of Reynolds stress turbulence model in generalized coordinates. In *American Institute of Aeronautics and Astronautics 22nd Fluid Dynamics, Plasmadynamics and Lasers Conference*, Honolulu, HI, 1991.
9. F. H. Champagne, V. G. Harris, and S. Corrsin. Experiments on nearly homogeneous shear flows. *Journal of Fluid Mechanics* 41: 81–140, 1970.
10. C. Chandrsuda and P. Bradshaw. Turbulence structure of a reattaching mixing layer. *Journal of Fluid Mechanics* 110: 171–94, 1980.
11. S. M. Chang. *Prediction of turbulent internal recirculating flows with k-ϵ models*. Ph.D. diss., University of Iowa, 1984.
12. C. J. Chen. *Prediction of turbulent flows*. Tech. Rep. 17, Central Research Institute of Electric Power Industry, Abiko Laboratory, Japan, November 1983.
13. C. J. Chen. Finite analytic method. In E. M. Sparrow and W. J. Minkowycz, ed., *Handbook of numerical heat transfer*, chap. 17, pp. 723–46. New York: Wiley-Interscience, 1986.
14. C. J. Chen and S. M. Chang. Prediction of turbulent flows in rectangular cavity with k-ϵ-A and k-ϵ-E models. In C. J. Chen, L. D. Chen, and F. M. Holly, Jr., eds., *Turbulence measurements and flow modeling*, 2nd International Symposium on Refined Flow Modeling and Turbulence Measurements, pp. 611–20, Iowa City, USA, September 16–18, 1985. Hemisphere Publishing.
15. C. J. Chen and C. H. Chen. On prediction and unified correlation for decay of vertical buoyant jets. *Journal of Heat Transfer* 101(3)(August): 532–37, 1979.

16. C. J. Chen and S. Y. Jaw. Present status and future approach of turbulence modeling. *Proceedings of JSCE (Japan Society of Civil Engineers)* 417/II–13, May 1990.
17. C. J. Chen, H. Naseri-Neshat, and K. S. Ho. Finite analytic numerical solution of heat transfer in two-dimensional cavity flow. *International Journal of Numerical Heat Transfer* 4: 179–97, 1981.
18. C. J. Chen and C. Nikitopoulos. On the near field characteristics of axisymmetric turbulent buoyant jets in a uniform environment. *International Journal of Heat and Mass Transfer* 22(1): 1–10, 1979.
19. C. J. Chen and W. Rodi. A mathematical model for stratified turbulent flows and its application to buoyant jets. In *XVIth Congress*, pp. 31–37, Sao Paulo, Brazil, July 27–August 1, 1975. International Association of Hydraulic Research.
20. C. J. Chen and W. Rodi. Vertical turbulent buoyant jets—A review of experimental data. *The science and application of heat and mass transfer series*. Heat Mass Transfer, Vol. 4, 1980.
21. C. J. Chen and K. Singh. Prediction of buoyant free shear flows by k-ϵ model based on two turbulence scale concept. *Proceedings International Symposium on Buoyant Flows*, pp. 26–36. Athens, Greece, September 1–5, 1986.
22. C. J. Chen and K. Singh. Development of a two-scale turbulence model and prediction of buoyant shear flows. In *American Institute of Aeronautics and Astronautics/American Society of Mechanical Engineering Thermophysics and Heat Transfer Conference*, 1990.
23. C. J. Chen, Y. H. Yoon, and C. H. Yu. *The finite analytic method*. Tech. Rep. IIHR. No. 232-VI, University of Iowa, 1983.
24. C. J. Chen, C. H. Yu, and K. B. Chandran. Finite analytic numerical solution of unsteady laminar flow past axisymmetric disc-valves. *Journal of Engineering Mechanics* 113: 1147–62(August), 1987.
25. C. J. Chen, C. H. Yu, and K. B. Chandran. Steady turbulent flow through a disc-type valve: 1. Finite analytic solution. *Journal of Engineering Mechanics* 114(5): 777–96, 1988.
26. C. P. Chen. *Confined swirling jet predictions using a multiple-scale turbulence model*. Tech. Rep. CR-178484, NASA (August), 1985.
27. H. C. Chen. *Calculations of submarine flows by a second-moment rans method*. Tech. Rep. COE No. 325, Texas A&M University Research Foundation, College Station, TX, 1992.
28. H. C. Chen and V. C. Patel. Near-wall turbulence models for complex flows including separation. *American Institute of Aeronautics and Astronautics Journal* 26(6): 641–48(June), 1988.
29. H. C. Chen, W. M. Lin, and K. M. Weems. Second-moment RANS calculations of viscous flow around ship hulls. *Proceedings of CFD Workshop Tokyo*, pp. 275–84. Japan, March 22–24, 1994.
30. K. Y. Chien. Predictions of channel and boundary-layer flows with a low-Reynolds-number turbulence model. *American Institute of Aeronautics and Astronautics Journal* 20(1): 33–8, 1982.
31. S. K. Choi and C. J. Chen. Finite analytic numerical solution of turbulent flow past axisymmetric bodies by zonal modeling approach. In *Symposium on Computational Fluid Dynamics*, Winter Annual Meeting of ASME, Chicago, IL, November 28–December 2, 1988.
32. S. K. Choi and C. J. Chen. A Navier–Stokes numerical analysis of a complete turbulent flow past finite axisymmetric bodies. *Journal of American Institute of Aeronautics and Astronautics* 29(6): 998–1001, 1991.
33. J. Cordes. *Entwicklung und anwendung eines zweischichten–turbulenz models fur abgeloste zweidimensionale stromungen* (Development and application of two layer turbulence models for stratified two-dimensional flows). Ph.D. diss., University of Karlsruhe, 1991.
34. D. E. Cormack, L. G. Leal, and J. H. Seinfeld. An evaluation of mean Reynolds stress turbulence models: The triple velocity correlation. *Journal of Fluids Engineering* 100: 47–54, 1978.
35. T. Craft, S. Fu, B. E. Launder, and D. P. Tselepidakis. *Developments in modeling the turbulent second-moment pressure correlations*. Tech. Rep. TFD/89/1, Mechanical Engineering Department, UMIST (The University of Manchester Institute of Science and Technology), Manchester, UK, 1989.
36. S. C. Crow. Viscoelastic properties of fine-grained incompressible turbulence. *Journal of Fluid Mechanics* 33: 1–20, 1968.
37. B. J. Daly and I. H. Harlow. Transport equations in turbulence. *Physics of Fluids* 13: 2634–49, 1970.
38. J. W. Deardorff. A numerical study of three-dimensional channel flow at large Reynolds numbers. *Journal of Fluid Mechanics* 41: 453–80, 1970.

39. C. Du P. Donaldson. *A computer study of an analytical model of boundary layer transition. American Institute of Aeronautics and Astronautics* paper, 1968.
40. C. Du P. Donaldson. Calculation of turbulent shear flows for atmospheric and vortex motions. *American Institute of Aeronautics and Astronautics Journal* 10: 4, 1972.
41. D. M. Driver and H. L. Seegmiller. Features of a reattaching turbulent shear layer in divergent channel flow. *American Institute of Aeronautics and Astronautics Journal* 23(February): 163–71, 1985.
42. P. A. Durbin. Near-wall turbulence modeling without damping functions. *Theoretical and Computational Fluid Dynamics* 3: 1–13, 1991.
43. P. A. Durbin. A Reynolds-stress model for near-wall turbulence. *Journal of Fluid Mechanics* 249: 465–98, 1993.
44. F. Durst and A. K. Rastogi. Turbulent flow over two-dimensional fences. In L. J. S. Bradbury, F. Durst, B. E. Launder, F. W. Schmidt, and J. H. Whitelaw, eds., *Turbulent shear flows: II*, pp. 218–32. Berlin: Springer-Verlag, 1980.
45. D. Dutoya and P. Michard. A program for calculating boundary layers along compressor and turbine blades. In R. W. Lewis, K. Morgan, and O. C. Zienkiewicz. *Numerical methods in heat transfer*. New York: Wiley, 1981.
46. L. Euler. Principes generaux du movement des fluides (General principle of fluid motion). *Hist. de l'Acad. de Berlin*, 1755.
47. J. B. Fourier. *The analytical theory of heat* (A. Freeman, Trans.) New York: Dover, 1955.
48. U. Frisch, P. L. Sulem, and M. Nelkim. A simple dynamical model of intermittent fully developed turbulence. *Journal of Fluid Mechanics* 87: 719–36, 1978.
49. S. Fu, B. E. Launder, and D. P. Tselepidakis. *Accommodating the effects of high strain rates in modeling the pressure-strain correlation*. Tech. Rep., Department of Mechanical Engineering Report TFD/87/5, UMIST (The University of Manchester Institute of Science and Technology), Manchester, 1987.
50. C. H. Gibson and W. H. Schwarz. The universal equilibrium spectra of turbulent velocity and scalar fields. *Journal of Fluid Mechanics* 16: 365, 1963.
51. M. M. Gibson and B. E. Launder. Ground effects on pressure fluctuations in the atmospheric boundary layers. *Journal of Fluid Mechanics* 86: 491–551, 1978.
52. J. Girard and R. Curlet. *Etude des courants de recirculation dans une cavity* (Study of recirculating flow in a cavity). Tech. Rep., Institute De Mecanique De Grenoble, Grenoble, France, 1975.
53. D. Grand. *Contribution a L'Etude des courants recirculation* (Contribution to the investigation of flow recirculation). Ph.D. diss., Universite de Grenoble, 17 April 1975.
54. T. Y. Han. *A Navier–Stokes analysis of three-dimensional turbulent flows around a bluff body in ground proximity*. Tech. Rep., General Motors Research Laboratory, Warren, Michigan, 1987.
55. K. Hanjalic and B. E. Launder. A Reynolds stress model of turbulence and its application to thin shear. *Journal of Fluid Mechanics* 52(4): 609–38, 1972.
56. K. Hanjalic and B. E. Launder. Contribution towards a Reynolds-stress closure for low-Reynolds-number turbulence. *Journal of Fluid Mechanics* 74(4): 593–610, 1976.
57. K. Hanjalic, B. E. Launder, and R. Schiestel. Multiple-time-scale concepts in turbulent transport modeling. In L. J. S. Bradbury, F. Durst, B. E. Launder, F. W. Schmidt, and J. H. Whitelaw, eds., *Turbulent shear flows: II*, pp. 36–49, Berlin: Springer-Verlag, 1980.
58. V. G. Harris, A. A. Graham, and S. Corrsin. Further experiments in nearly homogeneous turbulent shear flow. *Journal of Fluid Mechanics* 81: 657, 1977.
59. S. Hassid and M. Porch. A turbulent energy model for flow with drag reduction. *Transactions of the American Society of Mechanical Engineering: Journal of Fluid Engineering* 97: 234–41, 1975.
60. J. O. Hinze. *Turbulence*. New York: McGraw-Hill, 1975.
61. G. H. Hoffman. Improved form of the low Reynolds number k-ϵ turbulence model. *Physics of Fluids* 18: 309–12, 1975.
62. M. S. Hossain. *Mathematische Modellierung von Turbulenten Auftriebs-stromungen* (Mathematical modelling of turbulent Buoyant flow). Ph.D. diss., University of Karlsruhe, Germany, 1979.
63. T. T. Huang, N. C. Groves, and G. Belt. *Boundary layer flow on an axisymmetric body with an inflected stern*. Tech. Rep. DTNSRDC 80/064, David Taylor Naval Ship Research and Development Center, 1980.

64. T. T. Huang, N. Santelli, and G. Belt. Stern boundary layer flow on axisymmetric bodies. In *Proceedings of the 12th Symposium on Naval Hydrodynamics*, pp. 127–57, National Academy of Sciences, 1979.
65. H. Iacovides and B. E. Launder. The numerical simulation of flow and heat transfer in tubes in orthogonal rotation. In *Proceedings of the 6th Symposium on Turbulent Shear Flows*, Toulouse, France, 1990.
66. F. J. K. Ideriah. *Turbulent natural and forced convection in plumes and cavities*. Ph.D. diss., University of London, 1977.
67. S. Y. Jaw. *Development of an anisotropic turbulence model for prediction of complex flow*. Ph.D. diss., University of Iowa, 1991.
68. S. Y. Jaw and C. J. Chen. Development of turbulence model including fractal and Kolmogorov scale. In *Symposium on Advances and Applications in Computational Fluid Dynamics*, Winter Annual Meeting of ASME, Dallas, TX, November 25–30, 1990.
69. S. Y. Jaw and C. J. Chen. On determination of turbulence model coefficients for Reynolds stress closure model. Turbulence Session, *ASCE Engineering Mechanics Specialty Conference*, Columbus, Ohio, 1(May 19–22): 394–98, 1991.
70. W. P. Jones and B. E. Launder. The prediction of laminarization with a two-equation model of turbulence. *International Journal of Heat and Mass Transfer* 15: 301–14, 1972.
71. W. P. Jones and J. J. McGuirk. Computation of a round turbulent jet discharging into a confined cross flow. In L. J. S. Bradbury, F. Durst, B. E. Launder, F. W. Schmidt, and J. H. Whitelaw, eds., *Turbulent shear flows:* II, Berlin: Springer-Verlag, 1980.
72. W. M. Kays and M. E. Crawford. *Convective heat and mass transfer*. New York: McGraw-Hill, 1980.
73. J. Kim, S. J. Kline, and J. P. Johnston. *Investigations of separation and reattachment of a turbulent shear layer: Flow over a backward facing step*. Tech. Rep. MD-37, Thermal Sciences Division, Department of Mechanical Engineering, Stanford University, 1978.
74. J. Kim, P. Moin, and R. Moser. Turbulence statistics in fully developed channel flow at low Reynolds number. *Journal of Fluid Mechanics* 177: 133–66, 1987.
75. S.-W. Kim and C.-P. Chen. A multiple-time-scale turbulence model based on variable partitioning of the turbulent kinetic energy spectrum. *Numerical Heat Transfer* 16(B): 193–211, 1989.
76. S. J. Kline, B. J. Cantwell, and G. M. Lilley, eds. *AFOSR (Air Force Office of Scientific Research)–HTTM stanford conference on complex turbulent flows*. Stanford University, 1981.
77. A. N. Kolmogorov. Local structure of turbulence in an incompressible fluid at very high Reynolds numbers. *Dokl. Akad. Nauk (Report of academy of sciences) USSR* 30(4): 299–303, 1941.
78. A. N. Kolmogorov. Equations of turbulent motion of an incompressible fluid. *IZV Akad. Nauk. (Academy of science series) USSR, Ser. Phys.* 6: 56–58, 1942.
79. T. Komatsu and N. Matsunaga. Defect of k-ϵ turbulence model and its improvements. *Proceedings of 30th Japan Conference on Hydraulics*, pp. 529–34, February 1986.
80. Y. G. Lai and R. M. C. So. On near-wall turbulent flow modeling. *Journal of Fluid Mechanics* 221: 641–73, 1990.
81. C. K. G. Lam and K. Bremhorst. A modified form of the k-ϵ model for predicting wall turbulence. *Transactions of the American Society of Mechanical Engineering* 103(September): 456–60, 1981.
82. B. E. Launder. *Prediction methods for turbulent flows*. Tech. Rep. Lecture Series 76, von Kármán Institute for Fluid Dynamics, 1975.
83. B. E. Launder. A generalized algebraic stress transport hypothesis. *American Institute of Aeronautics and Astronautics Journal* 20(3): 436–37, 1982.
84. B. E. Launder. Low-Reynolds-number turbulence near walls. Tech. Rep. Department of Mechanical Engineering Rep. TFD/86/4, UMIST (The University of Manchester Institute of Science and Technology) Manchester, 1986.
85. B. E. Launder. Second-moment closure: Present and future? *International Journal of Heat and Fluid Flow* 10(4): 282–300, 1989.
86. B. E. Launder and A. Morse. Numerical prediction of axisymmetric free shear flows with a second-order Reynolds stress closure. In *Proceedings of Symposium on Turbulent Shear Flows: vol. 1*, Pennsylvania State University, April 1977.
87. B. E. Launder, A. Morse, W. Rodi, and D. B. Spalding. A comparison of performance of six turbulence models. In *Proceedings of NASA Conference on Free Shear Flows*, Langley, Virginia, 1972.

88. B. E. Launder, G. J. Reece, and W. Rodi. Progress in the development of a Reynolds-stress turbulence closure. *Journal of Fluid Mechanics* 68(3): 537–66, 1975.
89. B. E. Launder and W. C. Reynolds. Asymptotic near-wall stress distribution rates in turbulent flow. *Physics of Fluids* 26: 1157–58, 1983.
90. B. E. Launder and B. I. Sharma. Application of the energy-dissipation model of turbulence to the calculation of flow near a spinning disc. *Letters in Heat and Mass Transfer* 1: 131–38, 1974.
91. B. E. Launder and N. Shima. Second-moment closure for the near-wall sublayer: Development and application. *American Institute of Aeronautics and Astronautics Journal* 27(10): 1319–25, 1989.
92. B. E. Launder and D. B. Spalding. The numerical computation of turbulent flows. *Computer Methods in Applied Mechanics and Engineering* 3: 269–89, 1974.
93. M. Lee and W. C. Reynolds. On the structure of homogeneous turbulence. In *Proceedings of the 5th Symposium on Turbulent Shear Flows*, pp. 17.07–17.12, Cornell University, 1985.
94. M. A. Leschziner and W. Rodi. Calculation of strongly curved open channel flow. *Journal of the Hydraulics Division of ASCE (American Society of Civil Engineers)* 105(Hydraulic10): 1297–1314, 1979.
95. H. N. Liefmann and J. Laufer. *Investigation of free turbulent mixing*. Tech. Rep. 1257 NACA, 1957.
96. H. W. Liepmann. *The mechanics of turbulence*. New York: Gordon and Breach, 1964.
97. J. L. Lumley. Toward a turbulent constitutive relation. *Journal of Fluid Mechanics* 41(2): 413–34, 1970.
98. J. L. Lumley. Computational modeling of turbulent flows. *Advances in Applied Mechanics* 18: 123–76, 1978.
99. J. L. Lumley. Second order modeling of turbulent flows. In W. Kollman, ed., *Prediction methods for turbulent flows*, pp. 1–31. New York: Hemisphere Publishing, 1980.
100. J. L. Lumley and B. J. Khajeh-Nouri. *Modeling homogeneous deformation of turbulence*. Tech. Rep., Pennsylvania State University, 1973.
101. J. L. Lumley and B. J. Khajeh-Nouri. Computational modeling of turbulent transport. *Adv. Geophys.* 18A: 169–92, 1974.
102. N. N. Mansour, J. Kim, and P. Moin. Reynolds-stress and dissipation rate budgets in a turbulent channel flow. *Journal of Fluid Mechanics* 194(September): 15–44, 1988.
103. N. C. Markatos. The mathematical modeling of turbulent flows. *Applied Mathematical Modelling* 10(June): 190–220, 1986.
104. R. Martinuzzi and A. Pollard. Comparative study of turbulence models in predicting turbulent pipe flow: pt. 1, Algebraic stress and k-ϵ models. *American Institute of Aeronautics and Astronautics Journal* 27(1)(January): 29–36, 1989.
105. J. J. McGuirk and W. Rodi. A depth-averaged mathematical model for the near field of side discharges into open channel flow. *Journal of Fluid Mechanics* 86: 761–81, 1978.
106. J. J. McGuirk and D. B. Spalding. Mathematical modelling of thermal pollution in rivers. In C. A. Brebbia, ed., *Proceedings of the International Conference on Mathematical Models for Environmental Problems*, Southampton, England, September 8–12, 1975.
107. V. Michelassi and T.-H. Shih. *Low Reynolds number two-equation modeling of turbulent flows*. NASA Tech. Memo. 104368, May 1991.
108. R. D. Mills. On the closed motion of a fluid in a cavity. *Journal of Royal Aeronautical Society* 69(February): 116–20, 1965.
109. H. Ha Minh and P. Chassaing. Perturbations of turbulent pipe flow. In F. Durst, ed., *Turbulent shear flows*: I, pp. 178–97, Berlin: Springer-Verlag, 1979.
110. P. Moin and J. Kim. Numerical investigation of turbulent channel flow. *Journal of Fluid Mechanics* 118: 341–77, 1982.
111. R. D. Moser and P. Moin. Direct numerical simulation of curved channel flow. Tech. Rep. NASA TM 85974, NASA Ames Research Center, Moffett, CA, 1984.
112. Y. Nagano and M. Hishida. Improved form of the k-ϵ model for wall turbulence shear flows. *Transactions of the American Society of Mechanical Engineering: Journal of Fluids Engineering* 109(June): 1987.
113. D. Naot and W. Rodi. Calculation of secondary currents. *Journal of Hydraulic Division, ASCE (American Society of Civil Engineers)* 108(Hydraulic8)(August): 948–66, 1982.
114. M. Navier. *Memoire sur les Lois du Mouvement des Fluides (Memoir on the theory of fluid motion). Memoires Academie de Science (Memoir of academy of science) (Paris)* 6: 389–416, 1827.

115. N. H. Ng and D. B. Spalding. Some applications of a model of turbulence to boundary layers near walls. *Physics of Fluids* 15: 20, 1972.
116. J. Nikuradse. Gesetzmäßigkeit der turbulenten Strömung in glatten Rohren (Law of turbulent flow in smooth pipe). Forsch. Arb. Ing.-Wes. Research report of engineering and science. No. 356, 1932.
117. S. Nisizima and A. Yoshizawa. Turbulent channel and couette flows using an anisotropic k-ϵ model. *American Institute of Aeronautics and Astronautics Journal* 25(3): 414–20, 1987.
118. M. Normandin. *Etude experimentale de l'ecoulement Turbulent dans une cavite profonde Journal de Mécanique* 19(3): 540–60, 1980.
119. L. H. Norris and W. C. Reynolds. *Turbulent channel flow with a moving wavy boundary*, Tech. Rep. FM-10, Department of Mechanical Engineering, Stanford University, 1975.
120. S. V. Patankar, V. S. Pratap, and D. B. Spalding. Prediction of turbulent flow in curved pipes. *Journal of Fluid Mechanics* 67: 583–95, 1975.
121. V. C. Patel, W. Rodi, and G. Scheuerer. Turbulence models for near-wall and low Reynolds number flows: A review. *American Institute of Aeronautics and Astronautics Journal* 23(9): 1308–19, 1984.
122. V. C. Patel and G. Schueurer. Calculation of two-dimensional near and far wakes. *American Institute of Aeronautics and Astronautics Journal* 20(7): 900–07, 1982.
123. S. B. Pope. A more general effective-viscosity hypothesis. *Journal of Fluid Mechanics* 72: 331–40, 1975.
124. L. Prandtl. Uber die ausgebildete turbulenz (On the fully developed turbulence). *ZAMM (Journal of Applied Mathematics and Mechanics)* 5: 136–39, 1925.
125. L. Prandtl. Uber ein neues formelsystem fur die ausgebildete turbulenz (On a new formation for the fully developed turbulence). *Nachr. Akad. Wiss. (Report of academy of sciences) Gottingen*, p. 6, 1945.
126. G. J. Reece. *A generalized Reynolds-stress model of turbulence*. Ph.D. diss., University of London, England, 1977.
127. A. Reshko. Some measurements of flow in a rectangular cutout. NACA, (National Advisery Committee for Aeronautics) TN 3488, 1955.
128. O. Reynolds. On the dynamical theory of incompressible fluids and the determination of the criterion. *Philosophical Transactions of the Royal Society* 186, 1894.
129. W. C. Reynolds. *Computation of turbulent flows:* State-of-the-art. Tech. Rep., Report MD-27, Mechanical Engineering Department, Stanford University, 1970.
130. W. C. Reynolds. Computation of turbulent flows. AIAA paper 74–556, 1974.
131. J. P. Richter. *The Notebooks of Leonardo da Vinci*, Vols. 1 and 2. New York: Dover, 1970.
132. R. S. Rivlin. *Quarterly of Applied Mathematics*, Vol. 15, p. 212. Brown University, 1957.
133. W. Rodi. *On the equation governing the rate of turbulent energy dissipation*. Tech. Rep. TM/TN/A/14, Mechanical Engineering Department, Imperial College, London, 1971.
134. W. Rodi. *The prediction of free turbulent boundary layers by use of a two-equation model of turbulence*. Ph.D. diss. Department of Mechanical Engineering, Imperial College, London, 1972.
135. W. Rodi. *Turbulence models and their applications in hydraulics—A state of the art review*. Tech. Rep., University of Karlsruhe, Karlsruhe, Germany, 1980.
136. W. Rodi. Examples of turbulence models for incompressible flows. *American Institute of Aeronautics and Astronautics Journal* 20(7): 872–79, 1981.
137. W. Rodi. Recent developments in turbulence modeling. In Y. Iwasa, N. Tamai, and A. Wada, eds., *Proceedings 3rd International Symposium on Refined Flow Modeling and Turbulence Measurements*, Tokyo, Japan, July 26–28, 1988.
138. W. Rodi. Experience with two-layer models combining the k-ϵ model with a one-equation model near wall. In *29th Aerospace Science Meeting*, Vol. AIAA 91-0609, Reno, NV, 1991.
139. W. Rodi and G. Scheuerer. Scrutinizing the k-ϵ turbulence model under adverse pressure gradient conditions. *American Society of Mechanical Engineering Journal of Fluid Engineering* 108: 174–79, 1986.
140. J. J. Rohr, E. C. Itsweire, K. N. Helland, and C. W. Van Atta. An investigation of the growth of turbulence in a uniform-mean-shear flow. *Journal of Fluid Mechanics* 187: 1–33, 1988.
141. A. Roshko. *Some measurements of flow in a rectangular cutout*. NACA Tech. Note 3488, p. 21, 1955.
142. I. L. Rosovskii. *Flow of water in bends of open channels*. Academy of Sciences of the Ukrainian SSR, 1957. (Printed in Jerusalem by S. Manson).

143. J. Rotta. *Statistical theory of nonhomogeneous turbulence*. Tech. Rep. TWF/TN/38, TWF/TN/39, Mechanical Engineering Department, Imperial College, London, 1951.
144. J. Rotta. Turbulent boundary layers in incompressible flow. *Progress in Aero. Sci.*, Vol. 2, edited by Ferri, Kuchemann, and Sterne. 1962.
145. H. Rouse, C. S. Yih, and H. W. Humphreys. Gravitational convection from a boundary source. *Tellus* 4: 201–10, 1952.
146. M. Rowe. *Some secondary flow problems in fluid dynamics*. Ph.D. diss., Cambridge University, England, 1966.
147. R. Rubinstein and J. M. Barton. Nonlinear Reynolds stress models and the renormalization group. *Physics of Fluids A* 2(8)(August): 1472–76, 1990.
148. R. Schiestel. Multiple-time-scale modeling of turbulent flows in one point closures. *Physics of Fluids* 30(3)(March): 722–31, 1987.
149. H. Schlichting. *Boundary-layer theory* (J. Kestin, Trans.) New York: McGraw-Hill, 1979.
150. U. Schumann. Subgrid scale model for finite difference simulations of turbulent flows in plane channels and annuli. *Journal of Computational Physics* 18: 376–404, 1975.
151. T. H. Shih and J. L. Lumley. *Modeling of pressure correlation terms in Reynold stress and scalar flux equations*. Tech. Rep. FDA-85-3, Sibley School of Mechanical and Aeronautical Engineering, Cornell University, 1985.
152. T. H. Shih, J. L. Lumley, and J. Y. Chen. *Second-order modeling of boundary-free turbulent shear flows with a new form of pressure correlation*. Tech. Rep. FDA-85-7, Sibley School of Mechanical and Aeronautical Engineering, Cornell University, 1985.
153. T. H. Shih and N. N. Mansour. Modeling of near-wall turbulence. In *International Symposium on Engineering Turbulence Modeling and Measurements*, NASA Tech. Memo. 10322, ICOMP-90-0017, Dubrovnik, Yugoslavia, September 24–28, 1990.
154. N. Shima. A Reynolds-stress model for near-wall and low-Reynolds-number regions. *Transactions of the American Society of Mechanical Engineers, Journal of Fluid Engineering* 110(March): 38–44, 1988.
155. C. C. Shir. A preliminary numerical study of atmospheric turbulent flow in the idealized planetary boundary layer. *Journal of Atmospheric Sciences* 30: 1327, 1973.
156. L. M. Smith and W. C. Reynolds. On the Yakhot–Orszag renormalization group method for deriving turbulence statistics and models. *Physics of Fluids A* (4), p. 364, 1992.
157. R. Smyth. Turbulence flow over a plane symmetric sudden expansion. *Journal of Fluid Engineering* 101(September): 348–53, 1979.
158. H. J. Sneck and D. H. Brown. Plume rise from large thermal sources such as a cooling tower. *Transactions of the American Society of Mechanical Engineers: Journal of Heat Transfer* 96: 232–38, 1974.
159. R. M. C. So, H. S. Zhang, and C. G. Speziale. Near-wall modeling of the dissipation rate equation. *American Institute of Aeronautics and Astronautics Journal* 29(12): 2069–76, 1991.
160. R. M. C. So, Y. G. Lai, and B. C. Hwang. Near-wall turbulence closure for curved flows. *American Institute of Aeronautics and Astronautics Journal* 29: 1202–13, 1991.
161. R. M. C. So, Y. G. Lai, and H. S. Zhang. Second-order near-wall turbulence closure: A review. *American Institute of Aeronautics and Astronautics Journal* 29(11): 1819–35, 1991.
162. F. Sotiropoulos and V. C. Patel. Numerical calculation of turbulent flow through a circular-to-rectangular transition duct using advanced turbulence closures. *American Institute of Aeronautics and Astronautics* paper 93-3030, 1993.
163. F. Sotiropoulos and V. C. Patel. Second-moment modeling for shipstern and wake flows. In *Proceedings of CFD Workshop*, pp. 187–98, Tokyo, Japan, March 22–24, 1994.
164. P. R. Spalart. Numerical study of sink-flow boundary layers. *Journal of Fluid Mechanics* 172: 307–28, 1986.
165. D. B. Spalding. *The vorticity-fluctuations (kw) model of turbulence*. Tech. Rep. CFD Rep. CFD 82/17, Mechanical Engineering Department, Imperial College, London, 1982.
166. C. G. Speziale. On nonlinear k-l and k-ϵ models of turbulence. *Journal of Fluid Mechanics* 178: 459–75, 1987.
167. C. G. Speziale, T. B. Gatski, and N. Mac Giolla Mhuiris. A critical comparison of turbulence models

for homogeneous shear flows in a rotating frame. *Proc. 7th Symp. Turb. Shear Flows*, 2: 27.3.1–27.3.6. Stanford, Cali: Stanford University Press. 1989.

168. K. R. Sreenivasan and C. Meneveau. The fractal facets of turbulence. *Journal of Fluid Mechanics* 173: 357–86, 1986.
169. G. G. Stokes. On the theories of internal friction of fluids in motion. *Transactions of the Cambridge Philosophical Society* 8: 287–305, 1845.
170. D. J. Struik. *A concise history of mathematics*. New York: Dover Publications, 1948.
171. A. Tailland and J. Mathieu. Jet parietal. *Journal of Mechanics* 6: 103–31, 1967.
172. N. Takemitsu. A revised k-ϵ model. *Proceedings of the Journal of Mechanical Engineering* 53(494): 2928–36, 1986.
173. N. Takemitsu. An analytic study of the standard k-ϵ model. *Journal of Fluids Engineering* 112: 192–98, 1990.
174. S. Tavoularis and S. Corrsin. Experiments in nearly homogeneous turbulent shear flows with a uniform mean temperature gradient: pt. 1. *Journal of Fluid Mechanics* 104: 311, 1981.
175. M. M. M. Telbany and A. J. Reynolds. Turbulence in plane channel flows. *Journal of Fluid Mechanics* 111: 283–318, 1981.
176. A. A. Townsend. The uniform distortion of homogeneous turbulence. *Quarterly Journal of Mechanics and Applied Mathematics* 7: 104, 1954.
177. A. A. Townsend. *The structure of turbulent shear flow*. Cambridge, England: Cambridge University Press, 1976.
178. M. S. Uberoi. Equipartition of energy and local isotropy in turbulent flow. *Journal of Applied Physics* 28(10): 1165–70, 1957.
179. S. Ushijima, M. Kato, K. Fujimoto, and S. Moriya. *Application of some turbulence models—numerical analyses of two-dimensional circulation flows*. Civil Engineering Laboratory Rep. No. 385019 Central Research Institute of Electric Power Industry, Abiko-City, Japan, 1985.
180. T. von Kármán. Mechanische ahnlichkeit und turbulenz (Mechanical similarity and turbulence). *Math. Phys. Klasse (Mathematics and physics division)*, p. 58, 1930.
181. T. von Kármán. *TM 611*. Tech. Rep. NACA (National Advisory Committee for Aeronautics), 1931.
182. C. A. G. Webster. An experimental study of turbulence in a density stratified shear flow. *Journal of Fluid Mechanics* 19: 221–45, 1964.
183. J. Weinstock and S. Burk. Theoretical pressure-strain term, resistance to large anisotropies of stress and dissipation. In *Proceedings of the 5th Symposium on Turbulent Shear Flows*, pp. 12.13–18, Cornell, 1985.
184. D. C. Wilcox. Multiscale model for turbulent flows. *American Institute of Aeronautics and Astronautics Journal* 26(11): 1311–20, 1988.
185. D. C. Wilcox. Reassessment of the scale determining equation for advanced turbulence models. *American Institute of Aeronautics and Astronautics Journal* 26(11)(November): 1299–1310, 1988.
186. D. C. Wilcox and M. W. Rubesin. *Progress in turbulence modeling for complex flow fields including effects of compressibility*. Tech. Rep. Tech. Paper 1517, NASA, 1980.
187. M. Wolfshtein. The velocity and temperature distribution in one-dimensional flow with turbulence augmentation and pressure gradient. *International Journal of Heat and Mass Transfer* 12: 301–18, 1969.
188. I. Wygnanski and H. E. Fielder. The two-dimensional mixing region. *Journal of Fluid Mechanics* 41: 327–63, 1970.
189. V. Yakhot and S. A. Orszag. Renormalization group analysis of turbulence: 1. Basic theory. *Journal of Scientific Computing* 1(1): 3–51, 1986.
190. V. Yakhot, S. A. Orszag, S. Thangam, T. B. Gatski, and C. G. Speziale. Development of turbulence models for shear flows by a double expansion technique. *Physics of Fluids A*, 4(7): 1510–20, 1992.
191. A. Yoshizawa. Statistical modeling of a transport equation for the kinetic energy dissipation rate. *Physics of Fluid* 30(3): 628–31, 1987.
192. C. H. Yu, C. J. Chen, and K. B. Chandran. Steady turbulent flow through a disc-type valve: 2. Parametric study on disc size and position. *Journal of Engineering Mechanics* 114(5): 797–811, 1988.

INDEX

(*Continued on next page*)

(*Continued on next page*)

(*Continued on next page*)